NORTH-HOLLAND
MATHEMATICS STUDIES **27**

Notas de Matemática (63)

Editor: Leopoldo Nachbin

Universidade Federal do Rio de Janeiro
and University of Rochester

Functional Analysis: Surveys and Recent Results

Proceedings of the Conference on Functional Analysis
Paderborn, Germany, November 17-21, 1976

Edited by

KLAUS-DIETER BIERSTEDT

and

BENNO FUCHSSTEINER

University of Paderborn, Germany

1977

NORTH-HOLLAND PUBLISHING COMPANY - AMSTERDAM • NEW YORK • OXFORD

North-Holland ISBN: 0 444 85057 0

PUBLISHERS:
NORTH-HOLLAND PUBLISHING COMPANY
AMSTERDAM, NEW YORK, OXFORD

SOLE DISTRIBUTORS FOR THE U.S.A. AND CANADA:
ELSEVIER / NORTH HOLLAND, INC.
52 VANDERBILT AVENUE, NEW YORK, N.Y. 10017

Library of Congress Cataloging in Publication Data

Conference on Functional Analysis, Paderborn, 1976.
Functional analysis.

(Notas de matemática ; 63) (North-Holland mathematics studies ; 27)
Includes bibliographical references.
1. Functional analysis--Congresses. I. Bierstedt, Klaus-Dieter. II. Fuchssteiner, Benno. III. Title. IV. Series.
QA1.N86 no. 63 [QA320] 510'.8s [515.7] 77-20497
ISBN 0-444-85057-0

PRINTED IN THE NETHERLANDS

PREFACE

A Conference on Functional Analysis was held at the University of Paderborn (Gesamthochschule Paderborn) on November 17-21, 1976. This book contains the revised, updated and somewhat extended Proceedings of the conference. Organizers were K.-D. Bierstedt and B. Fuchssteiner who are also serving as editors of this volume.

At the conference, there were 17 invited lectures of either 50 or 90 minutes each on topics of current active research in functional analysis. Some blocks of lectures were centered around a main topic and studied related problems from different points of view. Many of the lectures started with a survey of the area before proceeding to recent contributions and new results.

The 18 articles in this book deal with applications of locally convex topological vector spaces and tensor products (B. Gramsch, W. Kaballo, D. Vogt); vector lattices (W. Hackenbroch, M. Wolff); spaces of continuous functions (J. Schmets, W. Ruess); measures (H. Buchwalter, S.D. Chatterji / V. Mandrekar) and distributions (R. Meise); complex analysis in infinite dimensions (M. Schottenloher); function algebras (H. König) and C*-algebras (J.D.M. Wright / M.A. Youngson); geometric aspects of Banach space theory (R.E. Huff, W. Lusky, E. Behrends); generalized spectral operators (E. Albrecht); and functional analysis in Solovay's model (J.D.M. Wright).

We would like to thank all those who participated in the conference for their interest and the stimulating discussions, above all the speakers and the chairmen of the sessions. We would also like to thank all contributors for the preparation of their manuscripts in time for this publication. We thank the analysis group of Gesamthochschule Paderborn for their help in organizing the meeting, in proofreading the manuscripts, and correcting some misprints. (We should mention R. Hollstein and W. Lusky in this connection.) Last, but not least, we thank Gesamthochschule Paderborn for providing the funds without which all this would not have been possible.

Paderborn, June 1977

K.-D. Bierstedt
B. Fuchssteiner

CONFERENCE ON FUNCTIONAL ANALYSIS, PADERBORN 1976

SCHEDULE OF LECTURES

Thursday, November 18, 1976:

Morning Session, Chairman: K.-D. Bierstedt

8.30 - 10.00 J. Schmets, Spaces of continuous functions

10.30 - 12.00 H. Buchwalter, Espaces de mesures sur un espace complètement régulier

Afternoon Session, Chairman: K. Floret

2.00 - 2.50 W. Ruess, Über die lokalkonvexe Struktur strikter Topologien

3.00 - 4.30 E. Albrecht, Verallgemeinerte Spektraloperatoren

5.00 - 6.30 M. Schottenloher, Reichhaltigkeit der Klasse holomorpher Funktionen auf einem lokalkonvexen Raum

Friday, November 19, 1976:

Morning Session, Chairman: G. Köthe

8.30 - 10.00 H. König, Abstrakte Theorie der Hardyschen Funktionenalgebren

10.30 - 12.00 B. Gramsch, Eine Fortsetzungsmethode der Dualitätstheorie lokalkonvexer Räume

Afternoon Session, Chairman: E. Behrends

2.00 - 2.50 W. Kaballo, ε-Tensorprodukte und Lifting vektorwertiger Funktionen

3.00 - 4.30 R. Meise, Darstellung von Distributionen und Ultradistributionen durch holomorphe Funktionen

5.00 - 5.50 D. Vogt, Unterräume und Quotientenräume von (s)

6.00 - 6.50 S.D. Chatterji, Singularity and absolute continuity of measures

Saturday, November 20, 1976:

Morning Session, Chairman: K. Kutzler

8.30 - 10.00 W. Hackenbroch, Darstellung von Vektorverbänden

10.30 - 12.00 M. Wolff, Über Korovkinsätze in lokalkonvexen Vektorverbänden

Afternoon Session, Chairman: B. Fuchssteiner

2.00 - 2.50 U. Schlotterbeck, Operatoren auf Banachverbänden

3.00 - 4.30 R.E. Huff, The Radon-Nikodým property for Banach spaces - A survey of geometric aspects

5.00 - 5.50 J.D.M. Wright, Functional analysis for the practical man

6.00 - 6.50 W. Lusky, Separable Lindenstrauss-Räume

CONFERENCE ON FUNCTIONAL ANALYSIS, PADERBORN 1976

PARTICIPANTS

E. Albrecht, Saarbrücken

B. Baumgarten, Darmstadt

E. Behrends, Berlin (F.U.)

H. Buchwalter, Lyon (France)

S.D. Chatterji, Lausanne (Suisse)

J. Cuntz, Berlin (T.U.)

B. Ernst, Paderborn

K. Floret, Kiel

H.-O. Flösser, Darmstadt

B. Gramsch, Kaiserslautern

P. Greim, Berlin (F.U.)

W. Hackenbroch, Regensburg

R. Hollstein, Frankfurt

R.E. Huff, Pennsylvania State University (U.S.A.) and Erlangen

W. Kaballo, Kaiserslautern

H. König, Saarbrücken

G. Köthe, Frankfurt

F. Krauß, Paderborn

K. Kutzler, Berlin (T.U.)

W. Lusky, Paderborn

E. Marschall, Münster

R. Meise, Düsseldorf

J. Michaliček, Hamburg

V. Osôrio, Darmstadt

H.-J. Petzsche, Düsseldorf

R. Riemer, Kaiserslautern

B. Rosenberger, Kaiserslautern

W. Ruess, Bonn

U. Schlotterbeck, Tübingen

J. Schmets, Liêge (Belgique)

M. Schottenloher, München

D. Vogt, Wuppertal

R. Wagner, Paderborn

L. Weis, Kaiserslautern

M. Wolff, Tübingen

J.D.M. Wright, Reading (England)

CONTRIBUTORS

E. Albrecht, Fachbereich 9 (Mathematik) der Universität des Saarlandes, Bau 27, D-6600 Saarbrücken 11, Germany

E. Behrends, I. Mathematisches Institut der Freien Universität, Hüttenweg 9, D-1000 Berlin 33 (West), Germany

H. Buchwalter, Département de Mathématiques, Université Claude-Bernard - Lyon I, 43, Boulevard du Onze Novembre 1918, F-69621 Villeurbanne, France

S. D. Chatterji, Département de Mathématiques, Ecole Polytechnique Fédérale de Lausanne, 61, Avenue de Cour, CH-1007 Lausanne, Switzerland

B. Gramsch, Fachbereich Mathematik der Universität, Bau 48, Pfaffenbergstr., Postfach 3049, D-6750 Kaiserslautern, Germany

W. Hackenbroch, Fachbereich Mathematik der Universität, Universitätsstr. 31, D-8400 Regensburg, Germany

R. E. Huff, Department of Mathematics, 215 McAllister Building, Pennsylvania State University, University Park, Pennsylvania 16802, U.S.A.

W. Kaballo, Fachbereich Mathematik der Universität, Bau 48, Pfaffenbergstr., Postfach 3049, D-6750 Kaiserslautern, Germany

H. König, Fachbereich Mathematik der Universität des Saarlandes, Bau 27, D-6600 Saarbrücken 11, Germany

W. Lusky, Fachbereich 17 der Gesamthochschule, Arbeitsstelle Mathematik, Warburger Str. 100, Postfach 1621, D-4790 Paderborn, Germany

R. Meise, Mathematisches Institut der Universität, Universitätsstr. 1, D-4000 Düsseldorf, Germany

W. Ruess, Institut für Angewandte Mathematik der Universität, Abt. Funktionalanalysis, Wegelerstr. 6, D-5300 Bonn, Germany

J. Schmets, Institut de Mathématique, Analyse Mathématique et Algèbre, Université de Liège, 15, Avenue des Tilleuls, B-4000 Liège, Belgium

M. Schottenloher, Mathematisches Institut der Universität, Theresienstr. 39, D-8000 München, Germany

D. Vogt, Fachbereich 7 (Mathematik) der Gesamthochschule Wuppertal, D-5600 Wuppertal, Germany

M. Wolff, Mathematisches Institut der Universität, Auf der Morgenstelle 10, D-7400 Tübingen, Germany

J. D. M. Wright, Department of Mathematics, University of Reading, Whiteknights, Reading, England

CONTENTS

K.-D. Bierstedt, B. Fuchssteiner (eds.)
Functional Analysis: Surveys and Recent Results

THE RADON-NIKODÝM PROPERTY FOR BANACH-SPACES — A SURVEY OF GEOMETRIC ASPECTS

R.E. Huff
Mathematisches Institut
der Universität Erlangen-Nürnberg
Erlangen, W. Germany*

INTRODUCTION

The theme for this talk is that the Radon-Nikodým property (RNP) gives a genuine dichotomy with respect to geometric properties invariant under isomorphisms of Banach spaces. Those spaces with the RNP are not only good, they are *very* good; those without it are not simply not good, they are bad.

I shall first remind you of some classical results and an analytic characterization. I will then try to indicate how bad spaces are without the RNP, and then run thru some results which show how good spaces are with the RNP with respect to extreme point phenomena and with respect to operators and functionals supporting sets. Finally, I will discuss the situation in dual spaces where the dichotomy is even more apparent.

The definition of the RNP is natural: a Banach space X (always over the real numbers $\mathbb{R}$) has it if and only if the classical Radon-Nikodým theorem holds for X-valued measures. More precisely,

DEFINITION 1: X *has the RNP if and only if for every σ-finite measure space (Ω,Σ,λ) and for every X-valued measure $\mu : \Sigma \to X$ which is of bounded variation and λ-absolutely continuous, there exists a λ-Bochner integrable function $g : \Omega \to X$ such that $\mu(E) = \int_E g d\lambda$ for every E in Σ.*

It is easy to prove that to check for the RNP one need consider only the case when $\lambda(\Omega) < \infty$, and $\|\mu(E)\| \leq \lambda(E)$ for all E in Σ. In that case, any derivative $g : \Omega \to X$ must be bounded. Thus,

DEFINITION 1′: X *has the RNP if and only if for every finite measure space (Ω, Σ,λ), either of the following equivalent conditions hold.*

(1) *For every additive $\mu : \Sigma \to X$ with $\|\mu(\cdot)\| \leq \lambda(\cdot)$, there exists a bounded measurable function $g : \Omega \to X$ such that $\mu(E) = \int_E g d\lambda$ for every E in Σ.*

---

*On leave of absence from the Department of Mathematics, Pennsylvania State University, University Park, Pa., 16802, USA. The author is grateful to the Alexander von Humboldt-Stiftung for support during his stay in Germany.

(2) *For every continuous linear operator* $T : L^1(\lambda) \to X$ *there exists a bounded measurable function* $g : \Omega \to X$ *such that* $T(f) = \int fg d\lambda$ *for every* f *in* $L^1(\lambda)$.

We shall see below that we can even reduce the criterion to the case $(\Omega,\Sigma,\lambda) =$ ([0,1], Borel sets, Lebesgue measure).

Some classical results are

(1) c_0 and $L^1[0,1]$ fail to have the RNP,

(2) (Phillips) reflexive spaces have the RNP [32],

(3) (Dunford-Pettis) separable dual spaces have the RNP [13].

Consider now a natural way of trying to find a derivative of a measure $\mu : \Sigma \to X$. Given a finite partition $\pi = (E_i)_{i=1}^n$ of Ω, let

$$g = \sum_{i=1}^{n} \frac{\mu(E_i)}{\lambda(E_i)} \chi_{E_i}.$$

Then g_π is a simple function such that $\int_E g_\pi d\lambda = \mu(E)$ for every E in the algebra generated by π. One would hope to find a derivative g for μ by taking some sort of limit of the g_π's as π ranges thru the net of all partitions. With some work on this idea, one obtains the following analytic characterization of the RNP.

THEOREM (Rønnow [36]). X *has the* RNP *if and only if for every sequential martingale* $(f_n)_{n=1}^\infty$ *of uniformly bounded* X-*valued simple functions*,

$$(*) \qquad \int \|f_n - f_m\| d\lambda \to 0 \quad \text{as} \quad m,n \to \infty.$$

I have chosen here a weak sufficient condition. (Using geometric considerations, we shall see below that (*) can be replaced by: for some t, $\{f_n(t)-f_m(t)\}_{m\neq n}$ has zero as a cluster point.) In the other direction, if X has the RNP, then very strong martingale convergence theorems hold. See Chatterji [7].

Two easy corollaries of the above characterization are

(1) *if* X *has the* RNP, *then so does every closed subspace of* X,

and (2) *if every closed separable subspace of* X *has the* RNP, *then so does* X.

The importance of (2) was first pointed out by Uhl [40]. Note that (2) combined with the Dunford-Pettis result gives

(3) *if every separable subspace of* X *is isomorphic to a subspace of a*

separable dual space, then X *has the* RNP.

Probably the most interesting open question concerning the RNP is that of Uhl: *does the converse of* (3) *hold*? One also obtains from (3) the widest general class of spaces known to have the RNP:

(4) (Kuo [23]). *If* $X \subset Y \subset Z$, Y *a dual space, and* Z *a weakly compactly generated* (WCG) *space, then* X *has the* RNP.

For a clever proof of (4) due to P. Morris, see [11, p. 36]. In connection with (4),

(i) ℓ^1 contains a subspace which has the RNP but is not isomorphic to a dual space [25], so that the first containment in (4) is useful,

(ii) there exists a dual space Y which is contained in a WCG space, but which is not itself WCG [37], so the second containment in (4) is useful, and

(iii) $\ell^1(\Gamma)$, for uncountable Γ, has the RNP (by (3)), but is not contained in any WCG space. Thus (4) does not give all dual spaces with the RNP.

We now turn to geometry and the RNP. A little history is of interest. When teaching a real variables class at Berkley in the mid 1960's, M.A. Rieffel decided to do vector-valued integration (i.e., the Bochner integral) rather than the simple Lebesgue integral. After all, the two theories are essentially the same; in one case one uses single bars $|\cdot|$, while in the other one uses double bars $\|\cdot\|$. Everything went fine in the class until they considered the Radon-Nikodým theorem. In an attempt to fit this result into his lectures, Rieffel [35] proved the following result.

THEOREM (Rieffel). *Call a bounded set* $A \subset X$ *non-dentable if there exists* $\varepsilon > 0$ *such that* $A \subset \overline{co}(A \backslash B_\varepsilon)$ *for every ball* B_ε *of radius* ε. ($\overline{co}$ = closed convex hull.)

If X *contains no non-dentable bounded sets, then* X *has the* RNP.

(It should be pointed out that this was not the first time that a sufficient geometric condition was given for the RNP. J.A. Clarkson [8] introduced the notion of uniform convexity in 1936 precisely as a condition which implies the RNP. Of course it was soon shown that uniformly convex spaces are reflexive and that all reflexive spaces have the RNP. Not all reflexive spaces are isomorphic to uniformly convex spaces, so Clarkson's condition is far from a necessary condition for the RNP. It is interesting however to note that it has recently been shown that X is isomorphic to a uniformly convex space iff X is "super-reflexive" iff X is "super-RNP" [33].)

It was not much beyond this stage of development that Joe Diestel, in a seminar at Kent State University, made the outlandish conjecture that the RNP is equivalent to the Krein-Milman property (KMP)! (X has the KMP provided every closed bounded convex set $K \subset X$ has an extreme point; equivalently, is the closed convex hull of its extreme points [26].) At this point, virtually all that was known about the KMP was that c_0 and $L^1[0,1]$ fail to have it, reflexive spaces and separable dual spaces have it [2]. As for the relationship between dentability and extreme points, examples were known of dentable sets without extreme points and of non-dentable sets which were the closed convex hulls of their extreme points [14]!

Another outlandish Diestel conjecture was that the RNP was equivalent to "Lindenstrauss' Property A" which we shall discuss below.

Much of the work in the past few years on the RNP has been motivated by these and other conjectures by Diestel. The two mentioned above remain open today, but have been proved to be true within a hair. They are discussed more fully below.

Joe Diestel deserves much credit as a prime moving force in research on the RNP, both for his own results and for his motivation of others.

SPACES WITHOUT THE RNP

We now will see how bad spaces are which fail the RNP. Thus assume

(B1) X *fails the* RNP.

Then by Rieffel's theorem

(B2) *there exists a bounded non-dentable subset of* X.

It is easy to see that a bounded set A is non-dentable iff there exists $\delta > 0$ such that for every $f \in X^*$ and for every $\alpha < \sup f(A)$, the set

$$S(A,f,\alpha) = \{x \in A : f(x) > \alpha\}$$

has diameter $\geqq \delta$. (Such a set $S(A,f,\alpha)$ is called a "slice" of A.) Since $\sup f(A) = \sup f(\overline{co}(A))$ and $S(A,f,\alpha) \subset S(\overline{co}(A),f,\alpha)$, it follows that $\overline{co}(A)$ is also non-dentable. Thus (B2) implies

(B3) *there exists a closed bounded convex non-dentable subset of* X.

Completely elementary methods can be used to prove that if A is non-dentable, then so is A+B for any set B (see [18]). In particular, A-A is non-dentable, and so is

$$B = \overline{co}[(A-A)+(\text{open unit ball of } X)].$$

This set B is the unit ball for a norm on X which is equivalent to the original norm on X. Thus [9], (B3) implies

(B4) *there exists an equivalent norm for* X, *whose open and closed balls are non-dentable*.

A result which is slightly more difficult than the above is the following. For a set A, let

$$B_\varepsilon(A) = A+\varepsilon\cdot(\text{closed unit ball of } X).$$

Let A^0 denote the interior of A.

THEOREM [21]. *If* K *is a closed bounded convex set which is non-dentable, then there exists* $\varepsilon > 0$ *such that*

$$K = \overline{co}(K\backslash B_\varepsilon(F))$$

and

$$K^0 = co(K^0\backslash B_\varepsilon(F)) \quad (co = \text{convex hull})$$

for every finite set $F \subset X$.

Using this, (B4) implies

(B5) *there exists a bounded non-void convex set* C *and* $\varepsilon > 0$ *such that*

$$C = co(C\backslash B_\varepsilon(F))$$

for every finite set $F \subset X$.

Now start in C with any point, write it as a convex combination of points ε-away from it, write each of these points as a convex combination of points ε-away from all the points used so far, and proceed by induction to get a "bush"

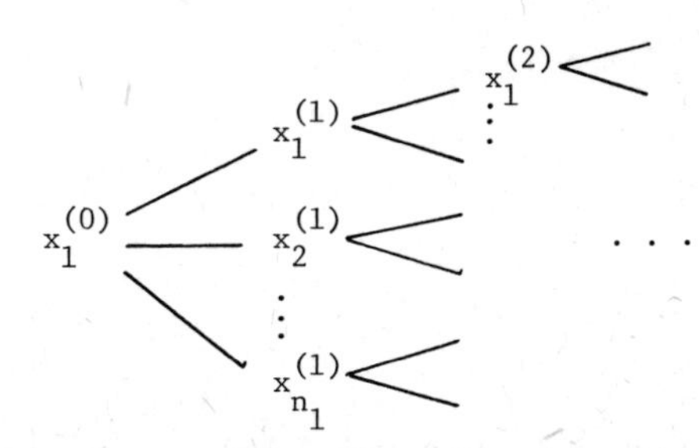

From this, one constructs a corresponding increasing sequence of partitions of [0,1] into intervals and a uniformly bounded martingale (f_n) of X-valued simple functions such that

$$\|f_n(t)-f_m(t)\| \geqq \varepsilon, \quad \text{all} \quad m \neq n, \quad \text{all} \quad t \quad \text{in} \quad [0,1]$$

(see [21]). Thus (B5) implies a rather dramatic failure of martingale convergence on (Ω,Σ,λ) = ([0,1], Borel sets, Lebesgue measure), and we see that the modification of Rønnow's theorem mentioned in the remark following its statement holds. We have now established the equivalence of (B1) thru (B5).

We now outline a construction from [21]. Suppose X fails the RNP, let Y, be a separable subspace without the RNP, and by renorming if necessary, assume that the open unit ball U of Y is non-dentable. Choose $\varepsilon > 0$ such that

$$U = co(U \setminus B_\varepsilon(F))$$

for every finite set $F \subset X$. Let $(y_n)_1^\infty$ be a dense sequence in U. Choose a sequence $F_1, F_2, \ldots$ of finite subsets of U such that

$$F_1 = \{y_1\}$$

$$\{y_n\} \cup F_n \subset co(F_{n+1})$$

$$dist(F_{n+1}, F_1 \cup \ldots \cup F_n) \geqq \varepsilon.$$

Let
$$A_n = F_n + (1-(n+1)^{-1})\frac{\varepsilon}{6}(\text{unit ball of } X),$$

and let
$$A = \bigcup_1^\infty A_n.$$

A is closed and A has no extreme points; i.e., every point of A is a convex combination of other points of A. Moreover, co(A) is dense in

$$W = U + \frac{\varepsilon}{6}\cdot(\text{open unit ball of } X)$$

which is the open unit ball for an equivalent norm for X. Note that if $0 \neq f \in X^*$, then

$$\sup f(A) = \sup f(co(A)) = \sup f(\overline{W})$$

and this cannot be attained on A. Thus in X,

(B6) <u>there exists a closed bounded set without extreme points,</u>

(B7) <u>there exists a closed bounded set contained in an open set</u> W <u>so that</u> $\overline{co}(A) = \overline{co}(W)$,

and (B8) <u>there exists a closed bounded set</u> A <u>such that the set</u>

$$P(A) = \{f \in X^* : f(x) = \sup f(A) \text{ for some } x \text{ in } A\}$$

<u>contains only the zero functional.</u>

We shall see below that these statements are also equivalent to (B1).

EXTREME POINTS

In this and the next section we assume

(G1) X <u>has the</u> RNP.

Soon after the characterization of the RNP in terms of dentability was established, Lindenstrauss gave the following easy proof that the RNP implies the KMP. Suppose (G1) holds and let K be a closed bounded convex set in X. Since K is dentable, there exist f in X^* and $\alpha < \sup f(K)$ such that $S(K,f,\alpha)$ has diameter < 1. By the Bishop-Phelps theorem (stated in the next section), there exists g in X^* which attains its supremum on K and is so close to f that

$$K_1 = \{x \in K : g(x) = \sup g(K)\} \subset S(K,f,\alpha).$$

Note that K_1 is extremal in K. Now repeat to get an extremal $K_2 \subset K_1$ with diam $K_2 < \frac{1}{2}$, and continue by induction. Then $\bigcap_1^\infty K_n$ is necessarily a singleton set whose only member is an extreme point of K.

Using considerably more tools, G.A. Edgar [15] proved that if X has the RNP, then for every separable closed bounded convex set K, the Choquet theorem holds. That is, for every point x in K there exists a probability measure μ supported on $\text{ext}(K)$ such that

$$(*) \qquad f(x) = \int_{\text{ext } K} f d\mu \, , \qquad \text{all } f \text{ in } X^*.$$

The idea of the proof was natural. If x is extreme we are done; if not write $x = \frac{1}{2}(y+z)$, $y \neq x \neq z$. If y and z are extreme we are done; if not write them as averages of other points. By using a selection theorem applied to the inverse of the "averaging map" from $(K \times K)\backslash(\text{diagonal})$ to $K\backslash(\text{ext } K)$ to perform the above for us, and by applying the martingale convergence theorem, one eventually finds x as an average of extreme points in the sense of (*). The execution of the idea was quite delicate; see [15]. (See also [16] and [28] for the non-separable case.)

The following questions remain open.

(a) _Does the KMP imply the RNP?_

(b) _Does the converse of Edgar's theorem hold?_

(c) _Does the KMP imply the conclusion of Edgar's theorem?_

Improving on Lindenstrauss' result, R.R. Phelps [31] was able to prove that (G1) implies

(G2) _Every closed bounded convex set in_ X _is the closed convex hull of its strongly exposed points._

(Recall that $x \in K$ is strongly exposed by $f \in X^*$ if $f(x) = \sup f(K)$ and

(#) $$x_n \in K, \quad f(x_n) \to f(x) \Rightarrow x_n \to x.$$

Note that (#) is equivalent to

$$\operatorname{diam} S(K,f,\alpha) \to 0 \quad \text{as} \quad \alpha \to \sup f(K).)$$

The proof of Phelps result is non-trivial, see [31], [4], or [18].)

If A is a closed set with $K = \overline{\text{co}}(A)$, then from (#) it is easy to see that every strongly exposed point of K must be in A. Hence (G2) implies

(G3) _Every closed bounded set_ $A \subset X$ _has extreme points; in fact,_

$$\overline{\text{co}}(A) = \overline{\text{co}}(A \cap \text{ext}\, \overline{\text{co}}(A)).$$

It follows that (B6), (B7), and (B8) are each equivalent to (B1), while (G1), (G2), and (G3) are mutually equivalent. Note that (G3) says that a Milman-type theorem holds in spaces with the RNP. (It should be pointed out that even for weakly closed bounded sets in spaces with the RNP, it is not always true that $A \supset \text{ext}(\overline{\text{co}}(A))$. (Let $(e_n)_1^\infty$ be the unit vector basis for ℓ^1 and let $A = \{e_n - e_{n+1}: n \in \mathbb{N}\}$. Then A is weakly closed and 0 is an extreme point of $\overline{\text{co}}(A)$.))

We say that X has the _strong_ - _KMP_ (SKMP) if (G3) holds. We have that SKMP $\Leftrightarrow$ RNP, and the question of equivalence of the KMP and the RNP is now the question: does KMP $\Rightarrow$ SKMP? We remark that in the space c of convergent real sequences with the supremum norm, the closed unit ball B is the closed convex hull of its extreme points while there exists a closed set $A \subset B^0$ with no extreme points and $\overline{co}(A) = B$ [21].

Finally, the RNP is separably determined, i.e., if every separable closed subspace of X has it, then so does X. It is open whether the KMP is separably determined or not.

PEAKING FUNCTIONALS AND OPERATORS

The characterization of spaces without the RNP given by (B8) is in contrast to the following well-known result.

THEOREM (Bishop-Phelps [3]). *For any Banach space* X, *for every closed bounded convex set* $K \subset X$, *the set*

$$P(K) = \{f \in X^* : f(x) = \sup f(K) \quad \text{for some} \quad x \in K\}$$

is norm dense in X^*.

Thus every Banach space is good in some sense, but RNP spaces are "very good": in [21] it is pointed out that Phelps proof of the equivalence of (G1) and (G2) contains implicitly a proof that (G1) implies

(G4) *for every closed bounded set* $A \subset X$, *the set of functionals which strongly expose* A *is a dense* G_δ *subset of* X^*.

In contrast to this is

THEOREM (Bourgain [6]). *If* A *is a separable closed bounded non-dentable set, then* $P(A)$ *is of* 1st *category in* X^*.

It is not known if separability can be removed from this result. However, we have immediately that (G1) is equivalent to

(G5) *For every closed bounded convex set* $A \subset X^*$, $P(A)$ *is of second category in* X^*.

(This equivalence is due independently to Bourgain [6] and to C. Stegall [39].)

When Bishop and Phelps proved their result stated above, they asked about possible extensions to more general operators than functionals.

To illustrate the state of ignorance here, the following question [22] is open. *Is it true that for every Banach space* X, *the set*

$$P(X, \mathbb{R}^2) = \{T \in L(X, \mathbb{R}^2) : \|Tx\| = \|T\| \quad \text{for some} \quad \|x\| = 1\}$$

is norm dense in $L(X,\mathbb{R}^2)$, *when* $\mathbb{R}^2$ *has the Euclidean norm?*

Some work has been done however on the general question [24], [22]. In [24], Lindenstrauss considered spaces with the following property (called *Property* A): for every Banach space Y the set $P(X,Y)$ is norm dense in $L(X,Y)$. Lindenstrauss showed that reflexive spaces have Property A. More recently, Bourgain [4] showed that the RNP implies Property A; i.e., (G1) implies

(G6) *Every space isomorphic to* X *has Property* A.

Building on a construction of Bourgain, one can show that (G6) is in fact equivalent to (G1) (see [19]). It is an open question whether or not Property A is invariant under isomorphisms.

Actually, for spaces with the RNP, Bourgain proved much more than (G6). Say that $T : X \to Y$ *strongly absolutely exposes* a bounded set $A \subset X$ if and only if there exists x in A with $\|Tx\| = \sup\|T(A)\|$, and such that if $x_n \in A$ and $\|Tx_n\| \to \|Tx\|$, then for some subsequence x_{n_k} we have either $x_{n_k} \to x$ or $x_{n_k} \to -x$. Then (G1) is equivalent to

(G7) *For every space* Y *and every closed bounded set* $A \subset X$, *the set of all* $T : X \to Y$ *which strongly absolutely expose* A *is a dense* G_δ *subset of* $L(X,Y)$.

DUAL SPACES

For Banach spaces which are dual spaces, many of the open problems mentioned above have been settled. The fundamental result here is a deep construction of C. Stegall [38]. Using it, Stegall showed that the statement

(D1) X^* *has the* RNP

is equivalent to each of the following.

(D2) *Every separable subspace* Y *of* X *has a separable dual*.

(D3) *Every separable subspace* Z *of* X^* *is isomorphic to a subspace of a separable dual space*.

Building on the Stegall construction, P. Morris and I were able to show [20] that (D1) is equivalent to

(D4) X^* *has the* KMP.

In [1] E. Asplund introduced and studied spaces he called *strong differentiability spaces*; they are now called *Asplund spaces* [30]. The definition is: X is an Asplund space if for every open convex set $U \subset X$, every continuous convex functional $\varphi : U \to \mathbb{R}$ is Fréchet differentiable at a dense G_δ set of points in U. It follows from Asplund's results and Rieffel's theorem that (D1) is implied by

(D5) X _is an Asplund space_.

Namioka and Phelps [30] showed that (D5) is equivalent to

(D6) _For every closed bounded set_ $A \subset X^*$, _the set_ $\{x \in X :$ _the evaluation functional_ $\hat{x}$ _strongly exposes_ $A\}$ _is a dense_ G_δ _subset of_ X.

Building on other results of Namioka and Phelps, Stegall [39] has shown that (D6) is equivalent to (D1). There is a rather easy corollary, the proof of which we omit, which says that if X _has an equivalent norm which is Fréchet differentiable everywhere except at zero, then_ (D1) _holds_. It is an open question whether or not the converse of this corollary holds.

SUMMARY AND ADDITIONAL REMARKS.

In summary, (B1) thru (B8) are mutually equivalent and indicate how far from finite dimensional the geometry of spaces without the RNP is; (G1) thru (G7) are equivalent and indicate how nicely the geometry behaves in spaces with the RNP; and (D1) thru (D6) are equivalent and indicate how much more is known for dual spaces.

Much recent work, primarily by C. Stegall and J. Bourgain (see [34]) has been devoted to _sets_ with the RNP (the definition of which can be taken to be that every bounded subset is dentable.) Many of the above equivalences have been shown to remain equivalent in this more general setting, and the proofs developed have often been even more elegant than the originals.

In this brief survey I have not attempted to always state results in their fullest generality, and I have not been able to give proper credit to all who have contributed to these results. Deserving of special mention is H. Maynard [29] who gave the first completely geometric characterization of the RNP, and thus initiated this area of research. To others whose work I have slighted, I apologize.

For further information, I refer to the references below, especially to the survey paper [11] and the text [12] of Diestel and Uhl.

REFERENCES

1. E. Asplund, Fréchet differentiability of convex functions, Acta Math. _121_ (1968), 31-49.
2. C. Bessaga and A. Pełczyński, On extreme points in separable conjugate spaces, Israel J. Math. _4_ (1966), 262-264.
3. E. Bishop and R.R. Phelps, The support functionals of a convex set, AMS Proc. of Symp. in Pure Math., Vol. VII, _Convexity_ (1963).

4. J. Bourgain, On dentability and the Bishop-Phelps property, Israel J. Math. (to appear).
5. __________, On strongly exposing functionals of convex sets, (preprint).
6. __________, Private communication.
7. S.D. Chatterji, Martingale convergence and the Radon-Nikodým theorem in Banach spaces, Math. Scand. 22(1968), 21-41.
8. J.A. Clarkson, Uniformly convex spaces, Trans. Amer. Math. Soc. 40(1936), 396-414.
9. W.J. Davis and R.R. Phelps, The Radon-Nikodým property and dentable sets in Banach spaces, Proc. Amer. Math. Soc. 45(1974), 119-122.
10. J. Diestel, Geometry of Banach spaces-selected topics, Lecture Notes in Math. #485, Springer-Verlag(1975).
11. __________ and J.J. Uhl, Jr., The Radon-Nikodým theorem for Banach space valued measures, Rocky Mountain J. Math. 6(1976), 1-46.
12. __________ and __________, The Theory of Vector Measures, Amer. Math. Soc. Surveys (to appear).
13. N. Dunford and B.J. Pettis, Linear operations on summable functions, Trans. Amer. Math. Soc. 47(1940), 323-392.
14. M. Edelstein, Concerning dentability, Pacific J. Math. 46(1973), 111-114.
15. G.A. Edgar, A non-compact Choquet theorem, Proc. Amer. Math. Soc. 48(1975), 354-358.
16. __________, Extremal integral representations, J. Functional Analysis 23 (1976), 145-161.
17. R.E. Huff, Dentability and the Radon-Nikodým property, Duke Math. J. 41(1974), 111-114.
18. __________, The Radon-Nikodým property for Banach spaces, Lecture Notes in Math #541, Measure Theory (Oberwolfach 1975), Springer-Verlag(1976), 229-242.
19. __________, On non-density of norm-attaining operators, (preprint).
20. __________ and P.D. Morris, Dual spaces with the Krein-Milman property have the Radon-Nikodým property, Proc. Amer. Math. Soc. 49(1975), 104-108.
21. __________ and __________, Geometric characterizations of the Radon-Nikodým property, Studia Math. 61(1976), 157-164.
22. J.A. Johnson and J. Wolfe, Norm attaining operators, Studia Math. (to appear).
23. T.-H. Kuo, On conjugate Banach spaces with the Radon-Nikodým property, Pacific J. Math. 59(1975), 497-503.
24. J. Lindenstrauss, On operators which attain their norm, Israel J. Math. 1(1963), 139-148.
25. __________, On certain subspaces of ℓ^1, Bull. Acad. Polon. Sci. 12 (1964), 539-542.

26. ______________, On extreme points in ℓ^1, Israel J. Math. 4(1966), 59-61.

27. P. Mankiewicz, On isometries in linear metric spaces, Studia Math. 55 (1976), 163-173.

28. ____________, A remark on Edgar's extremal integral representation theorem, (preprint).

29. H. Maynard, A geometric characterization of Banach spaces possessing the Radon-Nikodým property, Trans. Amer. Math. Soc. 185(1973), 493-500.

30. I. Namioka and R.R. Phelps, Banach spaces which are Asplund spaces, Duke Math. J. 42(1975), 735-750.

31. R.R. Phelps, Dentability and extreme points in Banach spaces, J. Functional Analysis 16(1974), 78-90.

32. R.S. Phillips, On weakly compact subsets of a Banach space, Amer. J. Math. 65(1943), 108-136.

33. G. Pisier, Martingales with values in uniformly convex spaces, Israel J. Math. 20(1975), 326-350.

34. J. Rainwater seminar notes, University of Washington, Seattle, Washington 1976-77.

35. M.A. Rieffel, Dentable subsets of Banach spaces, with applications to a Radon-Nikodým theorem, Proc. Conf. Functional Analysis, Thompson Book Co., Washington, D.C. (1967), 71-77.

36. U. Rønnow, On integral representation of vector-valued measures, Math. Scand. 21(1967), 45-53.

37. H.P. Rosenthal, The heredity problem for weakly compactly generated Banach spaces, Composito Math. 28(1974), 83-111.

38. C. Stegall, The Radon-Nikodým property in conjugate Banach spaces, Trans. Amer. Math. Soc. 206(1975), 213-223.

39. __________, Private communication.

40. J.J. Uhl, Jr., A note on the Radon-Nikodým property for Banach spaces, Rev. Roum. Math. 17(1972), 113-115.

41. ____________, Norm attaining operators in $L^1[0,1]$ and the Radon-Nikodým property, Pacific J. Math. 63(1976), 293-300.

K.-D. Bierstedt, B. Fuchssteiner (eds.)
Functional Analysis: Surveys and Recent Results
© North-Holland Publishing Company (1977)

SEPARABLE LINDENSTRAUSS SPACES

Wolfgang Lusky
Fachbereich Mathematik
Gesamthochschule Paderborn

One of the central theorems of functional analysis is the

THEOREM OF HAHN - BANACH :

Let $Y \subset W$ *be Banach spaces and let* $T : Y \to \mathbb{R}$ *be linear. Then there is a linear extension* $\tilde{T} : W \to \mathbb{R}$ *of* T *with* $\| \tilde{T} \| = \| T \|$.

It seems natural to replace $\mathbb{R}$ in the above theorem by some Banach space X and to ask which Banach spaces X can be characterized in terms of extensions of linear bounded operators. There are various ways to generalize the above Hahn - Banach condition.
Let us restrict ourselves in this article to real Banach spaces.

THEOREM 1 ([3], [9]) :

Let X *be a Banach space. Then the following are equivalent:*

(i) *If* Y *and* Z *are Banach spaces,* $Z \supset Y$, *and* $T : Y \to X^{**}$ *is a bounded linear operator, then there is a linear extension* $\tilde{T} : Z \to X^{**}$ *of* T *with* $\| \tilde{T} \| = \| T \|$.

(ii) $X^{**} \cong C(K)$, *the space of all continuous functions on a compact Hausdorff space.*

(iii) $X^* \cong L_1(\mu)$ *for some measure space* (Ω, Σ, μ).

(iv) *If* E *and* F *are finite dimensional Banach spaces,* $F \supset E$, $T : E \to X$ *is linear and* $\varepsilon > 0$, *then there is a linear extension* $\tilde{T} : F \to X$ *of* T *with* $\| \tilde{T} \| \le (1 + \varepsilon) \| T \|$.

Banach spaces X satisfying one of the above conditions are called <u>Lindenstrauss spaces</u>. Hence the class of Lindenstrauss spaces consists just of the preduals of L_1. It is somewhat surprising that this class contains Banach spaces with very different structures while the preduals of all other L_p - spaces, $1 < p \le \infty$, are always uniquely determined.

Examples of Lindenstrauss spaces ([8], [11]) :

1.) $C(K)$ - spaces, K compact Hausdorff space, which can be characterized as follows:
A Banach space X *is isometrically isomorphic to* $C(K)$ *if and only if* X *is a Lindenstrauss space, its closed unit ball has at least one extreme point and the set of all extreme points of the closed dual unit ball of* X *is* w*-*compact.*

2.) M - spaces, i.e. sublattices of $C(K)$. An M - space X turns out to be just a subspace of $C(K)$ for which there exists an index set A and triples $(k_\alpha^1, k_\alpha^2, \lambda_\alpha) \in K \times K \times \mathbb{R}_+$, $\alpha \in A$, so that X consists of the functions $f \in C(K)$ with $f(k_\alpha^1) = \lambda_\alpha f(k_\alpha^2)$, $\alpha \in A$.

3.) G - spaces, which are those subspaces of all functions $f \in C(K)$ satisfying $f(k_\alpha^1) = \lambda_\alpha f(k_\alpha^2)$ for a set of triples $(k_\alpha^1, k_\alpha^2, \lambda_\alpha) \in K \times K \times \mathbb{R}$.
G-spaces are generalizations of M-spaces which can be inferred also from the following:
A subspace X *of* $C(K)$ *is a* G - *space if and only if for any* $f_1, f_2, f_3 \in X$ *the function* $\max(f_1, f_2, f_3) + \min(f_1, f_2, f_3)$ *is in* X, *too.*

4.) $C_\delta(K)$ - and $C_\Sigma(K)$ - spaces, K compact:
$X \subset C(K)$ is said to be a $C_\delta(K)$ - space if there is an involutive homeomorphism $\delta : K \to K$ so that X consists of all $f \in C(K)$ with $f(k) = -f(\delta k)$, $k \in K$.
The definition for $C_\Sigma(K)$ - spaces is the same but we assume in addition that the homeomorphism here is fixpointfree.

5.) $A(S)$ - spaces, i.e. all continuous affine functions on a compact Choquet simplex S.
A Banach space X *is an* $A(S)$ - *space,* S *simplex, if and only if* X *is a Lindenstrauss space and its closed unit ball has at least one extreme point.*

We obtain the diagram:

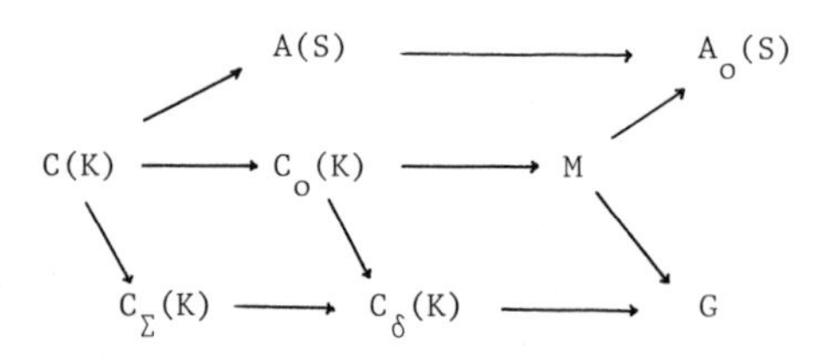

where $A_o(S) = \{ f \in A(S) \mid f(s_o) = 0 \}$, S simplex, s_o some extreme point of S and $C_o(K) = \{ f \in C(K) \mid f(k_o) = 0 \}$, k_o some element of K.
In this diagram the notation stands for classes; for example $C(K) \to A(S)$ means that the class of $C(K)$ - spaces is a subclass of all $A(S)$ - spaces.
In view of the preceding examples it is natural to ask if there are Lindenstrauss spaces which are essentially different from the just mentioned ones. It seems interesting to obtain more insight in the geometric nature of these spaces; in particular, a representation of a general Lindenstrauss space as a space of well behaved functions should be useful. Furthermore, since the duals of these spaces look like duals of $C(K)$, the question, how closely Lindenstrauss spaces are related to $C(K)$ - spaces, appears to be of some importance.
In the following we shall discuss these questions. However, later, we shall have to restrict ourselves to the separable case in order to obtain a satisfactory description of Lindenstrauss spaces since our main tool (Theorem 3) depends on the assumption of separability.
The next theorem, which is crucial for our following considerations, relates Lindenstrauss spaces to the theory of Choquet simplices and extends the well known Edwards separation theorem for simplices to dual unit balls of Lindenstrauss spaces. (We shall denote the closed unit ball of any Banach space Y by B(Y).)

THEOREM 2 ([8]) :

Let X *be a Lindenstrauss space and let* $g : B(X^*) \to (-\infty, \infty]$ *be concave and lower* w^* *- semicontinuous such that* $g(x^*) + g(-x^*) \geq 0$ *for all* $x^* \in B(X^*)$.
Then there is an affine, symmetric w^* *- continuous map* $f : B(X^*) \to \mathbb{R}$ *with* $-g(-x^*) \leq f(x^*) \leq g(x^*)$ *for all* $x^* \in B(X^*)$.

One can show that Theorem 2 implies the following

THEOREM 3 ([8]) :

Let X *be a separable Banach space. Then* X *is a Lindenstrauss space if and only if*

$$X = \overline{\bigcup_{n \in \mathbb{N}} E_n} \quad ; \; E_n \subset E_{n+1} \cong l_\infty^{n+1} \quad \textit{for all} \;\; n \in \mathbb{N}.$$

(By l_∞^n we mean $C(\{1,\ldots,n\})$.)

Thus Theorem 3 suggests to study isometric embeddings from l_∞^n into l_∞^{n+1} since apparently the individual character of any given separable Lindenstrauss space depends on them. For this purpose we introduce the notion of <u>admissible sets</u>, that means elements $e_1,\dots,e_n$ of a Banach space with

$$\left\| \sum_{i=1}^{n} \lambda_i e_i \right\| = \max_{i\le n} |\lambda_i| \quad \text{for all } \lambda_i \in \mathbb{R}.$$

Hence e_i can be identified with $(0,\dots,0,1,0,\dots,0)$ in l_∞^n .

THEOREM 4 ([16]) :

If l_∞^n *is embedded isometrically in* l_∞^{n+1} *and* $\{ e_{i,n} \mid i \le n \}$ *is an admissible basis of* l_∞^n *then there is an admissible basis* $\{ e_{i,n+1} \mid i \le n+1\}$ *of* l_∞^{n+1} *and there are real numbers* $a_{i,n}$; $i=1,\dots,n$; *such that*

$$(*) \quad \sum_{i=1}^{n} |a_{i,n}| \le 1$$

$$(**) \quad e_{i,n} = e_{i,n+1} + a_{i,n} e_{n+1,n+1} \quad \textit{for all } i.$$

Conversely, any set of n *real numbers,* $a_{i,n}$; $i=1,\dots,n$; *such that* (*) *holds, determines by* (**) *an isometric embedding of* l_∞^n *in* l_∞^{n+1} .

Hence all information about separable Lindenstrauss spaces is contained in an infinite triangular substochastic matrix $(a_{i,n})$ whose columns satisfy (*).

<u>Examples</u>:

1.) $A = (0)$ repesents c_o in the previously described manner.

(c_o = space of all sequences tending to 0 endowed with the sup-norm)

2.) The unit matrix, whose diagonal entries all are one and the entries off the diagonal are zero, represents c, the space of all converging sequences.

3.) More generally, for any C(K), where K is compact and totally disconnected, there is a representing matrix whose entries attain only zero and one ([8]).

In particular, all triangular matrices, in which every row contains zero and one infinitely many often and one appears in all columns, represent

$C(\Delta)$, Δ the Cantor set ([12]). For instance:

$$\begin{pmatrix} 1 & 0 & 1 & 0 & 0 & 1 & 0 & 0 & 0 & 1 & 0 & \dots & & \\ & 1 & 0 & 1 & 0 & 0 & 1 & 0 & 0 & 0 & 1 & 0 & \dots & \\ & & 0 & 0 & 1 & 0 & 0 & 1 & 0 & 0 & 0 & \dots & & \\ & & & 0 & 0 & 0 & 0 & 0 & 1 & 0 & 0 & \dots & & \\ & & & & \ddots & \ddots & \ddots & \ddots & \ddots & & & & & \end{pmatrix}$$

The matrix representation of a given Lindenstrauss space is not unique. But there are only few examples of Lindenstrauss spaces (including c_0 and c) where explicit descriptions of the set of all representing matrices are known ([10] , [12]).

$A(S)$ and $A_0(S)$ can be easily characterized as follows:

THEOREM 5 ([8]) :

Let X *be a separable Lindenstrauss space.*

(i) X *is an* $A_0(S)$ *- space if and only if there is a representing matrix* $(a_{i,n})$ *with* $a_{i,n} \geq 0$ *for all* $i, n \in \mathbb{N}$.

(ii) X *is an* $A(S)$ *- space if and only if there is a representing matrix* $(a_{i,n})$ *with* $\sum_{i=1}^{n} a_{i,n} = 1$ *for all* n.

Theorem 5 yields an answer to the previously mentioned question of a representation of a general Lindenstrauss space under the additional assumption of separability. Indeed, any infinite triangular substochastic matrix B can be changed to a matrix A by taking the moduli of all elements of B and inserting zeros between these elements. This can be done in such a way that the Lindenstrauss space, represented by B, is complemented in the newly constructed Lindenstrauss space, represented by A. The latter one is of type $A_0(S)$ by Theorem 5 (i). More precisely we obtain:

THEOREM 6 ([8]) :

Let X *be a separable Banach space.*
X *is a Lindenstrauss space if and only if* $X \subset A(S)$ *for some simplex* S *such that there is a contractive projection* $P : A(S) \to X$.

One might ask if it is possible to deduce simple conditions which are necessary and sufficient for a substochastic triangular matrix to represent an arbitrary $C(K)$ - space. In this connection it seems tempting to search for conditions which reflect special topological properties of K in the elements of such a matrix. This turns out to be considerably difficult except in the already mentioned case of $C(K)$ with totally disconnected K. Only partial and technically complicated results have been given up to now to clarify the connection between the topological structure of K and matrices representing $C(K)$ ([5]) .

The main advantage of the technique of representing matrices seems to be that we have a tool for easily constructing new Lindenstrauss spaces which was used already in the remark preceding Theorem 6 and which will become crucial for the following.

THEOREM 7 ([8] , [20]) :

(i) c_o *is "minimal" in the class of separable Lindenstrauss spaces in the following sense:*

If X *is separable,* $X^* \cong L_1(\mu)$, *then there is a subspace* Y *of* X, *which is isometrically isomorphic to* c_o , *together with a contractive projection* $P : X \to Y$.

(ii) $C(\Delta)$, Δ *the Cantor set, is "minimal" in the subclass of all separable Lindenstrauss spaces whose duals are non-separable. More precisely: If* X *is separable,* $X^* \cong L_1(\mu)$ *and* X^* *is non-separable, then there is a subspace* Y *of* X *with* $Y \cong C(\Delta)$ *and a contractive projection* $P : X \to Y$.

This suggests the existence of a "maximal" element in the class of all separable Lindenstrauss spaces. In other words, is there some separable Banach space G with $G^* \cong L_1(\mu)$ so that for every separable predual X of L_1 G contains a complemented subspace $\tilde{X}$ isometrically isomorphic to X?

In order to answer this question we consider the following conditions of substochastic triangular matrices $A = (a_{i,n})$:

(α) $(a_{1,n},\dots,a_{n,n},0,\dots)$; $n \in \mathbb{N}$; are dense in the unit ball of l_1 , the space of all absolutely summable sequences with the usual summation norm.

(β) $(a_{1,n},\dots,a_{n,n},0,\dots)$; $n \in \mathbb{N}$; are dense in $\{(b_1,\dots,b_n,\dots) \in l_1 \mid b_i \geq 0;\ i \in \mathbb{N};\ \sum_{i=1}^{\infty} b_i \leq 1\}$.

(γ) $(a_{1,n},\ldots,a_{n,n}0,\ldots)$; $n \in \mathbb{N}$; are dense in

$$\{(c_1,\ldots,c_n,\ldots) \in l_1 \mid c_i \geq 0 ; i \in \mathbb{N} ; \sum_{i=1}^{\infty} c_i = 1\}.$$

It is not difficult to show that the Lindenstrauss spaces represented by matrices of type (β) or (γ) are just $A_o(S)$ or $A(S)$, respectively, where S is a Poulsen simplex, i.e. a metrizable simplex whose extreme points ex S are dense in all S. (We thus can show by means of matrices the existence of such simplices). Furthermore a matrix of type (α) determines a Lindenstrauss space G with the following property:

For arbitrary finite dimensional Banach spaces $E \subset F$, $\varepsilon > 0$, *and any linear isometric operator* $T : E \to G$ *there is a linear extension* $\tilde{T} : F \to G$ *of* T *with*

$$(1-\varepsilon)\,\|x\| \leq \|\tilde{T}(x)\| \leq (1+\varepsilon)\,\|x\| \quad \textit{for all } x \in F.$$

Notice that hence G "possesses an almost isometric copy of every finite dimen= sional Banach space at almost every place in G", a very extensive property for a separable Banach space; but nevertheless the existence of such a space can be realized by exploiting matrices of type (α). The proof that these matrices represent such spaces reduces the above condition to the case where $E \cong l_\infty^n$; $F \cong l_\infty^{n+m}$. Then the embedding of E in F can be expressed by elements of m columns of a substochastic triangular matrix. These columns appear "almost exactly" infinitely many often in a matrix of type (α). Hence $T(E)$ is contained in a subspace $\tilde{F}$ of G which is an almost exact copy of F.
Spaces like G are called <u>Gurarij spaces</u>. The first proof for the existence of a separable Banach space having the above property, using a different approach, appeared in [4].

PROPOSITION 8 ([12] , [19]) :

For every separable Lindenstrauss space X *there is a Poulsen simplex* S *with* $A(S) \supset X$, *a Gurarij space* $G \supset X$ *and contractive projections* $P_1 : G \to X$, $P_2 : A(S) \to X$. *Furthermore* G *and* P_1 *can be chosen so that* $(\mathrm{id}-P_1)(G)$ *is a Gurarij space, too.*

PROOF:

Let the matrix A be of type (α), let B represent X, for instance

$$A = \begin{pmatrix} x & x & x & \ldots \\ & x & x & \ldots \\ & & x & \ldots \end{pmatrix} \qquad B = \begin{pmatrix} + & + & + & \ldots \\ & + & + & \ldots \\ & & + & \ldots \end{pmatrix}$$

We construct a new matrix

$$C = \begin{pmatrix} x & + & x & + & x & + & x & + & x & \cdots \\ & & 0 & x & 0 & x & 0 & x & 0 & x \cdots \\ & & & & 0 & + & x & + & x & + & x \cdots \\ & & & & & & 0 & 0 & 0 & x & 0 & x \cdots \\ & & & & & & & & 0 & + & 0 & + & x \cdots \\ & & & & & & & & & & 0 & 0 & 0 & 0 \cdots \\ & & & & & & & & & & & & 0 & + & 0 \cdots \\ & & & & & & & & & & & & & & 0 & 0 \cdots \end{pmatrix}$$

Again, the columns of C are dense in the unit ball of l_1, so C represents a Gurarij space, too. One can show that the special pattern of C yields the first assertion. The rest can be done similarly with Theorem 5 and 6.
A more careful exploitation of this technique yields the following dual statement of Proposition 8:

PROPOSITION 9 ([12]) :

For any compact metrizable simplex F *there is a Poulsen simplex* $S \supset F$ *such that* F *is a face of* S.

The last step to distinguish G and $A(S)$, S a Poulsen simplex, as maximal elements in the class of separable Lindenstrauss spaces (in the previously mentioned sense) is

THEOREM 10 ([10] , [13]) :

(i) *The (separable) Gurarij spaces are unique up to isometries.*
(ii) *The (metrizable) Poulsen simplices are unique up to affine homeomorphisms.*

Thus we obtain in addition that S is "maximal" in the class of all metrizable Choquet simplices, i.e. any metrizable simplex is affinely homeomorphic to a face of S. A consequence of Theorem 10 (ii) is furthermore that $A(S)$, S a Poulsen simplex, (and similarly $A_o(S)$) is unique up to isometries.
The proof of Theorem 10 is based on the following

LEMMA ([13] , [14]) :

(i) *Let* G *be a Gurarij space, let* $e_1,\ldots,e_n \in G$ *be admissible elements and consider* $r_1,\ldots,r_n \in \mathbb{R}$ *so that* $\sum_{i=1}^{n} |r_i| < 1$.

Then there is an extreme point Φ *of* $B(G^*)$ *with* $\Phi(e_i) = r_i$; $i=1,\ldots,n$.

(ii) *Let* S *be a Poulsen simplex; let* $e_1, \ldots e_n \in A_o(S)$ *be positive (in the natural pointwise order) admissible elements and consider real numbers*

$$0 < r_1, \ldots, r_n \quad \textit{with} \quad \sum_{i=1}^{n} r_i < 1.$$

Then there is an extreme point s *of* S *with* $e_i(s) = r_i$ *for all* i.

This implies that there are admissible elements $f_1, \ldots, f_{n+1} \in G$ satisfying the equation $e_i = f_i + r_i f_{n+1}$ for $i = 1, \ldots, n$; which enables us to show that every two Gurarij spaces G and $\tilde{G}$ admit a representation as in Theorem 3 so that the corresponding matrices coincide. The argumentation for $A_o(S)$ and $A_o(\tilde{S})$; S and $\tilde{S}$ Poulsen simplices; is similar which yields Theorem 10 (ii).

The above Lemma has some other consequences. It is not difficult to show that Lemma (i) is equivalent to

Let $E \cong l_\infty^n$; $F \cong l_\infty^{n+1}$; $E \subset F$, *so that* $\text{ex } B(E) \cap \text{ex } B(F) = \phi$. *Then any linear isometry* $T : E \to G$ *can be extended to an isometry* $\tilde{T} : F \to G$.

This strengthens the condition which defines a Gurarij space. On the other side the last assertion does not remain true any longer in general if B(E) and B(F) have a common extreme point ([13]). This shows that there is no separable Banach space where any linear isometry T from a finite dimensional Banach space E into G can be extended isometrically on any finite dimensional $F \supset E$.
Another consequence of Lemma (i) is that the extreme points of $B(G^*)$ are w^* - dense in the whole unit ball of G^* .
The assertion of the above Lemma suggests furthermore that there is much difference between C(K) - spaces and G and A(S) (S now always the Poulsen simplex), since nothing like the observation of the Lemma can be realized in any C(K) - space. Indeed, in [1] an example for a Banach space X was presented, whose dual is isometrically isomorphic to l_1, but X is not isomorphic to any complemented subspace of any C(K) - space. The maximality of G and A(S) immediately implies

THEOREM 11:

G *and* A(S) *are not isomorphic to any complemented subspace of any* C(K)*-space.*

On the other side G and A(S) are isomorphic to each other which follows from the maximality property ([12]).
Thus we have obtained Banach spaces which are opposite to the class of C(K) - spaces in some sense, although, surprisingly, their duals are isomorphic to the

dual of a $C(K)$ - space, namely $G^* \sim C(0,1)^*$. ([7]).
We can reformulate this assertion in terms of simplices:
The Poulsen simplex is opposite to the class of Bauer simplices, i.e. to those simplices F whose extreme points ex F are compact. (Recall $A(F) \tilde{=} C(\text{ex } F)$).
The difference between S and Bauer simplices is also reflected in different order structures: $C(K)$ is lattice ordered whereas $A(S)$ is an antilattice with respect to the natural order, i.e. $\max(a,b) \in A(S)$ for $a,b \in A(S)$ if and only if $a \leq b$ or $b \leq a$ ([10]) .

The preceding discussion indicates that it is of some relevance for the description of separable Lindenstrauss spaces as well as for metrizable Choquet simplices to know more facts about the geometric nature of G and the topological behaviour of S. We shall present some important topological properties of ex S and ex $B(G^*)$ as well as some assertions about rotations in G and $A(S)$ which reveal the intricate geometric structure of G and $A(S)$.

THEOREM 12 ([10]) :

Let S *be the Poulsen simplex.*

(i) *Any compact subset of* ex S *has empty interior (relative to* ex S).

(ii) *Every separable complete metric space is homeomorphic to a closed subset of* ex S.

(iii) *For each two homeomorphic compact subsets* K_1 *and* K_2 *of* ex S *there is an autohomeomorphism of* ex S *mapping* K_1 *onto* K_2.

(iv) ex S *is arcwise connected by simple arcs.*

PROOF:

(i) : If $K \subset \text{ex } S$ is compact and int $K \neq \phi$, then int $K = U \cap \text{ex } S$ for some open non-empty $U \subset S$. Hence int K is dense in U since ex S is dense in S. Thus the closed convex hull of K, F, would contain a non-empty open subset of S which is a contradiction because F is a face of S. (ii) is based on Proposition 9 and the fact that for any separable complete metric space H there is a metrizable simplex whose set of extreme points is homeomorphic to H.
(iii) is a consequence of the homogeneity property of S, given below, and (iv) follows immediately from (iii).

Theorem 12 can be strengthened considerably. Indeed, one can show that ex S as well as the extreme point set of the unit ball of G^* are homeomorphic to the pseudointerior of the Hilbert cube $[-1, 1]^{\mathbb{N}}$, that is the set

$\{ (x_1, x_2, \ldots) \mid |x_i| < 1 \,;\, i \in \mathbb{N} \}$ which is homeomorphic to the separable Hilbert space l_2 :

THEOREM 13 ([10]) :

(i) S *and* B(G*) *(with the restriction of the* w*- *topology) are homeomorphic to* $[-1, 1]^{\mathbb{N}}$.

(ii) ex S *and* ex B(G*) *are homeomorphic to the Hilbert space* l_2 .

There is an interesting rotation property of G which relates G to a problem about characterizations of separable Hilbert spaces posed by Mazur.
Mazur asked:

Let X *be a separable Banach space, so that for every* $x, y \in X$ *with* $\|x\| = \|y\| = 1$ *there exists a rotation* T *of* X *(i.e. a linear, isometric, surjective map) with* $T(x) = y$. *Is* X *then necessarily a Hilbert space?*

Mazur's problem for non-separable Banach spaces can be answered in the negative but the solution for the separable case still is lacking. However another consequence of the above Lemma is

THEOREM 14 ([13]) :

Let x *and* y *be smooth points of* G , *i.e.* $\|x\| = \|y\| = 1$ *and there is exactly one* $x^* \in B(G^*)$ *and one* $y^* \in B(G^*)$ *with* $x^*(x) = y^*(y) = 1$.
Then there is a rotation $T : G \to G$ *with* $T(x) = y$.

Notice that the set of smooth points of any separable Banach space X is dense in the unit sphere of X ([20]). Hence G has Mazur's rotation property on a dense subset without being a Hilbert space, indeed G is even not reflexive.
On the other side G cannot have this rotation property everywhere on the unit sphere, since the unit sphere of a separable Banach space, satisfying Mazur's property completely, consists evidently only of smooth points, but the extreme points of the unit ball of any l_∞^n - subspace of G, for instance, are not smooth.

THEOREM 15 ([14]) :

Let $f, g \in A(S)$ *and let* $0 \le f, g \le 1$ *so that* $f^{-1}(1), g^{-1}(1), f^{-1}(0), g^{-1}(0)$ *are singletons. Then there is a rotation* $T : A(S) \to A(S)$ *mapping* f *on* g.

Here one can show that the points f and g of the above statement are dense in $\{ h \in A(S) \mid 0 \le h \le 1 ; h^{-1}(1) \neq \phi \neq h^{-1}(0) \}$.

A more extensive property than that of Theorem 15 cannot be obtained in the case of $A(S)$. In particular, a rotation property as in Theorem 14 does not hold here. Indeed, the function which is 1 on S satisfies
$\max (\| 1 + f \| , \| 1 - f \|) = 1 + \| f \|$ for any $f \in A(S)$.
If $A(S)$ enjoyed a rotation property as G then for any element g of the unit sphere of $A(S)$ $\max(\| g + f \| , \| g - f \|) = 1 + \| f \|$ would hold for all $f \in A(S)$. This would imply that any element $g \in A(S)$ with $\| g \| = 1$ is an extreme point of $B(A(S))$ which is obviously false.

We want to conclude this article discussing some generalizations of the theory of separable Lindenstrauss space.
There are natural generalizations of the real theory to the complex case. Indeed, virtually everything remains true for complex Lindenstrauss spaces, i.e. complex Banach spaces whose duals are $L_1(\mu, \mathbb{C})$; although the proofs are no straightforward translations from the real case ([2], [6], [17]).

If we drop the assumption of separability then we cannot use Theorem 3 any more. It is unknown if for any Lindenstrauss space X there is a net T of subspaces $E \cong l_\infty^n$, directed by inclusion, such that $X = \overline{\bigcup_{E \in T} E}$. Thus the non-separable case is much more difficult to handle.
On the other side, it is possible, using different approaches, to introduce non-separable Gurarij spaces as well as non-metrizable Poulsen simplices. At least the assertions of Proposition 8 and Proposition 9 remain true in the non-separable case ([15]).

Very little is known about the isomorphic classification of Lindenstrauss spaces even in the separable case. There is an isomorphic version of Theorem 3:
A Banach space X is called a $\mathcal{L}_{\infty,\lambda}$ - space if there is a net T of subspaces E of X so that $\inf \{ \| T \| \, \| T^{-1} \| \mid T : E \to l_\infty^n$, isomorphic, $n = \dim E \} \le \lambda$, T is directed by inclusion and $X = \bigcup_{E \in T} E$.

It is not difficult to prove that if X is a $\mathcal{L}_{\infty,1+\varepsilon}$ - space for every $\varepsilon > 0$, then X is a Lindenstrauss space. One can describe $\mathcal{L}_{\infty,\lambda}$ - spaces by means of extensions of operators similar to those mentioned in Theorem 1. But it is still open what appears to be one of the main unsolved problems concerning this matter: Is every $\mathcal{L}_{\infty,\lambda}$ - space isomorphic to a Lindenstrauss space ?

REFERENCES

1. Y. Benyamini and J. Lindenstrauss (1972), A predual of l_1 which is not isomorphic to a C(K) - space, Israel J.Math. 13, 246 - 259
2. E. Effros (1974), On a class of complex Banach spaces, Illinois J. Math. 18, 48 - 59
3. A. Grothendieck (1955), Une caracterisation vectorielle métrique des espaces L^1, Canadian J. Math. 7, 552 - 561
4. V.I. Gurarij (1966), Space of universal disposition, isotopic spaces and the Mazur problem on rotations of Banach spaces, Sibirskii Mat. Zhurnal 7, 1002 - 1013
5. A.B. Hansen and Y. Sternfeld (1975), On the characterization of the dimension of a compact metric space K by representing matrices of C(K), Israel J. Math. 22, 148 - 167
6. B. Hirsberg and A. Lazar (1973), Complex Lindenstrauss spaces with extreme points, Trans. Amer. Soc. 459, 141 - 150
7. H.E. Lacey (1971), A note concerning $A^* = L_1(\mu)$, Proc. Amer. Math. Soc. 29, 525 - 528
8. A.J. Lazar and J. Lindenstrauss (1971), Banach spaces whose duals are L_1 spaces and their representing matrices, Acta Math. 126, 165 - 193
9. J. Lindenstrauss (1964), Extension of compact operators, Memoirs Amer.Math. Soc. 48
10. J. Lindenstrauss, G. Olsen and Y. Sternfeld, The Poulsen simplex, to appear in Ann. Inst. Fourier
11. J. Lindenstrauss and D. Wulbert (1969), On the classification of the Banach spaces whose duals are L_1-spaces, Journal of Funct. Anal. 4, 332 - 349
12. W. Lusky, On separable Lindenstrauss spaces, to appear in Journal of Funct. Anal.
13. W. Lusky (1976), The Gurarij spaces are unique, Arch. Math. 27, 627 - 635
14. W. Lusky, A note on the paper "The Poulsen simplex" of Lindenstrauss, Olsen and Sternfeld, to appear in Ann. Inst. Fourier
15. W. Lusky, On a construction of Lindenstrauss and Wulbert, Preprint
16. E. Michael and A. Pelczynski (1964), Separable Banach spaces which admit l_n^∞ - approximations, Israel J. Math. 4, 189 - 198
17. G.H. Olsen (1974), On the classification of complex Lindenstrauss spaces, Math. Scand. 35, 237 - 258
18. R.R. Phelps (1966), Lectures on Choquet's theorem, Van Nostrand, Princeton

19. P. Wojtaszczyk (1972), Some remarks on the Gurarij space, Studia Math. 47, 207 - 210

20. M. Zippin (1969), On some subspaces of Banach spaces whose duals are L_1-spaces Proc. Amer. Math. Soc. 23, 378 - 385

K.-D. Bierstedt, B. Fuchssteiner (eds.)
Functional Analysis: Surveys and Recent Results

AN APPLICATION OF M-STRUCTURE TO THEOREMS OF THE BANACH-STONE TYPE

Ehrhard Behrends
I. Mathematisches Institut
der Freien Universität Berlin

We investigate the structure of the M-ideals, the M-summands, and the centralizer of $C(K,X)$ (K a compact Hausdorff space, X a real Banach space, $C(K,X)$ the Banach space of continuous functions from K to X). For Banach spaces X, Y and compact Hausdorff spaces K, L, we get the following result as a corollary: If the centralizers of X and Y are one-dimensional (in particular, if X and Y have no nontrivial M-ideals), then the existence of an isometric isomorphism between $C(K,X)$ and $C(L,Y)$ implies that K and L are homeomorphic.

1. Introduction

The classical Banach-Stone theorem states that CK and CL are isometrically isomorphic iff K and L are homeomorphic (K, L compact Hausdorff spaces, CK and CL the respective Banach spaces of real-valued continuous functions).

The aim of this paper is to relate geometric properties of X (X a real Banach space) to the possibility of getting information on whether K and L are homeomorphic provided $C(K,X)$ and $C(L,X)$ are isometrically isomorphic (by $C(K,X)$ we mean the real Banach space of continuous functions from K to X; the norm of an element f of $C(K,X)$ is defined by $\|f\| := \sup_{k\in K} \|f(k)\|$).

1.1 Definition: Let X be a real Banach space.

(i) For $1\le p\le \infty$, a closed subspace J of X is called an L^p-summand, if there is a closed subspace $J^\perp$ in X such that algebraically $X = J \oplus J^\perp$, and $\|x+x^\perp\|^p = \|x\|^p + \|x^\perp\|^p$ (if $p = \infty$: $\|x+x^\perp\| = \max\{\|x\|, \|x^\perp\|\}$) for $x \in J$, $x^\perp \in J^\perp$. The (uniquely determined) space $J^\perp$ is called the L^p-summand complementary to J.

Projections onto L^p-summands with respect to these decompositions are called L^p-projections (for the basic properties of L^p-projections and L^p-summands see [B]). L^1-projections and L^1-summands (L^∞-projections and L^∞-summands) are called L-projections and L-summands (M-projections and M-summands), respectively.

(ii) A closed subspace J of X is called an M-ideal, if its annihilator in X' (the dual space of X) is an L-summand. The standard reference concerning properties of M-ideals is [AE].

(iii) Let T be a linear continuous operator from X to X. T is called M-bounded if there exists $\lambda > 0$ such that $Tx \in B$ for every closed (or every open) ball B with $\pm \lambda x \in B$. The set Z(X) of all M-bounded operators in X is called the centralizer of X. It is known that Z(X) is a commutative Banach algebra ([AE_2], theorem 4.8).

The main result of this paper is the characterization of the M-ideals (resp. M-summands) of C(K,X) by means of a type of lattice homomorphisms provided X contains only a finite number of M-ideals (resp. M-summands). We further describe the centralizer of C(K,X) if Z(X) is finite-dimensional. As a corollary we get the following theorem of the Banach-Stone type:

1.2 Theorem: Let X, Y be Banach spaces such that Z(X) and Z(Y) are finite-dimensional, n:= dim Z(X), m:= dim Z(Y). If K,L are compact Hausdorff spaces, then nK and mL are homeomorphic if C(K,X) and C(L,Y) are isometrically isomorphic. (nK is the topological sum, i. e. the disjoint union, of n copies of K.) In particular, if the centralizer of X is trivial (i.e. one-dimensional), then K is homeomorphic to L iff $C(K,X) \cong C(L,X)$.

In the second part of section 4 we relate our results to those of Jerison [J] who also gave conditions on X such that K may be reconstructed from C(K,X). It turns out that there are well-known spaces (e.g. L-spaces of dimension greater than two) which have a trivial centralizer but fail to satisfy the conditions of the Jerison-theorem. On the other hand we have not been able to determine whether the spaces which satisfy these conditions are a subclass of the Banach spaces with trivial centralizer (for a discussion of this problem see section 4). At the end of section 4 we consider examples of Banach spaces which satisfy the conditions of theorem 1.2.

2. Preliminaries

Banach spaces

We only consider real Banach spaces. To avoid trivialities we assume them to be nonzero.
By $S(x,r)$ we mean the open ball with center x and radius r in the Banach space X (so $S(f,r)$ is a ball in $C(K,X)$ whereas $S(f(k),r)$ is a ball in $X (k \in K)$). All norms (elements of X, functions in $C(K,X)$, operators) are denoted by the same symbol $\| \ \|$.
Let $(X_i)_{i=1,\ldots,n}$ be a finite family of Banach spaces, $1 \le p \le \infty$.
$\prod\limits_{i=1}^{n}{}^{p} X_i$ is the direct product of the X_i provided with the norm

$$\| (x_1,\ldots,x_n) \|_p := \sqrt[p]{\| x_1 \|^p + \ldots + \| x_n \|^p}$$

(if $p = \infty$: $\| (x_1,\ldots,x_n) \|_\infty := \max \{ \| x_1 \|, \ldots, \| x_n \| \}$).
If $X_1 = \ldots = X_n =: X$ we write $\prod\limits_{i=1}^{n}{}^{p} X$ instead of $\prod\limits_{i=1}^{n}{}^{p} X_i$.

M-ideals

It is known that the set $\underline{M}(X)$ (resp. $\underline{M}^{\infty}(X)$) of the M-ideals (resp. M-summands) in X is a lattice (resp. a Boolean algebra); cf. [AE]. The sup of two M-ideals is their sum, and the inf is their intersection.
Every M-summand is an M-ideal, but the converse is not true in general.
The subspaces X, $\{0\}$ are always M-ideals (resp. L^p-summands) in X. They are called the <u>trivial M-ideals</u> (resp. L^p-summands).

Cunningham algebra

$C(X)$ (= the Cunningham algebra of X) is the Banach algebra which is generated by the L-projections on X. It is commutative by [AE_2], 1.7.

Centralizer

The centralizer is isometrically isomorphic to the space of continuous functions on some compact Hausdorff space. The operators in the centralizer are in one-to-one correspondence with the space of structurally continuous functions on the space of primitive M-ideals (Dauns-Hofmann theorem, [AE_2] 4.9). It follows that $Z(X)$ is trivial, i.e. consists only of the multiples of the identity operator, if X has no nontrivial M-ideals.
We need the following relation between $\|T\|$ and the set of λ for which 1.1(iii) is valid ($T \in Z(X)$):-

2.1 Lemma: For $T \in Z(X)$, define $S_T := \left\{\lambda \,\middle|\, \begin{array}{l} Tx \in B \text{ for all balls} \\ B \text{ with } \pm \lambda x \in B \end{array}\right\}$.

(i) $\lambda \in S_T$, $\lambda' \geq \lambda \Rightarrow \lambda' \in S_T$; $S_T = S_{-T}$.

(ii) For $T \geq 0$ in $Z(X)$, $\|T\| \in S_T$ (the order being defined by $Z^+(X)$; see [AE_2], section 4).

(iii) For $T \in Z(X)$, $2\|T\| \in S_T$.

(iv) If G is a norm bounded family in $Z(X)$, there is a λ such that 1.1(iii) is true for all T in G.

Proof:

(i) obvious (note that balls are convex).

(ii) By [AE_2], 4.8, the norm of $Z(X)$ is the order unit norm with respect to the cone $Z^+(X)$ and the identity operator. Thus, for $T \in Z^+(X)$, $T <_M \|T\|\, Id$ (i.e. $Tx <_M \|T\|\, x$ for x in X). But $x <_M y$ is equivalent to the condition that x is in every closed ball containing 0 and y (lemma 4.2 in [AE_2]). It follows that $\|T\| \in S_T$.

(iii) For $T \in Z(X)$, we have $\pm T + \|T\|\, Id \in Z^+(X)$, i.e. $\|\pm T + \|T\|\, Id\| \in S_{(\pm T + \|T\|\, Id)}$. By (i), $2\|T\| \in$

$S_{(T + \|T\|\, Id)} \cap S_{(T - \|T\|\, Id)}$. Now let B be a closed ball

containing $\pm 2\|T\|\, x$. It follows that $Tx \pm \|T\|\, x \in B$ and therefore $Tx \in B$.

(iv) is a direct consequence of (i) and (iii).

Isomorphisms

We write $K \cong L$ if K and L are homeomorphic. $X \cong Y$ means that X and Y are isometrically isomorphic.

Functions

If x is in X (and K is understood), $\underline{x}$ denotes the constant function on K which attains the value x for every k in K. In particular, $\underline{1} \in CK$ is the usual order unit of CK.

Topology

Let K be a compact Hausdorff space. We already noted that nK means the n-fold topological sum of n copies of K ($n \in \mathbb{N}$). t.d.(K) denotes the totally disconnected compact Hausdorff space associated to K, i.e. the quotient of K with respect to the relation " $k \sim l$ iff k and l lie in the same component of K". It is easy to see that there is a one-to-one correspondence between the clopen subsets of K and those of t.d.(K) (which is, in fact, an isomorphism of Boolean algebras).

3. M-ideals, M-summands, and the centralizer of C(K,X)

Let K be a compact Hausdorff space, X a real Banach space. At first we prove that multiplication by continuous functions defines operators in the centralizer of C(K,X). As a corollary we are able to show that M-ideals in C(K,X) are CK-modules, a result which will be very important for the investigation of the M-structure of C(K,X).

3.1 Proposition: For $h \in CK$, the multiplication operator $M_h: C(K,X) \to C(K,X)$, defined by $M_h f := hf$, is in the centralizer of C(K,X).

Proof: It is clear that M_h is linear and continuous. We prove that M_h satisfies the condition of definition 1.1 (iii) with $\lambda = \| h \|$. Let $S(f,r)$ be an open ball containing λg and $-\lambda g$ ($f,g \in C(K,X)$, $r > 0$), that is $\| f \pm \lambda g \| < r$. It follows that $\| f(k) \pm \lambda g(k) \| < r$ and thus, because the open ball $S(f(k),r)$ in X is convex, that $\| f(k) - h(k)g(k) \| < r$ (all k in K). But then $\| f - hg \| < r$, that is $M_h g \in S(f,r)$.

3.2 Corollary: If J is an M-ideal in C(K,X), then $hf \in J$ for $f \in J$, $h \in CK$.

Proof: By [AE_2], theorem 4.8, M_h' is in the Cunningham algebra of $(C(K,X))'$. It therefore commutes with E, the L^1-projection onto the annihilator of J. A standard application of the Hahn-Banach theorem gives $M_h J \subset J$, that is $hf \in J$ for f in J.
We can now characterize the M-ideals and the M-summands of C(K,X).

3.3 Theorem: Let J be a closed subspace of C(K,X), $X_k := \{f(k) \mid f \in J\}$ for k in K.

(i) J is an M-ideal iff J is a CK-module (i.e. $M_h J \subset J$ for h in CK) and every X_k is an M-ideal in X.

(ii) If J is an M-summand, then J is a CK-module and every X_k is an M-summand in X. The converse is not true.

Proof:

(i) Let J be an M-ideal, $k \in K$. J is a CK-module by 3.2, so we only have to show that

1) X_k is closed,

2) X_k satisfies the conditions of the Alfsen-Effros criterion for M-ideals ([AE_1], 5.9).

ad 1): Let x_o be a point in X_k^-, $\|x_o\| = 1$. Construct a sequence $(f_n)_{n\in\mathbb{N}}$ in J as follows:

I. Choose $f_1 \in J$ such that $\|x_o - f_1(k)\| < 2^{-1}$.
Because of $\|f_1(k)\| < 2^{-0} + 2^{-1}$ and corollary 3.2 we may assume that $\|f_1\| < 2^{-0} + 2^{-1}$ (we have $\|f(l)\| < 2^{-0} + 2^{-1}$ for l in some neighbourhood U of k; replace, if necessary, f_1 by hf_1, where $h \in CK$, $h|_{K \setminus U} = 0$, $\|h\| = 1$, $h(k) = 1$).

II. Suppose f_n has been chosen with $\|x_o - (f_1 + \ldots + f_n)(k)\| < 2^{-n}$, $\|f_n\| < 2^{-(n-1)} + 2^{-n}$.

III. Choose $f_{n+1} \in J$ such that $\|x_o - (f_1 + \ldots + f_{n+1})(k)\| < 2^{-(n+1)}$ (note that $x_o - (f_1 + \ldots + f_n)(k) \in X_k^-$) and $\|f_{n+1}\| < 2^{-(n+1)} + 2^{-n}$ (the existence of such a function may be verified as in I.).

For $f := \sum_{n=1}^{\infty} f_n \in J$ we then have $f(k) = x_o$, that is $x_o \in X_k$.
(Note that we only used the fact that J is a closed CK-module in $C(K,X)$).

ad 2): Let $S(x_i,r_i)$ $(i=1,2,3, r_i > 0, x_i \in X)$ be three open balls in X having the following intersection property:

a)$_{X_k}$ $\bigcap_{i=1}^{3} S(x_i,r_i) \neq \emptyset$,

b)$_{X_k}$ $S(x_i,r_i) \cap X_k \neq \emptyset$ for $i = 1,2,3$.

We have to show that $X_k \cap \bigcap_{i=1}^{3} S(x_i,r_i) \neq \emptyset$.

By b)$_{X_k}$, there are functions f_1, f_2, f_3 in J with $\|f_i(k) - x_i\| < r_i$ $(i=1,2,3)$. Let U be a closed neighbourhood of k such that $\|f_i(l) - x_i\| < r_i$ for $i = 1,2,3$, $l \in U$. Choose an Urysohn function $h \in CK$, $\underline{0} \leq h \leq \underline{1}$, $h|_{K \setminus U} = 0$, $h(k) = 1$. Define $f_i' := hf_i$, $g := h\underline{x}_o$ (where $x_o \in \bigcap_{i=1}^{3} S_i(x_i,r_i)$), $g_i := h\underline{x}_i$ $(i=1,2,3)$. Then the balls $S(g_i,r_i)$ satisfy

a)$_J$ $\bigcap_{i=1}^{3} S(g_i,r_i) \neq \emptyset$ (since $\|g_i - g\| < r_i$),

b)$_J$ $S(g_i,r_i) \cap J \neq \emptyset$ (since $\|g_i - f_i'\| < r_i$; note that $f_i' \in J$ by 3.2).

Therefore there is an $f \in J$, $f \in \bigcap_{i=1}^{3} S(g_i,r_i)$ ([AE_1], 5.9). In particular, $f(k) \in X_k \cap \bigcap_{i=1}^{3} S(x_i,r_i)$.

Now, suppose that J is a CK-module such that all X_k are M-ideals. Let $S(f_i,r_i)$ $(i=1,2,3,\ r_i > 0,\ f_i \in C(K,X))$ be three open balls with $\bigcap_{i=1}^{3} S(f_i,r_i) \neq \emptyset$, $S(f_i,r_i) \cap J \neq \emptyset$ for $i = 1,2,3$. For $k \in K$ we therefore have $\bigcap_{i=1}^{3} S(f_i(k),r_i) \neq \emptyset$, $S(f_i(k),r_i) \cap X_k \neq \emptyset$.
Because X_k is an M-ideal, there is an $x_k \in X_k \cap \bigcap_{i=1}^{3} S(f_i(k),r_i)$.

Choose $f_k \in J$ such that $f_k(k) = x_k$. For a suitable open neighbourhood of k, say U_k, we have $f_k(l) \in \bigcap_{i=1}^{3} S(f_i(l),r_i)$ (all l in U_k). Select $U_{k_1}, \dots, U_{k_n}$ with $U_{k_1} \cup \dots \cup U_{k_n} = K$. Let $h_1, \dots, h_n$ be a partition of unity for the U_{k_i}, i.e. $\underline{0} \le h_i \le \underline{1}$, $h_i|_{K \setminus U_{k_i}} = 0$, $\sum_{i=1}^{n} h_i = \underline{1}$. If we now define $f := \sum_{i=1}^{n} h_i f_{k_i}$, we have $f \in J$, and $\| f_j(l) - f_{k_i}(l) \| < r_j$ (for l in U_{k_i}) implies $\| h_i f_j - h_i f_{k_i} \| < r_i$ $(j=1,2,3;\ i=1,\dots,n)$. By addition we obtain
$\| (\sum_{i=1}^{n} h_i) f_j - \sum_{i=1}^{n} h_i f_{k_i} \| = \| f_j - f \| < r_j$ $(j=1,2,3)$, i.e.
$f \in J \cap \bigcap_{i=1}^{3} S(f_i,r_i)$.

(ii) If J is an M-summand with complementary M-summand $J^{\perp}$, X_k and $X_k^{\perp} := \{f(k) \mid f \in J^{\perp}\}$ are M-ideals by the first part of the proof. The equality $X_k + X_k^{\perp} = X$ is an immediate consequence of the condition $J + J^{\perp} = C(K,X)$. Suppose $X_k \cap X_k^{\perp} \neq \{0\}$. We choose an $x \in X_k \cap X_k^{\perp}$, $\|x\| = 1$. For suitable $f \in J$, $f^{\perp} \in J^{\perp}$ we have $f(k) = f^{\perp}(k) = x$. Let U be a neighbourhood of k such that $\| f(l) - f^{\perp}(l) \| < 1/2$ for l in U, h an Urysohn function for U which attains the value 1 at k. Then $hf \in J$, $hf^{\perp} \in J^{\perp}$, $\| hf \|$, $\| hf^{\perp} \| \ge \| x \| = 1$, $\| hf - hf^{\perp} \| < 1/2$. The last inequality contradicts the norm condition for complementary M-summands, so that $X_k \cap X_k^{\perp} = \{0\}$. We thus have proved that X_k and $X_k^{\perp}$ are complementary M-ideals which already implies that they are in fact M-summands ([AE_2], corollary 2.9 ;note that it is possible to verify the norm condition for $X_k, X_k^{\perp}$ directly:

if $\| x_k + x_k^\perp \| \neq \max\{\| x_k\|, \| x_k^\perp\|\}$ for some $x_k \in X_k$, $x_k^\perp \in X_k^\perp$, choose $f \in J$, $f^\perp \in J^\perp$ with $f(k) = x_k$, $f^\perp(k) = x_k^\perp$. It follows that $\| hf + hf^\perp \| \neq \max\{\| hf\|, \| hf^\perp\|\}$ for a suitably chosen Urysohn function h in contradiction to the norm condition for $J, J^\perp$).
The converse of (ii) is not true, even for elementary situations. Let L be a compact Hausdorff space, $A \subset L$ a closed set which is not open. It is known ([AE_2], prop. 6.18; see also the discussion below) that $J_A \subset CL$, $J_A := \{f \mid f \in CL, f|_A = 0\}$, is an M-ideal but not an M-summand. But all X_k are M-summands:
we have $X_k = \{0\}$ or $X_k = X (= \mathbb{R})$ according to whether k is in A or k is not in A.

Note: We note that the spaces X_k determine J: if J, J* are M-ideals of C(K,X), X_k, X_k^* the corresponding families of M-ideals in X, then $X_k = X_k^*$ (all k in K) implies $J = J^*$.

(Let f be a function in J, $\varepsilon > 0$. For $k \in K$, determine $f_k \in J^*$, $f_k(k) = f(k)$. Select a finite subcover $U_{k_1}, \ldots, U_{k_n}$ of the family $\{U_k \mid k \in K\}$, where $U_k := \{l \mid \| f(l) - f_k(l)\| < \varepsilon\}$. If $h_1, \ldots, h_n$ is a corresponding partition of unity, $f^* := \sum_{i=1}^{n} h_i f_{k_i}$ belongs to J*, and $\| f - f^*\| < \varepsilon$. We therefore have $f \in (J^*)^- = J^*$ so that $J \subset J^*$. The reverse inclusion is valid by symmetry).

For the centralizer of C(K,X) we have a similar localization principle: -

3.4 Theorem: Let $T: C(K,X) \to C(K,X)$ be a continuous linear operator. For k in K, define $T_k : X \to X$ by $T_k x := (T\underline{x})(k)$ (obviously, the T_k are linear continuous operators).
T is in the centralizer of C(K,X) iff T commutes with all M_h (see prop. 3.1) and all T_k are in the centralizer of X.

Proof: Let T be in the centralizer of C(K,X). For $h \in CK$, $M_h T = TM_h$, because the centralizer is a commutative Banach algebra ([AE_2],4.8). Choose λ for T as in def. 1.1(iii). We prove that all T_k satisfy the condition of 1.1(iii), too. Let $x,y \in X$, $r > 0$, $\pm \lambda x \in S(y,r)$. It follows that $\pm \lambda \underline{x} \in S(\underline{y},r)$ and therefore that $T\underline{x} \in S(\underline{y},r)$, in particular that $T_k x \in S(y,r)$.
Conversely, let T be a continuous linear operator such that all T_k are in the centralizer of X and $M_h T = TM_h$ for all h in CK. We first note that $\| T_k\| \leq \| T\|$ (all $k \in K$). Consequently, by 2.1 (iv), there is a $\lambda > 0$ such that all T_k satisfy 1.1(iii) for this λ. Let f,

$g \in C(K,X)$, $r > 0$, $\pm \lambda f \in S(g,r)$. It follows that $\pm \lambda f(k) \in S(g(k),r)$ and thus that $T_k(f(k)) \in S(g(k),r)$ $(k \in K)$. We only have to show that $T_k(f(k)) = (Tf)(k)$, i.e. $(T(\underline{f(k)}))(k) = (Tf)(k)$ or $[T(f - \underline{f(k)})](k) = 0$. For k fixed, $\varepsilon > 0$, let U be a neighbourhood of k such that $\| f(l) - f(k) \| < \varepsilon$ for l in U. Choose $h \in CK$, $\underline{0} \le h \le \underline{1}$, $h(k) = 1$, $h|_{K \setminus U} = 0$.

Because of $TM_h = M_h T$ we have
$$\begin{aligned} \| [T(f - \underline{f(k)})](k) \| &= \\ &= \| [hT(f - \underline{f(k)})](k) \| \\ &= \| [T(h(f - \underline{f(k)}))](k) \| \\ &\le \| T(h(f - \underline{f(k)})) \| \\ &\le \| T \| \, \| h(f - \underline{f(k)}) \| \le \| T \| \, \varepsilon . \end{aligned}$$

Hence $T(\underline{f(k)})(k) = (Tf)(k)$. (It follows that $f(k) = g(k)$ implies $(Tf)(k) = (Tg)(k)$, i.e. the value of Tf at k is determined by the value of f at k.)

We are now able to give a complete description of the M-ideals, the M-summands, and the centralizer of $C(K,X)$ provided the respective structure of X is not too complicated.
There seem to be no similar results for the general situation.

<u>3.5 Definition</u>: Let $\underline{M}(X)$ be the set of all M-ideals of X, ordered by inclusion. By $\underline{M}^{\infty}(X)$ we mean the subset of M-summands (see section 2). A map $\varphi : \underline{M}(X) \to \underline{A}(K)$ (:= the lattice of all closed subsets of K) is called an <u>M-homomorphism</u>, if the following conditions are satisfied:

(i) $\varphi(M_1 \cap M_2) = \varphi(M_1) \cap \varphi(M_2)$ (all M_1, M_2 in $\underline{M}(X)$),

(ii) $\varphi(X) = K$.

$\varphi : \underline{M}^{\infty}(X) \to \underline{A}^{\infty}(K)$ (: = the Boolean algebra of the clopen subsets of K) is called an <u>M^{∞}-homomorphism</u> if (i) and (ii) are valid for $M_1, M_2 \in \underline{M}^{\infty}(X)$.

<u>3.6 Lemma</u>: Let J be an M-ideal in $C(K,X)$. We define

$\varphi_J : \underline{M}(X) \to \underline{A}(K)$ by $\varphi_J(M) := \{k \mid X_k \subset M\}$ $(X_k := \{f(k) \mid f \in J\}$, cf. 3.3).

(i) φ_J is an M-homomorphism.

(ii) If J is an M-summand and X contains only a finite number of M-summands, then the restriction of φ_J to $\underline{M}^{\infty}(X)$ (which we denote by φ_J, too) is an M^{∞}-homomorphism.

<u>Proof</u>:

(i) For $k \notin \varphi_J(M)$, $X_k \not\subset M$. Choose an $x \in X_k$, $x \notin M$. Since M is closed, x has a positive distance to M. Therefore $f(l) \notin M$ in a suitable neighbourhood of k where f is chosen in J with $f(k) = x$. This neighbourhood is therefore contained in the complement of $\varphi_J(M)$. Thus φ_J is well-defined. It is easily verified that φ_J satisfies (i) and (ii) of 3.3.

(ii) We only have to show that $\varphi_J(M)$ is clopen in K for M in $\underline{M}^\infty(X)$. By (i), $\varphi_{J^\perp}(M^*) := \{k \mid k \in K, \{f(k) \mid f \in J^\perp\} \subset M^*\}$ is closed for $M^* \in \underline{M}^\infty(X)$. It follows that $\{k \mid X_k = M\} = \varphi_J(M) \cap \varphi_{J^\perp}(M^\perp)$ is closed (note that $X_k = M$ iff $X_k \subset M$ and $X_k^\perp \subset M^\perp$). Thus $\{\{k \mid X_k = M\} \mid M \in \underline{M}^\infty(X)\}$ is a finite disjoint partition of X into closed subsets so that all $\{k \mid X_k = M\}$ are clopen. Hence $\varphi_J(M)$ is clopen as a finite union of clopen sets.

3.7 Lemma:

(i) Let X be a Banach space having only a finite number of M-ideals. For $\varphi : \underline{M}(X) \to \underline{A}(K)$, φ an M-homomorphism, define
$J_\varphi := \{f \mid f \in C(K,X),\ f(k) \in M \text{ for all } M \in \underline{M}(X) \text{ with } k \in \varphi(M)\}$.
J_φ is an M-ideal in C(K,X).

(ii) If $\underline{M}^\infty(X)$ is finite and $\varphi : \underline{M}^\infty(X) \to \underline{A}^\infty(K)$ is an M^∞-homomorphism, $J_\varphi := \{f \mid f \in C(K,X),\ f(k) \in M \text{ for all } M \in M^\infty(X) \text{ with } k \in \varphi(M)\}$ is an M-summand in C(K,X).

Proof:

(i) It is clear that J_φ is a closed subspace of C(K,X) which is invariant with respect to all M_h, $h \in CK$. We thus only have to show that $X_k := \{f(k) \mid f \in J_\varphi\}$ is an M-ideal for k in K (cf. 3.3(i)). We prove that $X_k = \bigcap_{\substack{k \in \varphi(M) \\ M \in \underline{M}(X)}} M$.

The inclusion $X_k \subset \bigcap_{\substack{k \in \varphi(M) \\ M \in \underline{M}(X)}} M$ is trivially satisfied. Conversely, let $x \in \bigcap_{\substack{k \in \varphi(M) \\ M \in \underline{M}(X)}} M$. The set $\bigcup_{\substack{k \notin \varphi(M) \\ M \in \underline{M}(X)}} \varphi(M)$ is closed as a finite union of closed sets. Choose $h \in CK$, $h(k) = 1$, $h|_{\varphi(M)} = 0$ for all $M \in \underline{M}(X)$ with $k \notin \varphi(M)$, and define $f := h\underline{x}$. It is easy to see that $f \in J_\varphi$ so that $x \in X_k$.
It follows that X_k is an M-ideal as the finite intersection of M-ideals.

(ii) As in (i), J_φ is a closed subspace of C(K,X) with $M_h J_\varphi \subset J_\varphi$ for all h in CK. $X_k = \bigcap_{\substack{k\in\varphi(M)\\ M\in\underline{M}^\infty(X)}} M$ is an M-summand as the finite intersection of M-summands ($k \in K$). It follows that $X_1 = \bigcap_{\substack{k\in\varphi(M)\\ M\in\underline{M}^\infty(X)}} M$ for $1 \in \bigcap_{k\in\varphi(M)} \varphi(M)$.

Therefore all sets $A_M := \{k \mid X_k = M\}$ are clopen in K, and $\{A_M \mid M \in \underline{M}^\infty(X)\}$ is a disjoint partition of K. It is now easy to see that $J_\varphi = \{f \mid f(k) \in M \text{ for } k \in A_M\}$, and that $J^\perp := \{f \mid f(k) \in M^\perp,\ k \in A_M\}$ is an M-summand complementary to J_φ.

3.8 Theorem:

(i) Let X be a Banach space such that $\underline{M}(X)$ is finite.

a) For every M-ideal $J \subset C(K,X)$ we have $J_{(\varphi_J)} = J$.

b) For every M-homomorphism φ we have $\varphi_{(J_\varphi)} = \varphi$.

Thus there is a one-to-one correspondence between the set of M-ideals of C(K,X) and the set of M-homomorphisms. With respect to this correspondence the inclusion order for M-ideals induces the order

$$\varphi_1 \le \varphi_2 \Leftrightarrow \varphi_1(M) \supset \varphi_2(M) \quad (\text{all } M \text{ in } \underline{M}(X))$$

for M-homomorphisms.

(ii) Let X be a Banach space such that $\underline{M}^\infty(X)$ is finite.

a) $J_{(\varphi_J)} = J$ for every M-summand J in C(K,X).

b) $\varphi_{(J_\varphi)} = \varphi$ for every M^∞-homomorphism.

Proof:

(i) ad a): Let $X_k^* := \{f(k) \mid f \in J_{(\varphi_J)}\}$, $X_k := \{f(k) \mid f \in J\}$ for k in K. We have verified in the proof of 3.7 that $X_k^* = \bigcap_{\substack{k\in\varphi_J(M)\\ M\in\underline{M}(X)}} M$. But $k \in \varphi_J(M)$ is equivalent to $X_k \subset M$ so that $X_k^* = \bigcap_{\substack{M\in\underline{M}(X)\\ M\supset\overline{X}_k}} M = X_k$. By the note at the end of 3.3 it follows that $J = J_{(\varphi_J)}$.

ad b): Let φ be an M-homomorphism, $M_o \in \underline{M}(X)$. If X_k ($k \in K$) denotes the M-ideal $\{f(k) \mid f \in J_\varphi\}$ we know that $X_k = \bigcap_{\substack{k\in\varphi(M)\\ M\in\underline{M}(X)}} M$.

The equation $\varphi(M_1\cap\ldots\cap M_r) = \varphi(M_1)\cap\ldots\cap\varphi(M_r)$, applied to the finite family of those M for which $k \in \varphi(M)$, implies $k \in \varphi(X_k)$. We now are able to prove that the closed sets $\varphi(M_o)$ and $\varphi_{(J_\varphi)}(M_o)$ are identical. For $k \in \varphi(M_o)$, the above formula for X_k gives $X_k \subset M_o$. But then, by

definition, $k \in \varphi_{(J_\varphi)}(M_o)$.
Conversely, $k \in \varphi_{(J_\varphi)}(M_o)$ implies $X_k \subset M_o$. φ is monotone (an easy consequence of 3.5(i)), so that $k \in \varphi(X_k) \subset \varphi(M_o)$. We have proved that φ and $\varphi_{(J_\varphi)}$ coincide for every M-ideal, i.e. $\varphi = \varphi_{(J_\varphi)}$.
It is obvious that $\varphi_1 \le \varphi_2$ implies $J_{\varphi_1} \subset J_{\varphi_2}$ and that $J_1 \subset J_2$ yields $\varphi_{J_1} \le \varphi_{J_2}$. By the first part of the proof it follows that $\varphi_1 \le \varphi_2$ iff $J_{\varphi_1} \subset J_{\varphi_2}$.
(ii) is proved analogously.

<u>3.9 Corollary</u>:

(i) If X has no nontrivial M-ideals, the lattices $\underline{M}(C(K,X))$ and $\underline{A}(K)$ are anti-isomorphic. The M-ideals in $C(K,X)$ are exactly the subspaces of the form $J_A := \{f \mid f|_A = 0\}$ for A closed in K.

(ii) If X and $\{0\}$ are the only M-summands in X, the Boolean algebras $\underline{M}^\infty(C(K,X))$ and $\underline{A}^\infty(K)$ are isomorphic. A subspace $J \subset C(K,X)$ is an M-summand iff
$J = J_A := \{f \mid f|_A = 0\}$ for some clopen subset A of K.

<u>Proof</u>:

(i): If $\underline{M}(X) = \{\{0\}, X\}$, the M-homomorphisms are determined by their value at the M-ideal $\{0\}$. For $A \in \underline{A}(K)$, we define φ_A by $\varphi_A(\{0\}) = A$, $\varphi_A(X) = K$. It is clear that φ_A is an M-homomorphism and that $\varphi_A \le \varphi_{A^*}$ iff $A \supset A^*$. The equation $J_{\varphi_A} = J_A$ proves the second part of the assertion.

(ii): We omit the proof because it is similar to the proof of (i). The anti-isomorphism $A \mapsto J_A$ of (i) yields an isomorphism for Boolean algebras $A \mapsto J_{K \setminus A}$ $(A \in \underline{A}^\infty(K))$.

<u>3.10 Corollary</u>: If*) Cz $K > 1$, then $C(K,X)$ has no nontrivial L^p-summands for $1 < p < \infty$. If, in addition, $\dim C(K,X) > 2$, $C(K,X)$ has no nontrivial L-summands.
<u>Proof</u>: For $A \subset K$ closed, $\emptyset \ne A \ne K$, J_A is a nontrivial M-ideal in $C(K,X)$. By [B], section 5, the L^p-summands ($1 < p < \infty; 1 \le p < \infty$ if $\dim C(K,X) > 2$) must be trivial.
The following investigations of the centralizer are of great interest in view of the applications of our theory to theorems of the Banach-Stone type in the next chapter.
<u>3.11 Theorem</u>: Let X be a Banach space such that $Z(X)$ is n-dimensional ($n \in \mathbb{N}$). Then the centralizer of $C(K,X)$ is isometrically isomor-

*) CzK denotes the cardinality of K.

phic to $\prod_{i=1}^{n} {}^{\infty} CK$.

Proof: $Z(X)$ is an n-dimensional space of continuous functions. Therefore $Z(X)$ is isometrically isomorphic to $(\mathbb{R}^n, \| \ \|_\infty)$. For $(a_1,\ldots,a_n) \in \mathbb{R}^n$, we denote the corresponding operator by $T_{[a_1,\ldots,a_n]}$, and for $T \in Z(X)$, $(a_1(T),\ldots,a_n(T))$ is the associated n-tuple. As $Z(X)$ is n-dimensional, there are $x_1,\ldots,x_r$ $(r \le n)$ in X such that $\|T\|^* := \sup_{i=1,\ldots,r} \|Tx_i\|$ is a norm for $Z(X)$. (For $T \in Z(X)$, $T \neq 0$, there is $x_1 \in X$ such that $Tx_1 \neq 0$. If $Z_1 := \{S \mid S \in Z(X), Sx_1 = 0\}$ is $\{0\}$, put $r = 1$. Otherwise, there is $S \in Z_1$, $S \neq 0$. For a suitable vector x_2, we have $Sx_2 \neq 0$. Define $Z_2 := \{P \mid P \in Z_1, Px_2 = 0\}$. Continuing this construction we get $Z(X) \supsetneqq Z_1 \supsetneqq Z_2 \supsetneqq \ldots$ so that $Z_{r+1} = \{0\}$ for some $r \le n$. The vectors $x_1,\ldots,x_r$ have the desired properties.) We fix these vectors for the rest of the proof and note that there is an $R > 0$ such that $\|T\| \le R \|T\|^*$ for $T \in Z(X)$.

Now let $(h_1,\ldots,h_n)$ be an element of $\prod_{i=1}^{n} {}^{\infty} CK$. We define

$$(T_{[h_1,\ldots,h_n]} f)(k) := T_{[h_1(k),\ldots,h_n(k)]} f(k) \quad \text{for } f \in C(K,X),\ k \in K.$$

It will be proved that $(h_1,\ldots,h_n) \mapsto T_{[h_1,\ldots,h_n]}$ is an isometric isomorphism from $\prod_{i=1}^{n} {}^{\infty} CK$ onto $Z(C(K,X))$.

1) $T_{[h_1,\ldots,h_n]} f : K \to X$ is continuous as the composition of the continuous maps

$k \mapsto (h_1(k),\ldots,h_n(k), f(k))$ (from K to $\mathbb{R}^n \times X$)

$(a_1,\ldots,a_n, x) \mapsto T_{[a_1,\ldots,a_n]}, x)$ (from $\mathbb{R}^n \times X$ to $Z(X) \times X$)

$(T,x) \mapsto Tx$ (from $Z(X) \times X$ to X).

It follows that $T_{[h_1,\ldots,h_n]} : C(K,X) \to C(K,X)$ is welldefined.

2) $T_{[h_1,\ldots,h_n]}$ is a linear continuous operator with

$\|T_{[h_1,\ldots,h_n]}\| = \|(h_1,\ldots,h_n)\|_\infty$:

The linearity of $T_{[h_1,\ldots,h_n]}$ is easily verified. For $f \in C(K,X)$, we have

$$\|(T_{[h_1,\ldots,h_n]} f)(k)\| \le \|T_{[h_1(k),\ldots,h_n(k)]}\| \, \|f(k)\|$$
$$= \|(h_1(k),\ldots,h_n(k))\|_\infty \|f(k)\|$$

and therefore $\| T_{[h_1,\dots,h_n]} f \| \le \| (h_1,\dots,h_n) \|_\infty \| f \|$.

If $k \in K$ is chosen with $\| (h_1,\dots,h_n) \|_\infty = \| (h_1(k),\dots,h_n(k)) \|_\infty$, it follows that

$$
\begin{aligned}
\| T_{[h_1,\dots,h_n]} \| &\ge \sup_{\substack{x\in X\\ \|x\|=1}} \| T_{[h_1,\dots,h_n]} \underline{x} \| \\
&\ge \sup_{\substack{x\in X\\ \|x\|=1}} \| (T_{[h_1,\dots,h_n]} \underline{x})(k) \| \\
&= \sup_{\substack{x\in X\\ \|x\|=1}} \| T_{[h_1(k),\dots,h_n(k)]} x \| \\
&= \| T_{[h_1(k),\dots,h_n(k)]} \| \\
&= \| (h_1(k),\dots,h_n(k)) \|_\infty = \| (h_1,\dots,h_n) \|_\infty .
\end{aligned}
$$

Hence $\| T_{[h_1,\dots,h_n]} \| = \| (h_1,\dots,h_n) \|_\infty$.

3) $T_{[h_1,\dots,h_n]}$ commutes with M_h ($h \in CK$, cf. 3.1).

This is a direct consequence of the linearity of the mapping $(a_1,\dots,a_n) \mapsto T_{[a_1,\dots,a_n]}$.

4) $T_{[h_1,\dots,h_n]}$ is in the centralizer of $C(K,X)$:

By 3.4, we have only to show that the mapping $T_k : X \to X$, $x \mapsto (T_{[h_1,\dots,h_n]}\underline{x})(k)$, is in the centralizer of X for k in K.

But $T_k = T_{[h_1(k),\dots,h_n(k)]}$ by definition.

5) We thus have proved that $(h_1,\dots,h_n) \mapsto T_{[h_1,\dots,h_n]}$ is an isometric mapping from $\prod_{i=1}^{n\,\infty} CK$ into $Z(C(K,X))$. This mapping is obviously linear. It remains to show that it is onto.

6) For $T \in Z(C(K,X))$ there is $(h_1,\dots,h_n) \in \prod_{i=1}^{n\,\infty} CK$ such that $T = T_{[h_1,\dots,h_n]}$:

For k in K, the mapping $x \mapsto (T\underline{x})(k)$ is in $Z(X)$ (3.4) and thus has the form $T_{[a_1^k,\dots,a_n^k]}$. Define $h_i : K \to \mathbb{R}$ by $k \mapsto a_i^k$ $(i=1,\dots,n)$. We have to show that

a) $(h_1,\dots,h_n) \in \prod_{i=1}^{n\,\infty} CK$

b) $T_{[h_1,\dots,h_n]} = T$.

<u>ad a)</u>: Let $k_o \in K$ be given, $\varepsilon > 0$. The functions $T\underline{x}_1, \ldots, T\underline{x}_r$ ($x_1, \ldots, x_r$ being the fixed vectors chosen at the beginning of this proof) are continuous at k_o. Thus there is a neighbourhood U of k_o such that $\| (T\underline{x}_i)(k_o) - (T\underline{x}_i)(l) \| \le \varepsilon$ for $i = 1, \ldots, r$, $l \in U$. It follows that

$$\| T_{[h_1(k_o), \ldots, h_n(k_o)]}\, x_i - T_{[h_1(l), \ldots, h_n(l)]}\, x_i \| \le \varepsilon$$

($i=1, \ldots, r$, $l \in U$), i.e.

$$\| T_{[h_1(k_o), \ldots, h_n(k_o)]} - T_{[h_1(l), \ldots, h_n(l)]} \|^* \le \varepsilon \text{ for } l \text{ in } U.$$

Hence $\| (h_1(k_o), \ldots, h_n(k_o)) - (h_1(l), \ldots h_n(l)) \|_\infty =$

$= \| T_{[h_1(k_o), \ldots, h_n(k_o)]} - T_{[h_1(l), \ldots, h_n(l)]} \| \le R\varepsilon$ which implies

$\| h_i(k_o) - h_i(l) \| \le R\varepsilon$ for $l \in U$, $i=1, \ldots, n$. We thus have proved the continuity of $h_1, \ldots, h_n$ at k_o, i.e. $(h_1, \ldots, h_n) \in \prod_{i=1}^{n} {}^{\infty}\, CK$ since k_o was arbitrary.

<u>ad b)</u>: We already noted in the proof of 3.4 that $(Tf)(k) = (T\underline{f(k)})(k)$ for $T \in Z(C(K,X))$, $f \in C(K,X)$, $k \in K$. Hence

$$\begin{aligned} (T_{[h_1, \ldots, h_n]} f)(k) &= T_{[h_1(k), \ldots, h_n(k)]}(f(k)) \\ &= (T(\underline{f(k)}))(k) = (Tf)(k), \text{ i.e.} \end{aligned}$$

$$T_{[h_1, \ldots, h_n]} = T$$

<u>3.12 Corollary</u>: If X is a Banach space such that Z(X) is trivial (i.e. one-dimensional) then the centralizer of C(K,X) is isometrically isomorphic to CK. The operators in the centralizer are exactly the operators of the form M_h, $h \in CK$.

4. Applications to theorems of the Banach-Stone type

<u>4.1 Theorem</u>:

(i) Let X,Y be Banach spaces such that X and Y have no nontrivial M-summands. If K, L are compact Hausdorff spaces, then $C(K,X) \cong C(L,Y)$ implies $\text{t.d.}(K) \cong \text{t.d.}(L)$. In particular, if K and L are totally disconnected, $K \cong L$ is a consequence of $C(K,X) \cong C(L,Y)$.

(ii) Let X, Y be Banach spaces such that the centralizers Z(X),

$Z(Y)$ are finite dimensional, say $\dim Z(X) = n$, $\dim Z(Y) = m$. If K and L are compact Hausdorff spaces such that $C(K,X)$ and $C(L,Y)$ are isometrically isomorphic, then $nK \cong mL$. In particular, if $Z(X)$ and $Z(Y)$ are one-dimensional (e.g. when X and Y have no nontrivial M-ideals), $C(K,X) \cong C(L,Y)$ implies $K \cong L$.

Proof:

(i) $\underline{A}^{\infty}(\text{t.d.}(K)) \underset{B}{\cong} \underline{A}^{\infty}(K) \underset{B}{\cong} \underline{M}^{\infty}(C(K,X)) \underset{B}{\cong} \underline{M}^{\infty}(C(L,Y)) \underset{B}{\cong} \underline{A}^{\infty}(L) \underset{B}{\cong}$ $\underline{A}^{\infty}(\text{t.d.}(L))$, where "$\underset{B}{\cong}$" means "isomorphism for Boolean algebras" (the second and the fourth isomorphisms are considered in 3.9(ii)). Thus the Boolean algebras of the clopen sets in t.d.(K) and t.d.(L) are isomorphic which has $\text{t.d.}(K) \cong \text{t.d.}(L)$ as a consequence ([L], p. 118)

(ii) $C(nK) \cong \prod_{i=1}^{n}{}^{\infty}\, CK \cong Z(C(K,X)) \cong Z((CL,Y)) \cong \prod_{i=1}^{m}{}^{\infty}\, CL \cong C(mL)$

(the second and the fourth isometry are established in 3.11, the others are easily constructed). By the classical Banach-Stone theorem it follows that $nK \cong mL$.

We now want to compare our results with those of Jerison [J] who also gave conditions on X such that $C(K,X) \cong C(L,Y)$ implies $K \cong L$.

4.2 Definition: Let X be a Banach space. A T-set in X is a maximal subset for which the norm is additive. Two T-sets, T_1 and T_2, are called discrepant if either

1) $T_1 \cap T_2 = \{0\}$

or 2) there is a T-set T such that $T \cap T_1 = T \cap T_2 = \{0\}$.

4.3 Theorem: (Jerison) Let X be a Banach space for which every two T-sets are discrepant. Then $K \cong L$ iff $C(K,X) \cong C(L,X)$.

Proof: [J], Th. 5.2

Note: In fact, with the same proof one has the following stronger result: If X, Y are Banach spaces such that every two T-sets in X (resp. Y) are discrepant, then $C(K,X) \cong C(L,Y)$ implies $K \cong L$.

4.4 Definition: By $\underline{J}$ we denote the class of Banach spaces X such that every pair of T-sets in X is descrepant. $\underline{M}_o$ (resp. $\underline{Z}_1$) is the class of Banach spaces for which there are no nontrivial M-ideals (resp. the centralizer is trivial).

Note that $\underline{M}_o \subset \underline{Z}_1$ and that for $X \in \underline{Z}_1$ $K \cong L$ iff $C(K,X) \cong C(L,X)$.

<u>4.5 Lemma</u>: Let $X = \prod_{i=1}^{2} {}_{\infty} X_i$ be the direct product of two nonzero Banach spaces X_1, X_2, $\|(x_1,x_2)\| = \max\{\|x_1\|, \|x_2\|\}$.

(i) If T_1 is a T-set in X_1, then

$(T_1,X_2) := \{(x_1,x_2) \mid x_1 \in T_1,\ x_2 \in X_2,\ \|x_2\| \le \|x_1\|\}$

is a T-set in X

(ii) Similarly, (X_1,T_2) is a T-set in X for every T-set T_2 of X_2.

(iii) Every T-set of X is either of the form (T_1,X_2) (T_1 a T-set of X_1) or (X_1,T_2) (T_2 a T-set of X_2).

(iv) Two T-sets of X are discrepant iff both are of the form (i) (resp. (ii)) and the associated T-sets in X_1 (resp. X_2) are discrepant.

It follows that X is not in <u>J</u> (note that there are T-sets in X_1 and X_2 since both are assumed to be nontrivial).

<u>Proof</u>:

(i): The norm is additive in (T_1,X_2):

$$\begin{aligned}\|(x_1,x_2) + (x_1',x_2')\| &= \max\{\|x_1+x_1'\|, \|x_2+x_2'\|\}\\ &= \max\{\|x_1\| + \|x_1'\|, \|x_2+x_2'\|\}\\ &= \|x_1\| + \|x_1'\|\\ &= \|(x_1,x_2)\| + \|(x_1',x_2')\|\end{aligned}$$

since $\|x_2+x_2'\| \le \|x_2\| + \|x_2'\| \le \|x_1\| + \|x_1'\|$.

(T_1,X_2) is maximal with this property for let (x_1',x_2') be in X such that the norm is additive on $(T_1,X_2) \cup \{(x_1',x_2')\}$. For $x_1 \in T_1$, $x_1 \ne 0$, we have $a(x_1,0) \in (T_1,X_2)$ ($a \ge 0$ arbitrary) and consequently $\|(x_1',x_2') + a(x_1,0)\| = \max\{\|x_1'+ax_1\|, \|x_2'\|\}$ $= \max\{\|x_1'\|, \|x_2'\|\} + a\|x_1\|$. This implies $\|x_1'+ax_1\| \ge \|x_2'\|$ for all $a \ge 0$ (otherwise, $\max\{\|x_1'+ax_1\|, \|x_2'\|\}$ would not be an affine function of a as it must be) and thus $\|x_1'+ax_1\| = \|x_1'\| + a\|x_1\|$. It follows that the norm is additive on $T_1 \cup \{x_1'\}$, hence $x_1' \in T_1$ as T_1 is a T-set. Thus, because of $\|x_1'\| \ge \|x_2'\|$, $(x_1',x_2') \in (T_1,X_2)$.

(iii) Let T be a T-set in X. We claim that either $T_1 := \{x_1 \mid (x_1,x_2) \in T\}$ or $T_2 := \{x_2 \mid (x_1,x_2) \in T\}$ is a T-set. To this end let S be the set $\{(x_1,x_2) \mid (x_1,x_2) \in T, \|(x_1,x_2)\| = 1\}$. The T-set property of T implies that S in convex (note in particular that T-sets are convex [L] , p. 78) so that

$S_1 := \{x_1 \mid (x_1,x_2) \in S\}$ (resp. $S_2 := \{x_2 \mid (x_1,x_2) \in S\}$)

is a convex subset of X_1 (resp. X_2), too. If S_1 and S_2 contain elements of norm less than one, say x_1 and x_2', there are $x_1' \in X_1, x_2 \in X_2$, such that $(x_1,x_2) \in S$, $(x_1',x_2') \in S$. But then $\|1/2(x_1,x_2) + 1/2(x_1',x_2')\| < 1$ in contradiction to the convexity of S. Thus either S_1 or S_2 contains only elements of norm one. Suppose that $\|x_1\| = 1$ for all $x_1 \in S_1$. It follows that $\|x_1\| \geq \|x_2\|$ for all $(x_1,x_2) \in T$ and that T_1 is a set for which the norm is additive. Since (T_1,X_2) contains T and the norm is additive on (T_1,X_2) by (i), we necessarily have $(T_1,X_2) = T$, and T_1 must be a T-set.

(iv) The proof is elementary. Note that the T-sets (T_1,X_2), (X_1,T_2) (T_1 and T_2 T-sets in X_1 and X_2, respectively) always have nonzero intersection.

Remark: A special case of this situation has been considered in more detail in [S]: Cylindrical Banach spaces have exactly two nontrivial M-summands.

We are now able to compare the classes $\underline{J}$ and $\underline{M}_o$, at least for reflexive spaces.

4.6 Proposition: If $X \in \underline{J}$ and X is reflexive (in particular if X is finite dimensional), then $X \in \underline{M}_o$.

Proof: For reflexive spaces, any M-ideal is an M-summand ([AE_2],2.8). The assertion is therefore a direct consequence of the preceding lemma.

We have not been able to construct a Banach space X which is in $\underline{J}$ but not in $\underline{Z}_1$. The following difficulties arise:

1. The typical examples for spaces that are in $\underline{J}$ (e.g. L^p-spaces $1 < p < \infty$) are also in $\underline{Z}_1$ (cf. the following theorem).

2. Some important classes of classical Banach spaces, e.g. the spaces of the form CK, K a compact Hausdoff space with Cz K > 1, are neither in $\underline{J}$ norin $\underline{Z}_1$.
3. The space in question therefore must be a nonreflexive space which is not a CK-space, with nontrivial centralizer but without nontrivial M-summands, for which it is possible to determine all T-sets and to decide whether two of them are discrepant.

Finally we give some examples of Banach spaces in $\underline{Z}_1$. Maybe it is interesting to note that the L-spaces of dimension greater than two are in $\underline{M}_o$ ($\subset \underline{Z}_1$) but not in $\underline{J}$ so that $\underline{Z}_1$ is not a subclass of $\underline{J}$ (it is easy to check that T is a T-set in

$L^1 = L^1(S,\Sigma,m)$ iff $T = \{f \mid f \in L^1,\ f|_A \geq 0 \text{ m-a.e.}, f|_{S\setminus A} \leq 0 \text{ m-a.e.}\}$

for $A \in \Sigma$ so that only T and -T are discrepant; the measure space is assumed to be strictly localizable).

4.7 Theorem: Let X be a Banach space. Each of the following conditions implies that X is in $\underline{Z}_1$:

1) X is reflexive and $X \in \underline{J}$,
2) X is reflexive and rotund,
3) X is uniformly smooth or uniformly rotund,
4) X, or X', or any other iterated dual space $X^{(n)}$ of X contains a nontrivial L^p-summand for $1 < p < \infty$,
5) dim X > 2, and X' (or X''', or $X^{(5)}$,...) has a nontrivial M-ideal,
6) dim X > 2, and X (or X'', or $X^{(4)}$,...) has a nontrivial L-summand,
7) X is an abstract L^p-space, $1 < p < \infty$ (cf. [L], § 15),
8) dim X > 2, and X is an abstract L-space,
9) $X = AK := \{f \mid f: K \to \mathbb{R},$ f affine and continuous$\}$ (sup-norm), where K is a compact convex set having no proper closed split-faces,
10) $X = Y_{sa}$, where Y_{sa} is the self-adjoint part of a C*-algebra Y with trivial center.

Proof:

1) see 4.6.
2) It is obvious that rotund spaces are in $\underline{J}$.
3) Because X is reflexive we have only to show that X contains no nontrivial M-summands. But this is easily verified for rotund

or smooth spaces.

4) -8) cf. [B], section 5.

9) The nontrivial M-ideals of AK are in one-to-one correspondence with the proper closed split faces of K ([AE_2], p. 164).

10) The operators in the centralizer of X correspond to the self-adjoint elements in the centre of Y ([AE_2], 6.17).

In view of theorem 4.1(iii) it is interesting to describe $\underline{Z}_n$, the class of Banach spaces for which the centralizer is n-dimensional. The following proposition, together with the preceding theorem, provides us with some concrete nontrivial elements of $\underline{Z}_n$.

<u>4.8 Proposition</u>: Let $X_1, \ldots X_n$ be Banach spaces such that the centralizer of X_i is n_i-dimensional for $i = 1,\ldots,n$. Then the centralizer of $X := \prod_{i=1}^{n} {}^{\infty} X_i$ is isometrically isomorphic to $\prod_{i=1}^{n} {}^{\infty} Z(X_i)$ so that $\dim Z(X) = \sum_{i=1}^{n} n_i$. Thus, if all X_i are in $\underline{Z}_1$, then $X \in \underline{Z}_n$.

<u>Note</u>: For the general case of infinitely many summands, see [G].

<u>Proof</u>: Define $\phi : \prod_{i=1}^{n} {}^{\infty} Z(X_i) \rightarrow Z(X)$ by

$$\phi(T_1,\ldots,T_n)(x_1,\ldots,x_n) := (T_1x_1,\ldots,T_nx_n).\ \phi(T_1,\ldots,T_n)$$

is a linear continuous operator with norm $\|(T_1,\ldots,T_n)\|_\infty$ which satisfies the condition for centralizer operators 1.1 (iii) for $\lambda \geq \sup\{\lambda_1,\ldots,\lambda_n\}$ (where T_i satisfies 1.1 (iii) for λ_i). If $T \in Z(X)$, let $T_i := P_i \circ T \circ I_i$ (I_i = the i-th canonical injection, P_i = the canonical projection onto X_i, $i=1,\ldots,n$) which is in $Z(X_i)$. (Note that T leaves the M-summands X_i invariant.) It can easily be shown that $\phi(T_1,\ldots,T_n) = T$.

References

[AE$_1$] E.M. Alfsen-E.G. Effros: Structure in real Banach spaces I
Ann. of Math. 96 (1972), 98-128

[AE$_2$] E.M. Alfsen-E.G. Effros: Structure in real Banach spaces II
Ann. of Math. 96 (1972), 129-173

[B] E. Behrends: L^p-Struktur in Banachräumen
Studia Math. 55 (1975), 71-85

[G] S. Göbel: Über die L^p-Struktur in Produkten und den Zentralisator reeller Banachräume
Dissertation FU Berlin, 1975

[J] M. Jerison: The space of bounded maps into a Banach space
Ann. of Math. 52 (1950), 309-327

[L] H.E. Lacey: The isometrical theory of classical Banach spaces
Springer Verlag, Grundlehren 208, 1974

[S] K. Sundaresan: Spaces of continuous functions into Banach spaces
Studia Math. 48 (1973), 15-22

Ehrhard Behrends
I. Mathematisches Institut
Hüttenweg 9
D 1000 Berlin 33

K.-D. Bierstedt, B. Fuchssteiner (eds.)
Functional Analysis: Surveys and Recent Results

REPRESENTATION OF VECTOR LATTICES BY SPACES OF REAL FUNCTIONS

WOLFGANG HACKENBROCH

Dept. of Mathematics
University of Regensburg
84 Regensburg, Germany

INTRODUCTION

The general method of representation theory consists in a decomposition of the structure under consideration into components for which sufficently concrete models exist, and a device for re-integration of these components. As the title of this exposition indicates, the structure to be represented is that of a (real) vector lattice; the components may be various classes of ideals, Boolean algebras of projections or more general algebras of operators. Basic tools for the recombination process are topologies or just more or less classical integration procedures. The final aim consists in establishing vector lattice isomorphisms with well-behaved lattices of real functions like $\mathcal{C}(\Omega)$ (the continuous functions on some compact space Ω), $L^1(\mu)$ (the equivalence classes of μ-integrable functions with respect to some measure μ) or sublattices and appropriate generalizations of such spaces.
Given a vector lattice E there will usually be several ways of realizing E as a vector lattice $\hat{E}$ of functions; the "quality" of the representation will depend on the explicit description of $\hat{E}$ as well as on the way how additional structures (e.g. the norm structure if E is a Banach lattice) are reflected in $\hat{E}$.
In these notes we shall discuss a number of - partly classical - representation theorems from a uniform (functional analytic) point of view, starting with order unit spaces and stressing integration procedures. Thus e.g. representations by continuous (extended real valued) functions are derived from more detailed "integral representations" by a (strong) lifting argument. Throughout, the vector lattice E is seen together with its order center $\mathcal{Z}(E)$, and the main representations turn out to be spectral representations of Z(E). This aspect gives a better understanding also for many classical approaches from the Japanese school in the early forties starting out with the Stone representation of various Boolean algebras of projections.

1. ORDER UNIT SPACES AND THEIR REPRESENTATION

In this section we briefly fix notations and collect some fundamental facts on order unit spaces, using freely basic notions like "ordered vector space", "vector lattice" (abbreviated by v.l.), "order ideal"

and so on as presented e.g. in the first paragraphs of chapter II of H.H.Schaefer's monograph [20].
The letter E always denotes a real ordered vector space with order relation $\leq$ and positive cone E_+; if E is even a v.l., the symbols $\vee, \wedge, |\,|$ denote the lattice supremum, infimum and absolute value respectively. To avoid pathologies we shall always assume (without essential loss of generality) the ordering to be *Archimedean*, i.e. $x,y \in E$ with $nx \leq y$ for all $n \in \mathbb{N}$ implies $x \leq 0$.
For a fixed $u \in E_+$ let

$$E_u = \{x \in E:\ \pm x \leq \lambda u \text{ for some } \lambda \in \mathbb{R}_+\}$$

denote the order ideal generated by u. In view of the Archimedicity, $\|\ \|_u : \|x\|_u = \inf\{\lambda \in \mathbb{R}_+ : \pm x \leq \lambda u\}$ is a norm on E_u. If $E_u = E$, u is called an *order unit* for E and the pair (E,u) endowed with the norm $\|\ \|_u$ an *order unit space*. Thus in particular (E_u, u) (with its relative structure) is always an order unit space.
We denote by

$$S_u = \{\varphi \in E_u^*:\ \varphi \geq 0 \text{ and } \varphi(u) = 1\} = \{\varphi \in E_u^*:\ \|\varphi\| = \varphi(u) = 1\}$$

(E_u^* the algebraic dual of E_u; $\|\varphi\| := \sup\{|\varphi(x)| : \|x\|_u \leq 1\}$) the *state space* of (E_u, u) and observe, that S_u is a convex and weak*compact subset of E_u^*. It is not hard to see that the positive cone $E_u \cap E_+$ of E_u is closed with respect to the $\|\ \|_u$-topology. Therefore (by the Hahn-Banach theorem for $x \in E_u$ we have $x \geq 0$ iff $\varphi(x) \geq 0$ for all $\varphi \in E_u \cap E_+$. Thus the canonical evaluation mapping $x \mapsto \hat{x}$: $\hat{x}(\varphi) = \varphi(x)$ defines a bi-positive linear mapping from E_u into the order unit space $\mathcal{C}(S_u)$ (with the constant function 1 as order unit and the sup-norm $\|\ \| = \|\ \|_1$) of all continuous real valued functions on S_u, sending u onto 1. Finally, this situation can only be improved by restricting the functions $\hat{x}$ to the Silov boundary $\Omega_u = \overline{\text{ex}S_u}$ (weak*closure of the set $\text{ex}S_u$ of extremal points of S_u): by the boundary-property we still have a bi-positive linear mapping $x \mapsto \hat{x}|\Omega_u$ from (E_u, u) onto some (point separating) subspace of $(\mathcal{C}(\Omega_u), 1)$, sending u onto 1 (*Kadison's representation*). Such a mapping is automatically continuous with respect to the order unit norms $\|\ \|_u$ in E_u and $\|\ \|$ in $\mathcal{C}(\Omega_u)$. The details of this chain of arguments can be found in Alfsen's book [1], chapt. II §1.

In this context the next two propositions are of obvious interest. (As before, E is an Archimedean ordered vector space and u any element in E_+).

PROPOSITION 1: E_u is complete with respect to $\| \ \|_u$ iff the supremum in E_u, $\bigvee_{n=1}^{\infty} \sum_{k=1}^{n} x_k$, exists for any sequence (x_n) in E_u such that $0 \leq x_n \leq \lambda_n u$ with $\lambda_n \in \mathbb{R}_+$: $\sum_1^\infty \lambda_n < \infty$.
The proof is contained in [19] p.231.

PROPOSITION 2: i) Let E_u be a lattice in its relative ordering. Then

$$\Omega_u = \mathrm{ex}S_u = \{\varphi \in S_u \colon \varphi(x \vee y) = \max\{\varphi(x), \varphi(y)\}, \text{ all } x,y \in E_u\}.$$

ii) Let a bilinear multiplication "$\cdot$" on E_u be given such that $u \cdot x = x \cdot u = x$ for all $x \in E_u$ and $x \cdot y \geq 0$ whenever $x \geq 0$ and $y \geq 0$. Then

$$\Omega_u = \mathrm{ex}S_u = \{\varphi \in S_u \colon \varphi(x \cdot y) = \varphi(x)\varphi(y), \text{ all } x,y \in E_u\}.$$

PROOF: Since the subsets of lattice-preserving as well as of multiplicative states are closed in S_u it is enough to prove the second equality in i) and ii) respectively. For i) we refer to Alfsen's book [1], p.75. ii) Let $\varphi \in \mathrm{ex}S_u$ be given. Then for $0 \leq a \leq u$ $\varphi_a : x \mapsto \varphi(xa)$ is a linear functional on E_u with $0 \leq \varphi_a(x) \leq \varphi(x \cdot u) = \varphi(x)$ for every $0 \leq x \in E_u$, since $x \cdot (u-a) \geq 0$ as a product of non-negative elements. As $\varphi \in \mathrm{ex}S_u$, we have $\varphi_a = \lambda(a) \cdot \varphi$ for some $\lambda(a) \in \mathbb{R}_+$; evaluation at $x=u$ gives $\lambda(a) = \lambda(a)\varphi(u) = \varphi(u \cdot a) = \varphi(a)$. Therefore $\varphi(xa) = \varphi(a)\varphi(x)$ for all $x \geq 0$ and every $a : 0 \leq a \leq u$ and thus by linearity for all $x, a \in E_u$.
Conversely let $\varphi \in S_u$ be multiplicative and consider, for arbitrary $0 \leq a \in E_u$, the continuous affine mapping $l_a : S_u \to \mathbb{R}$ defined by $l_a(\psi) = \psi(a \cdot a) - 2\psi(a)\varphi(a) + \varphi(a \cdot a)$ ($= (\psi(a) - \varphi(a))^2$ if also ψ happens to be multiplicative, in particular for $\psi \in \mathrm{ex}S_u$). Then $l_a(\psi) \geq 0$ for $\psi \in \mathrm{ex}S_u$ and therefore for all $\psi \in S_u$ = convex closure of $\mathrm{ex}S_u$. Defining now
$l := \sum_{a : 0 \leq a \in E_u} l_a : S_u \to \overline{\mathbb{R}}_+$ (pointwise supremum over all finite partial sums), we obtain a lower semicontinuous affine function with $l(\varphi) = 0$. By Bauer's minimum principle (cf. [1], p.46) the minimum value 0 is attained at some point $\psi \in \mathrm{ex}S_u$. So $0 = l_a(\psi) = (\psi(a) - \varphi(a))^2$ for all $0 \leq a \in E_u$ and thus $\varphi = \psi \in \mathrm{ex}S_u$. ■

As special cases of prop.1 we notice

COROLLARY 3: E is relatively *uniformly complete* (i.e. E_u is complete with respect to $\| \ \|_u$ for all $u \geqq 0$) if

either i) E is *monotone σ-complete* (i.e. for every increasing majorized sequence (x_n) in E_+ the supremum $\overset{\infty}{\underset{1}{\vee}} x_n$ exists in E)

or ii) E is a Banach lattice.

As a very important example for the situation described in prop.2. ii) we mention the *(order-) center* of an ordered vector space E, i.e. the order ideal $\mathcal{Z}(E)$ in the ordered vector space L(E) of linear endomorphisms of E (with positive cone $L(E)_+ = \{T \in L(E): Tx \geqq 0$ for all $x \geqq 0\}$), generated by the identity operator 1:

$$\mathcal{Z}(E) = \{T \in L(E): \pm T \leqq \lambda 1 \text{ for some } \lambda \in \mathbb{R}_+\}.$$

<u>COROLLARY 4</u>: If E is relatively uniformly complete, then also $\mathcal{Z}(E)$ is complete with respect to its order unit norm $\| \ \|_1$.

<u>PROOF</u>: According to prop.1 take any sequence (T_n) in $\mathcal{Z}(E)$ such that $0 \leqq T_n \leqq \lambda_n 1$ for some summable sequence (λ_n) in $\mathbb{R}_+$. By prop.1, for every $u \in E_+$ the supremum

$$Su := \bigvee_{n=1}^{\infty} \sum_{k=1}^{n} T_k u \ \left(\leqq \left(\sum_1^{\infty} {}_n \right) \cdot u \right)$$

exists. Obviously, the mapping $S: u \longmapsto Su$ is additive and positively homogeneous on E_+ and therefore extends to a unique linear mapping $S: E \to E$. It is immediate from the definition that $0 \leqq S \leqq \left(\sum_1^{\infty} \lambda_n \right) 1$ (and thus $S \in \mathcal{Z}(E)$) and also that S is the smallest upper bound for all finite partial sums $\sum_1^n T_k$ in $\mathcal{Z}(E)$. Another application of prop.1 gives the desired result. ■

Combining prop.2 with Kadison's representation and the Stone-Weierstrass-theorem, we obtain in particular for any (Archimedean) v.l. E the following canonical isomorphisms, isometrical with respect to the corresponding order unit norms:

i) For every $u \geqq 0$, E_u is lattice isomorphic to a dense sublattice $\hat{E}_u$ of $\mathcal{C}(\Omega_u)$ (Ω_u compact Hausdorff).

ii) $\mathcal{Z}(E)$ is algebraically and order isomorphic to a dense subalgebra $\hat{\mathcal{Z}} \subset \mathcal{C}(\Omega)$ (Ω compact Hausdorff); in particular $\mathcal{Z}(E)$ is a commutative normed algebra over the real field.

iii) If E is in addition relatively uniformly complete, we have $E_u = \mathcal{C}(\Omega_u)$ and (by cor.4) $\hat{\mathcal{Z}}=\mathcal{C}(\Omega)$; in particular, $\mathcal{Z}(E)$ is then also a lattice.

In the derivation of these representations, the compact spaces Ω_u and Ω respectively arose as the extremal sets of the corresponding state spaces. They are in fact intrinsically characterized as such and in particular determined up to homeomorphism by the following simple result:

PROPOSITION 5: Let Ω be a compact Hausdorff space and $F\subset\mathcal{C}(\Omega)$ a norm-dense linear subspace containing the constants; consider $(F,1)$ as an order unit space with state space S (endowed with its compact weak* topology). Then

$$\begin{array}{l}\delta:\Omega\to S\\ \quad t\mapsto\delta_t\end{array} \quad : \delta_t f := f(t),\ f\in F$$

maps Ω homeomorphically onto exS.

PROOF: If $F=\mathcal{C}(\Omega)$ this is a well-known fact. But since otherwise F is assumed to be dense in $\mathcal{C}(\Omega)$ we may identify (even topologically) S with the state space of $(\mathcal{C}(\Omega),1)$ by continuously extending the functionals in S to functionals on $\mathcal{C}(\Omega)$. (For a slightly more general result compare theorem II 2.1 in [1]). ■

EXAMPLE: Let Ω_o be a locally compact Hausdorff space and consider the (relatively uniformly complete) v.l. $E=\mathcal{K}(\Omega_o)$ of all continuous functions with compact support. Then $\mathcal{Z}(E)$ is algebraically and order isomorphic (in particular isometric) to the space $\mathcal{C}_b(\Omega_o)$ of all bounded continuous real valued functions on Ω_o by means of the mapping from $\mathcal{C}_b(\Omega_o)$ to $\mathcal{Z}(E)$ sending τ onto $[\tau]$, the operator of multiplication by τ.
(For, evidently $\tau\mapsto[\tau]$ is a bi-positive and also multiplicative linear mapping from $\mathcal{C}_b(\Omega_o)$ into $\mathcal{Z}(E)$, sending 1 to the identity operator. Conversely, given $T\in\mathcal{Z}(E)$, for any fixed $t_o\in\Omega_o$ the Radon measure $u\mapsto(Tu)(t_o)$ is estimated according to $|(Tu)(t_o)|\leq\|T\|_1|u(t_o)|$, all $u\in\mathcal{K}(\Omega_o)$. So its support is contained in $\{t_o\}$, i.e. $(Tu)(t_o) = \tau(t_o)u(t_o)$, $u\in\mathcal{K}(\Omega_o)$, with some function $\tau:\Omega_o\to\mathbb{R}$ bounded by $\|T\|_1$. Since Tu is continuous for every $u\in\mathcal{K}(\Omega_o)$, by Urysohn's lemma it is easily seen that also τ has to be continuous).

2. INTEGRATION PROCEDURES

We consider an Archimedean v.l. E. In section 1. it was shown that for every $u \in E_+$ we have a v.l. isomorphism of the ideal E_u generated by u onto a dense sublattice $\hat{E}_u \subset \mathfrak{C}(\Omega_u)$ for some compact space Ω_u. We want to "integrate" these isomorphisms to a representation of all of E. Although more general situations are conceivable and partly carried out (e.g. in [9]) we shall restrict ourselves for simplicity mostly to the simplest cases where E is either a Banach lattice or σ-complete (without any topology). By cor.3 of section 1 in both cases we have $\hat{E}_u = \mathfrak{C}(\Omega_u)$.

Unless u is an order unit for E, E_u will be a proper ideal of E. Closing with respect to order limits, we arrive at the *band* generated by u, which (in view of the Archimedicity of E) is equal to $u^{\perp\perp} := (u^\perp)^\perp$; here the superscript ⊥ denotes the *orthogonal complement* (i.e. $M^\perp =$ $= \{x \in E : |x| \wedge |y| = 0 \text{ for all } y \in M\}$, defined for any subset M of E; cf. [20] p.61). If $u^{\perp\perp} = E$ we call u a *weak order unit* for E; if E is a topological v.l. ([19] p.235) and the topological closure $\overline{E_u}$ of E_u equals E, we call u a *topological order unit* for E. Evidently, order units are topological as well as weak order units, and topological order units are still weak order units (by the continuity of the lattice operations). For $E = \mathfrak{C}([0,1])$, $u : u(t) = t$ is an example for a weak order unit which is not a topological order unit.
Instead of considering just one $u \in E_+$ we can look at maximal subsets $U \subset E_+$ of pairwise orthogonal elements. Then the order ideal E_U generated by U is the order direct sum of the E_u, $u \in U$, and the band $U^{\perp\perp}$ generated by U equals E (by maximality). Again, if E is a topological v.l., U is called a *topological orthogonal system* whenever $\overline{E_U} = E$. This will e.g. certainly be the case when order convergence implies topological convergence (thus in particular for the important class of Banach lattices with order continuous norm, studied in chapt. II §5 of [20]). But there are Banach lattices without any topological orthogonal system ([20], p.170).

If E is relatively uniformly complete and we choose (by Zorn's lemma) any maximal system U of pairwise orthogonal positive elements (briefly m.o.s.) in E, by adding up the canonical v.l. isomorphisms $E_u \cong \mathfrak{C}(\Omega_u)$, $u \in U$, we obtain a v.l. isomorphism

$$(*) \qquad T_U : E_U = \bigoplus_{u\in U} E_u \to \bigoplus_{u\in U} \mathcal{C}(\Omega_u) = \mathcal{K}(\Omega),$$

where the locally compact space Ω is the topological sum of the compact spaces Ω_u, $u\in U$, and where an arbitrary element $(f_{u_1},\dots,f_{u_r}) \in \bigoplus_{u\in U} \mathcal{C}(\Omega_u)$ has been identified with $f\in\mathcal{K}(\Omega)$: $f(t) = f_{u_i}(t)$ for $t\in\Omega_{u_i}$ and $f(t)=0$ otherwise. $\mathcal{K}(\Omega)$ again denotes the v.l. of all continuous functions with compact support on Ω.
By (*), every positive linear functional φ on E corresponds to a positive Radon measure $\varphi\circ T_U^{-1}$ on Ω. Since locally convex v.l. topologies are induced by families of positive linear functionals, we are quite naturally led to study the following situation: Let Ω be a locally compact space and M a family of positive Radon (= regular Borel = Baire) measures on Ω; for measurable $f:\Omega\to\mathbb{R}$ put $\|f\|_M := \sup_{\mu\in M} \int |f|\,d\mu$ and consider the space

$$\mathcal{L}^1(M) := \{f:\ f:\Omega\to\mathbb{R} \text{ measurable and } \|f\|_M<\infty\}.$$

Evidently $\|\ \|_M$ is a seminorm on $\mathcal{L}^1(M)$; by $L^1(M)$ we denote the corresponding normed quotient of $\mathcal{L}^1(M)$.

REMARKS: i) $L^1(M)$ is a Banach lattice (with respect to the quotient structure induced from pointwise v.l. operations in $\mathcal{L}^1(M)$).

(For, only completeness requires a proof. Let (f_n) in $\mathcal{L}^1(M)$ be $\|\ \|_M$-Cauchy and consider a subsequence $\left(f_{n_k}\right)_k$ such that $\sum_{k=1}^{\infty} \|f_{n_{k+1}}-f_{n_k}\|_M<\infty$. Then there is some measurable $f:\Omega\to\mathbb{R}$ such that $f_{n_k}\to f$ μ-a.e. for every $\mu\in M$. To see that $f\in\mathcal{L}^1(M)$ and $\|f_{n_k}-f\|_M\to 0$ as $k\to\infty$ just notice that for all $\mu\in M$

$$\|f-f_{n_k}\|_{L^1(\mu)} \leq \sum_{i=k}^{\infty} \|f_{n_{i+1}}-f_{n_i}\|_{L^1(\mu)} \leq \sum_{i=k}^{\infty} \|f_{n_{i+1}}-f_{n_i}\|_M).$$

ii) If M consists of only one element μ, obviously $L^1(M)=L^1(\mu)$; if on the other hand $M=\{\delta_t;t\in\Omega\}$ (δ_t = Dirac measure for t), we have $\mathcal{L}^1(M)=L^1(M)$ = set of all bounded measurable functions. The general situation is a mixture of these two extremal cases, which shows that we cannot expect to have nice properties like dominated convergence for $L^1(M)$. Nevertheless the notion seems to be useful also otherwise (see e.g. [14] and [17]).

If $M=\{\mu\}$, $\mathcal{K}(\Omega)$ is dense in $L^1(M)$; for general M with $\mathcal{K}(\Omega)\subset\mathcal{L}^1(M)$ let $L^1(M;\mathcal{K})$ denote the closure of $\mathcal{K}(\Omega)$ in $L^1(M)$. The next theorem reveals the general character of these Banach lattices.

THEOREM 1 ([21]): Let E be a Banach lattice possessing a topological orthogonal system. Then there exists a locally compact space Ω, direct topological sum of compact spaces, and a uniquely determined smallest vaguely compact family M of positive Radon measures, such that E is isometrically v.l. isomorphic to $L^1(M;\mathcal{K})$ and $\mathcal{K}(\Omega)$ is "contained" in $L^1(M;\mathcal{K})$ as a (norm dense) ideal.

PROOF: Let U be a topological orthogonal system and $T_U:E_U\to\mathcal{K}(\Omega)$ the corresponding v.l. isomorphism described in (*) above. Let N denote the unique smallest weak*compact subset of $\{\varphi\in E'_+:\|\varphi\|\leq 1\}$ (namely its Silov boundary with respect to the cone E_+) which determines the norm of E according to $\|x\| = \sup_{\varphi\in N} \varphi(|x|)$. Since E_U is dense in E, $\sigma(E',E)$ and $\sigma(E',E_U)$ coincide on N and therefore $M:=\{\varphi\circ T_U^{-1}:\varphi\in N\}$ is a vaguely compact set of Radon measures on Ω such that $\mathcal{K}(\Omega)\subset\mathcal{L}^1(M)$ and T_U is isometric with respect to the corresponding norm $\| \ \|_M$ on $\mathcal{K}(\Omega)$. The continuous extension of T_U to all of E is then an isometrical lattice isomorphism from E onto $L^1(M;\mathcal{K})$. Also, by the uniqueness of the Silov boundary N, it is clear that M is the unique smallest vaguely compact set of positive Radon measures with the stated properties. ■

For further discussion of this result we refer the reader to chapt. III §5 of [20]. It should be obvious that it interpolates the classical representations of Kakutani [11] for *abstract L-spaces* (i.e. Banach lattices whose norm is additive on the positive cone) and *abstract M-spaces* with unit (i.e. Banach lattices with order unit whose norm is the order unit norm) in the same way as $L^1(M;\mathcal{K})$ interpolates between $L^1(\mu)$ and $\mathcal{C}(\Omega)$.—For abstract M-spaces without unit, representations as spaces of "φ-dominated" continuous functions obtained by Goullet de Rugy [6] are particularly worthwhile mentioning.

So far, the possibility for completing the representation (*) of a dense ideal of E (arbitrary relatively uniformly complete v.l.) to a representation of all of E rested upon the use of positive Radon measures, induced by positive linear functionals on E. But, if no v.l. topology on E is given, there is no guarantee for the existence of such functionals. Nevertheless, under reasonable assumptions the lat-

tice E itself contains a certain abstract (vector) measure space which allows us again to build up E by an integration procedure as the corresponding space of integrable real valued functions.
Instead of repeating the proof of the general result given in [9], let us study in some detail only how the vector measure emerges, leaving aside all purely technical considerations of the corresponding integration theory. The notion of integrability and the integral for real valued functions with respect to a measure taking values in the positive cone of a monotone σ-complete ordered vector space has been developed - in complete analogy to the order-oriented approach in real analysis - by J.D.M. Wright ([22]). In view of the results of section 1 on the representation of order unit spaces we shall consider this situation:
Let Ω be a compact Hausdorff space and $F\subset\mathcal{C}(\Omega)$ a point separating sublattice, containing the constants. Let us recall (cf.[20] p.64) that for $u\in F_+$ the band $u^{\perp\perp}$ generated by u is a *projection band* (i.e. F is the direct sum of $u^{\perp}$ and $u^{\perp\perp}$) iff for every $f\in F_+$ the supremum $f_u :=$ $:= \bigvee_{n=1}^{\infty} (f\wedge nu)$ exists in F; if so, f_u is the component of f in $u^{\perp\perp}$ of the direct decomposition $F=u^{\perp\perp}\oplus u^{\perp}$. F is said to have the *principal projection property* if every principal band $u^{\perp\perp}$ is a projection band. This will, by the above criterion, certainly be the case whenever F is a σ-complete v.l..
Now we have the following series of simple remarks:

i) The sets $[f>0]:=\{t\in\Omega : f(t)>0\}$, $f\in F_+$, form a basis for the topology of Ω (since Ω is compact and $F\subset\mathcal{C}(\Omega)$ a dense sublattice).

ii) If $u^{\perp\perp}$ is a projection band ($u\in F_+$), the component f_u of any $f\in F_+$ is given by $f_u = \chi_{\overline{[u>0]}}\cdot f$.
(For, by the above formula for f_u, $\chi_{[u>0]}\cdot f\leqq f_u$ is obvious. On the other hand, we have $f_u\leqq\chi_{\overline{[u>0]}}\cdot f$ (which proves our claim), because otherwise we would have $f_u(t_o)>0$ for some $t_o\notin\overline{[u>0]}$ and thus $f_u(t)\geqq$ $\geqq\delta>0$ on some open set V in the complement of $\overline{[u>0]}$, containing t_o. By i) choose $h\in F_+$ with $0\leqq h\leq 1=h(t_o)$ and support in V. Then $f_u \gneqq f_u-\delta\cdot h$ $\geqq f\wedge nu$ for all n, a contradiction).

iii) Let $\mathfrak{a}_o$ denote the Boolean algebra of all sets $[f>0]$ for $f\in F_+$ with $0\leqq f=f^2$, and consider

$$\mathfrak{a} = \{A\in\mathrm{Borel}(\Omega) : A\Delta A' \text{ is meager for some } A'\in\mathfrak{a}_o\},$$

the Boolean algebra of all Borel sets in Ω differing from some set in $\mathfrak{A}_o$ only by a meager set (a set is *meager* if it is a countable union of sets whose closure has empty interior). Since the sets in $\mathfrak{A}_o$ must be clopen, $\mathfrak{A}_o$ is trivial when Ω is connected. On the other hand, if $F=\mathcal{C}(\Omega)$ is σ-complete or even Dedekind complete, it is well-known (cf. [10] §23) that $\mathfrak{A}$ is the collection of all Baire- or Borel sets respectively. In general, the appropriately adjusted assumption seems to be that F have the principal projection property. Then every $f\in F$ is $\mathfrak{A}$-measurable.

(For, $[f\geq 0]=[|f|-f=0]$, so that, since F is a sublattice of $\mathcal{C}(\Omega)$, it is enough to show $[f>0]\in\mathfrak{A}$ for $f\in F_+$. But $\overline{[f>0]}\setminus[f>0]$ is meager, and $\overline{[f>0]} = [\chi_{\overline{[f>0]}} > 0]\in\mathfrak{A}_o$, since by assumption $f^{\perp\perp}$ is a projection band and so by ii) $\chi_{\overline{[f>0]}} = 1_f\in F$).

iv) Finally let us note that the $A'\in\mathfrak{A}_o$ belonging to $A\in\mathfrak{A}$ are uniquely determined (since, in a compact space, every open meager set must be empty). Therefore we can define the mapping $\pi:\mathfrak{A}\to F_+$ sending A to $\chi_{A'}$. We claim that π is *σ-additive*, i.e. additve and $A_n\searrow\emptyset$ in $\mathfrak{A}$ $\Rightarrow$ $\bigwedge_1^\infty\pi(A_n)=0$ in F.

(Additivity is obvious. If $A_n\searrow\emptyset$ in $\mathfrak{A}$, there are meager sets N_n such that $A_n'\subset A_n\cup N_n$ and thus $\bigcap_1^\infty A_n' \subset (\bigcap_1^\infty A_n)\cup(\bigcup_1^\infty N_n) = \bigcup_1^\infty N_n$. Again, since $\bigcup_1^\infty N_n$ is meager, the interior of $\bigcap_1^\infty A_n'$ must be empty, which implies $\bigwedge_1^\infty\pi(A_n)=0$).

It is rather obvious that, for any reasonable notion of an integral with respect to the measure π contructed above, we should have the tautology $f=\int f d\pi$ for all $f\in F$. In fact, as shown in [9] even under the weaker hypothesis of the principal projection property, much more is true:

Let E be an arbitrary σ-complete v.l. and fix a maximal system U of positive elements in E. For each $u\in U$, let $\pi_u:\mathrm{Baire}(\Omega_u)\to E_+$ denote the vector measure constructed above (for $F=E_u\cong\hat{E}_u=\mathcal{C}(\Omega_u)$; $\mathfrak{A}=\mathrm{Baire}(\Omega_u)$). On the topological direct sum Ω of the Ω_u let $\pi:\mathrm{Baire}(\Omega)\to\overline{E_+}$ be the direct sum of the π_u (with $\pi(A):=\infty$ whenever for some Baire set $A\subset\Omega$ the set of finite partial sums $\Sigma\pi_u(A\cap\Omega_u)$ does not have a supremum in E). This is a Baire measure in the sense of [22]; let $L^1(\pi)$ denote the corresponding space of π-integrable real functions on Ω, identifying functions equal π-locally almost everywhere (cf. [7],[9]). If $x\mapsto\hat{x}$ denotes the representation T_U defined in (*), by the very contruction of π we have $x=\int\hat{x}d\pi$ for all $x\in E_U$. But at the same time π gives the desired extension of T_U to a representation of all of E, namely

THEOREM 2 ([7],[9]): $L^1(\pi)$ is a σ-complete v.l. isomorphic to E. An isomorphism is given explicitly by the integration mapping $f \mapsto \int f d\pi$.

The theorem essentially states that an arbitrary σ-complete v.l. is built up by its principal-band projections in much the same way as any L^1-space is built up by its integrable sets. It should be mentioned also that the π-integral enjoys most of the nice properties (like continuity with respect to pointwise dominated convergence) familiar from classical Lebesgue integration.
To close this section let us briefly describe two typical applications of theorem 2. The first one shows that it is essentially a general Freudenthal spectral theorem, related to the classical Freudenthal theorem (cf. [15] chapt. 6) in the same way as the Gelfand isomorphism for a commutative C*-algebra is related to the spectral theorem for a single normal operator.

COROLLARY 3 ([7]): Let E be a σ-complete v.l. and $u \in E_+$ be given. For every x in the band $u^{\perp\perp}$ generated by u there is exactly one σ-additive ρ:Borel$(\mathbb{R}) \to E_+$ which is *spectral* (i.e. $\rho(A \cap B) = \rho(A) \wedge \rho(B)$), such that

$$\rho(\mathbb{R}) = u \quad \text{and} \quad x = \int \lambda \rho(d\lambda).$$

In fact, the distribution of ρ is given by *Freudenthal's formula*

$$\rho(]-\infty,\lambda[) = \bigvee_{n=1}^{\infty} (u \wedge n(\lambda u - x)_+), \ \lambda \in \mathbb{R}.$$

PROOF: Applying theorem 2 to the σ-complete v.l. $E = u^{\perp\perp}$, we obtain an E_+-valued measure π on the Baire sets of some compact space Ω (namely π_u on Ω_u) such that $\pi(\Omega) = u$ and π-integration is a v.l. isomorphism from $L^1(\pi)$ onto E. Let the π-integrable function f be a representative for x, i.e. $\int f d\pi = x$, and take for ρ the image measure $A \mapsto \pi(f^{-1}(A))$. This measure evidently has the desired properties; in particular, Freudenthal's formula immediately follows from $\chi_{]-\infty,\lambda[} = \sup_{n \in \mathbb{N}} \inf\{1, n(\lambda - id)_+\}$ together with the Beppo Levi theorem for monotone sequences. ∎

The second application combines the abstract integration procedure with additive linear functionals to obtain a classical integral; for p=1 it is Kakutani's well-known representation of abstract L-spaces, and our way of proof is much in the spirit of Kakutani's original paper [11].

COROLLARY 4 (cf. [12], p.112): Let the Banch lattice E be an *abstract* *L^p-space*, i.e. $\|x+y\|^p=\|x\|^p+\|y\|^p$ for $x \wedge y = 0$ ($1 \leq p<\infty$). Then there is a positive Radon measure μ on some locally compact space Ω such that $E \cong L^p(\mu)$ isometrically lattice isomorphic.

PROOF: Using a lemma by Meyer-Nieberg ([16] p.p.643 and 645) it is easily seen that every increasing majorized sequence (x_n) in E_+ is Cauchy. Therefore, (x_n) converges to its supremum, and theorem 2 can be applied, giving a vector measure π such that π-integration is a v.l. isomorphism from $L^1(\pi)$ to E. But then $\mu=\|\pi(\cdot)\|^p$ is a non-negative Baire measure (to be canonically identified with a Radon measure) having the desired properties, since for every π-integrable simple function $f=\Sigma\alpha_i\chi_{A_i}$ (A_i disjoint Baire sets; $\alpha_i \geq 0$)

$$\|\int f d\pi\|^p = \Sigma\|\alpha_i\pi(A_i)\|^p = \Sigma|\alpha_i|^p\mu(A_i) = \int|f|^p d\mu = \|f\|^p_{L^p(\mu)} . \blacksquare$$

3. REPRESENTATION BY CONTINUOUS FUNCTIONS

The family M of Radon measures in theorem 1 and in the same way the vector measure π in theorem 2 of section 2 determine the "shape" of the function space $\hat{E}$ representing a v.l. E, defining the right kind of closure for $\mathcal{K}(\Omega)$ as well as describing which functions in $\hat{E}$ have to be identified. In addition, M-integration gives us the norm of E if E is a Banach lattice, and π-integration explicitly represents the v.l. isomorphism $\hat{E}\to E$.

In dealing with equivalence classes M-a.e. or π-a.e. respectively, it will be convenient to know that each such class has exactly one continuous representative if we are ready to admit the values $\pm\infty$. We shall now derive this "strong lifting" property, thereby becoming able to compare the results of section 2 with several classical v.l. representations obtained already in the early forties mostly by the Japanese school. (For a discussion of these more historical aspects and also for many related results, chapt.7 of Luxemburg-Zaanen's monograph [15] is highly recommended).

According to the situation encountered in theorems 1 and 2 of section 2, let us consider a locally compact space Ω and an ideal $\mathfrak{n}$ of subsets of Ω, each $N\in\mathfrak{n}$ having empty interior. Denote by "$\sim$" the equivalence relation "$f \sim g$ iff $[f \neq g] \in \mathfrak{n}$" on the set $\mathbb{R}^{\Omega}$ of all real functions on Ω. As usual then the quotient set $\mathbb{R}^{\Omega}/_{\sim}$ is a linear space and a vector lattice, the v.l. operations being defined pointwise for representatives of the corresponding equivalence classes. Note that we may consider $\mathcal{K}(\Omega)$ as a sublattice of $\mathbb{R}^{\Omega}/_{\sim}$, since for $f,g \in \mathcal{K}(\Omega)$ with $f \sim g$ the set $[f \neq g] \in \mathfrak{n}$, is open and thus empty by assumption, so that $f=g$. Denoting

$$\mathcal{C}^{\infty}(\Omega) := \{f: f:\Omega \to \overline{\mathbb{R}} \text{ continuous with } [f=\pm\infty] \in \mathfrak{n}\},$$

we have the following

PROPOSITION 1: Let $F \subset \mathbb{R}^{\Omega}/_{\sim}$ be a sublattice containing $\mathcal{K}(\Omega)$ as an ideal and assume that for every class $\tilde{f} \in F$ and every $h \in \mathcal{K}(\Omega)_{+}$ the class of $h \cdot \frac{f}{1+|f|}$ again belongs to F. Then every $\tilde{f} \in F$ has exactly one representative $f \in \mathcal{C}^{\infty}_{\mathfrak{n}}(\Omega)$.

PROOF: Since every $N \in \mathfrak{n}$ has empty interior there is at most one such representative. By the same reason it is also sufficient, for a given class $\tilde{f} \in F$ belonging to the real valued function $f:\Omega \to \mathbb{R}$, to find for every compact subset $K \subset \Omega$ a function $f_K \in \mathcal{C}^{\infty}_{\mathfrak{n}}(\Omega)$ such that $[f \neq f_K] \cap K \in \mathfrak{n}$. So let f and K be given and fix any $k \in \mathcal{K}(\Omega)$ such that $0 \leq k \leq 1$ and $k(t)=1$ for $t \in K$. Consider the bijective, continuous mapping $\sigma:\overline{\mathbb{R}} \to [-1,1]$ with $\sigma(t) = \frac{t}{1+|t|}$ for $t \in \mathbb{R}$, $\sigma(\pm\infty)=\pm 1$, and its continuous inverse $\sigma^{-1}:[-1,1] \to \overline{\mathbb{R}}$ sending $t \to \frac{t}{1-|t|}$ for $-1<t<1$ and ± 1 to $\pm\infty$ respectively. By hypothesis the class $\tilde{g}$ of $g:=(\sigma \circ f) \cdot k$ is in F. Since $\mathcal{K}(\Omega)$ is an ideal in F, there is a representative $h \in \mathcal{K}(\Omega)$ for $\tilde{g}$. Then necessarily $-1 \leq h \leq 1$; put $f_K :=$ $:= \sigma^{-1} \circ h$. Then $f_K:\Omega \to \overline{\mathbb{R}}$ is continuous and we have, with $[h \neq g] =: N \in \mathfrak{n}$,

$$[f_K \neq \pm\infty] = [h=\pm 1] \subset [g=\pm 1] \cup N \subset [f=\pm\infty] \cup N = N$$

i.e. $f_K \in \mathcal{C}^{\infty}_{\mathfrak{n}}(\Omega)$, and furthermore

$$K \cap [f_K \neq f] = K \cap [h \neq \sigma \circ f] = K \cap N \in \mathfrak{n}. \quad \blacksquare$$

REMARK: It is clear that the "lifting" mapping which assigns to every $\tilde{f} \in F$ its extended real valued representative is a v.l. isomorphism from F onto some v.l. $\hat{F} \subset \mathcal{C}^{\infty}_{\mathfrak{n}}(\Omega)$. Here the v.l. operations in $\hat{F}$ are defined pointwise on the (open dense) set where the functions involved have finite values.

The next result is classical (see [4],[13],[21],[7];[15] chapt.7 §51). That it can also be deduced easily from the integral representations of section 2 indicates the significance of integration procedures in representation theory.

COROLLARY 2: Let E be either a Banach lattice with topological orthogonal system or any σ-complete v.l.. Then E is lattice isomorphic to a v.l. $\hat{E}$ of continuous extended real valued functions on some locally compact space Ω (direct topological sum of compact spaces), containing $\mathcal{K}(\Omega)$ as an ideal which is topologically or order dense respectively. The v.l. operations in $\hat{E}$ are defined pointwise on common sets of finiteness of the functions involved.

PROOF: We start with the integral representations given in theorems 1 and 2 of the last section. In both cases we have a locally compact space Ω (direct topological sum of compact spaces) and v. lattices $L^1(M;\mathcal{K})$ and $L^1(\pi)$ of equivalence classes of real functions, v.l. isomorphic to E. They contain $\mathcal{K}(\Omega)$ as a dense ideal (in such a way that different functions in $\mathcal{K}(\Omega)$ determine different classes). In particular, the common M-null sets as well as the local π-null sets have to have empty interior. Taking for $\mathfrak{n}$ the ideal of common M-null sets we obviously have $L^1(M;\mathcal{K}) \subset \mathbb{R}^{\Omega}/_{\sim}$; in the same way $L^1(\pi) \subset \mathbb{R}^{\Omega}/_{\sim}$ if we take $\mathfrak{n}$ to be the ideal of local π-null sets. By the "remark" above it is therefore enough to show the existence of a lifting from $L^1(M;\mathcal{K})$ and $L^1(\pi)$ respectively into $\mathcal{C}^{\infty}_{\mathfrak{n}}(\Omega)$. But this is done by prop.1: That for $f\in\mathcal{L}^1(\pi)$ and $h\in\mathcal{K}(\Omega)_+$ we have $\frac{f}{1+|f|}\cdot h \in \mathcal{L}^1(\pi)$ is immediate from the definition of π-integrability. On the other hand for $\tilde{f}\in L^1(M;\mathcal{K})$ choose a representative f and a sequence (f_n) in $\mathcal{K}(\Omega)$ tending to f with respect to $\| \; \|_M$. Since for all $s,t\in\mathbb{R}$ we have $\left|\frac{s}{1+|s|} - \frac{t}{1+|t|}\right| \leqq |s-t|$, we obtain, for arbitrary $h\in\mathcal{K}(\Omega)_+$,

$$\left\|\frac{f}{1+|f|}\cdot h - \frac{f_n}{1+|f_n|}\cdot h\right\|_M \leqq \|f-f_n\|_M \cdot \|h\| \to 0.$$

Therefore again the class of $\frac{f}{1+|f|}\cdot h$ is in $L^1(M;\mathcal{K})$ as required in prop.1.

■

4. CONNECTIONS WITH SPECTRAL THEORY AND RELATED REPRESENTATIONS

In spectral theory one is given a Boolean algebra of projections or, more generally, any commutative algebra $\mathcal{Z}$ of operators acting on a given vector space E. The intention of *spectral representation* theory is to find a linear isomorphism R from E onto some function space $\hat{E}$ which "trivializes" $\mathcal{Z}$ in the sense that the operators RTR^{-1}, $T\in\mathcal{Z}$, become multiplications by scalar valued functions. The most familiar example is possibly von Neumann's spectral representation of commutative C*-algebras of operators on a Hilbert space. This example already shows that in general, for $\hat{E}$ we have to take into account also vector valued functions (viz. direct integrals of Hilbert spaces in the case of C*-algebras). It is not hard to imagine that, the better $\mathcal{Z}$ corresponds to the structure of E, the simpler the construction of $\hat{E}$ will be.

In slight difference to the situation in spectral theory, in our case a v.l. E is given and we are free to choose a commutative operator algebra $\mathcal{Z}$ on E matching the structure of E in an optimal way. If E is a purely algebraic v.l., the order center $\mathbf{Z}(E)$ discussed in section 1 will be a "canonical" candidate, as we shall presently see. But already in the case of Banach lattices, in view of the additional norm structure other algebras might be of similar importance, in particular those investigated by Alfsen-Effros [2] revealing L- and M-structure in normed spaces. We shall not pursue this interesting aspect further here but rather refer to the recent paper [5] by Gierz and to the forthcoming lecture notes [3] by Behrends et al. Also the work of Portenier [18] should be mentioned in this context.

In view of the fact, that the "continuous" representation of cor.2 in section 3 of a v.l. E is obtained by lifting the "integral" representations of theorems 1 and 2 in section 2, the next theorem shows that all those representations (in the σ-complete but not complete case at least when the locally compact space Ω is a countable union of compact subsets) are in fact spectral representations of the center of E. One has only to note that, for any v.l. isomorphism $R:E\to\hat{E}$; evidently $RZ(E)R^{-1} = Z(\hat{E})$.

To formulate the theorem, let us call an ideal F in the v.l. E [σ-] *order dense* whenever to each $x\in E_+$ there is an increasing net [sequence] $(x_i)_{i\in T}$ in F_+ with lattice supremum $\bigvee_{i\in I} x_i = x$. In particular, if U is any maximal orthogonal system of positive elements in E, we have (cf. chapt. II, prop.2.11 of [20]), for any $x \in E_+$

$$x = \vee\left\{x \wedge \left(n \sum_{u \in V} u\right) : n \in \mathbb{N};\ V \subset U \text{ finite.}\right\};$$

since each $x \wedge (n \sum_{u \in V} u)$ is in the ideal E_U generated by U, we see that E_U is order dense and, for countable U, even σ-order dense.

Before stating the theorem recall the situation in cor.2 of section 3: We consider a locally compact space Ω and an ideal $\mathfrak{n}$ of subsets of Ω with empty interior. Let $F \subset \mathcal{C}_{\mathfrak{n}}^{\infty}(\Omega)$ denote any v.l. (with v.l. operations pointwise on sets of finiteness, as explained in section 3), containing $\mathcal{K}(\Omega)$ as an ideal. Note that $\mathcal{K}(\Omega)$ is necessarily order dense in F (since, by Urysohn's lemma, $\mathcal{K}(\Omega)^{\perp} = \{0\}$); for σ-compact Ω, $\mathcal{K}(\Omega)$ is even σ-order dense. Let us say a bounded continuous $\tau : \Omega \to R$ *multiplies* F whenever for each $f \in F$ the continuous real function $t \mapsto \tau(t) f(t)$, defined on $[f \neq \pm\infty]$, extends to a (unique) function in F, denoted by $[\tau]f$. With these conventions we have

<u>THEOREM 1</u>: i) The set $\mathcal{C}_b^F(\Omega) := \{\tau \in \mathcal{C}_b(\Omega) : \tau \text{ multiplies } F\}$ is a subalgebra of $\mathcal{C}_b(\Omega)$, and the mapping $\tau \mapsto [\tau]$ describes a bi-positive algebra isomorphism from $\mathcal{C}_b^F(\Omega)$ onto $\mathcal{Z}(F)$.

ii) We have $\mathcal{C}_b^F(\Omega) = \mathcal{C}_b(\Omega)$ if either the v.l. F is complete or if F is σ-complete and Ω is σ-compact or if F is a Banach lattice and $\mathcal{K}(\Omega)$ is norm dense in F.

<u>PROOF</u>: i) It is obvious that $\mathcal{C}_b^F(\Omega)$ is a subalgebra of $\mathcal{C}_b(\Omega)$ and also that $\tau \mapsto [\tau]$ is a bi-positive algebra isomorphism from $\mathcal{C}_b^F(\Omega)$ into $\mathcal{Z}(F)$. To show its surjectivity, let $T \in \mathcal{Z}(F)$ be given, without loss of generality $T \geqq 0$. $\mathcal{K}(\Omega)$ being an ideal in F is invariant under T, so that the restriction of T to $\mathcal{K}(\Omega)$ is in $\mathcal{Z}(\mathcal{K}(\Omega))$. By the example at the end of section 1 we have $T|\mathcal{K}(\Omega) = [\tau]$ for some non negative $\tau \in \mathcal{C}_b(\Omega)$. To show $\tau \in \mathcal{C}_b^F(\Omega)$ and $Tf = [\tau]f$ for any $f \in F_+$ consider an increasing net $(f_i)_{i \in I}$ in $\mathcal{K}(\Omega)_+$ with $\vee_{i \in I} f_i = f$. Since $0 \leqq T \leqq \text{const.}1$ we have

$$\vee_{i \in I} \tau f_i = \vee_{i \in I} T f_i = Tf.$$

From this the desired result is clear in view of the following general

<u>REMARK</u>: Let $0 \leqq \tau \in \mathcal{C}_b(\Omega)$ and $f \in F_+$ be given For an increasing net $(f_i)_{i \in I}$

in $\mathcal{K}(\Omega)_+$ with $\bigvee_{i\in I} f_i = f$ assume that also $\bigvee_{i\in I} \tau f_i =: g$ exists in F. Then $\tau(t_o)f(t_o)=g(t_o)$ for all $t_o\in[f\neq\infty]$.
To prove this remark assume first $\tau(t_o)f(t_o)<g(t_o)$. Then, by continuity and Urysohn's lemma, there would exist a neighbourhood V of t_o and some $0\neq h\in \mathcal{K}(\Omega)_+$ with support in V such that $\tau(t)f(t)\leqq g(t)-h(t)$ for $t\in V$ and so $\tau f_i\leqq g-h$ for all $i\in I$ which contradicts the definition of g since $\mathcal{K}(\Omega)\subset F$. Conversely suppose $\tau(t_o)f(t_o)>g(t_o)$ and therefore also $\tau(t_o)>0$ (since $f(t_o)<\infty$). Again by continuity and Urysohn's lemma this would imply $\tau(t)f(t)-h(t)\geqq g(t)$ and $\tau(t)>0$ on a neighbourhood V of t_o with some $0\neq h\in \mathcal{K}(\Omega)_+$ having its support in V (so that $\frac{h}{\tau}$ is a well-defined element $\neq 0$ in $\mathcal{K}(\Omega)$). But then

$$f_i(t) \leqq \frac{g(t)}{\tau(t)} \leqq f(t) - \frac{h(t)}{\tau(t)}, \quad t\in V$$

and thus $f_i \leqq f - \frac{h}{\tau}$, contradictory to $f = \bigvee_{i\in I} f_i$.

ii) Let $0\leqq\tau\in\mathcal{C}_b(\Omega)$ be given and fix $f\in F_+$. We must show that the function $t\longrightarrow\tau(t)f(t)$ on $[f\neq\infty]$ extends to an element of F. Since $\mathcal{K}(\Omega)$ is order dense there is an increasing net $(f_i)_{i\in I}$ in $\mathcal{K}(\Omega)_+$ with $\bigvee_{i\in I} f_i = f$; if F is σ-complete and not complete or if F is a Banach lattice, by hypothesis we may even assume $I=\mathbb{N}$; also $f_i\to f$ in the normed case. But then, since τ is bounded, $\bigvee_{i\in I}\tau f_i=:g$ exists in F in view of the corresponding completeness assumptions for F. Thus again the above remark finishes the proof. ∎

<u>COROLLARY 2</u> ([21], [8]): Let E be either a Banach lattice with a topological orthogonal system or any [σ-]complete v.l. and consider the locally compact space Ω arising in the representations of E given in section 2 (theorems 1 and 2) and section 3 (cor.2); in the σ-complete but not complete case assume in addition Ω to be σ-compact. Then the Stone-Čech-compactification $\beta\Omega$ is uniquely determined up to homeomorphism (note that Ω itself, being locally compact, is always at least a dense open subset of $\beta\Omega$).

<u>PROOF:</u> Since our representations are spectral representations for $Z(E)$, we know by theorem 1 that $Z(E)$ is algebraically and order isomorphic to $\mathcal{C}(\beta\Omega)$, with identity operator corresponding to the constant function 1. Therefore, as already remarked at the end of section 1., $\beta\Omega$ is determined up to homeomorphism. ∎

In the σ-complete case of cor.2 also $Z(E)\cong\mathcal{C}(\beta\Omega)$ is σ-complete. Therefore $\beta\Omega$ is totally disconnected (an easy consequence of remarks i) and ii) preceding theorem 2 in section 2) and thus the Stone space of the Boolean algebra of its clopen sets which, by the above isomorphism, correspond to the idempotent elements $P\in Z(E)$. These are exactly the *band projections* of E, i.e. the linear endomorphisms P of E with $0\leqq P=P^2\leqq 1$. Therefore in this case our "continuous" representation is that given already by *Nakano* (1941; cf. [15] chapt. 7 §51). Note also that the null sets of π (= local null sets since Ω is supposed to be σ-compact) are, acording to our construction exactly the meager Baire sets and therefore the ideal $\mathfrak{n}$ defining $\mathcal{C}^{\infty}_{\mathfrak{n}}(\Omega)$ is that of all meager sets in Ω. In this case $\mathcal{C}^{\infty}_{\mathfrak{n}}(\Omega)$ is usually denoted as $\mathcal{C}^{\infty}(\Omega)$.

The representation theorems discussed so far rested on completeness assumptions for the v.l. E. It is interesting to note that one can use the Dedekind completion (cf. [15] p. 191) to derive the classical Ogasavara-Maeda theorem (1942; cf. [15] chapt. 7 §49), a representation theorem valid for any Archimedean v.l. E (but with a "bad" representing function space $\hat{E}$, corresponding to the weak hypotheses about E). Since the Boolean algebra P(E) of all projection bands in the non-complete case may be trivial, Ogasawara-Maeda started with the set B(E) of all *bands* (i.e. ideals $I\subset E$ such that $\bigvee_{\alpha\in A} x_\alpha\in I$ whenever $(x_\alpha)_{\alpha\in A}$ is an increasing net in $I_+:=I\cap E_+$). We shall deduce their result in a series of remarks:

REMARKS: Let E be any Archimedean v.l.

i) If I is any ideal, $I^{\perp\perp}$ is the smallest band containing I.

ii) Let $F\subset E$ be a sublattice. Then $I=I^{\perp\perp}\cap F$ for any $I\in B(F)$ (with orthocomplements taken in E). If the *inclusion* $F\subset E$ is *order continuous* (i.e. for every increasing net $(x_\alpha)_{\alpha\in A}$ in F for which there exists a least upper bound x in F, x is also the supremum $\bigvee_{\alpha\in A} x_\alpha$ in E), we have also conversely $I\cap F\in B(F)$ for each $I\in B(E)$.
(The second statement is obvious; the first is shown by

$$I^{\perp\perp}\cap F\subset(I^{\perp}\cap F)^{\perp}\cap F=I=I\cap F\subset I^{\perp\perp}\cap F).$$

iii) Let either F be an order dense (i.e. $F^{\perp\perp}=E$) ideal in E with continuous inclusion $F\subset E$ or let F be any Archimedean v.l. and E its Dedekind completion. In both cases the mapping $I\longmapsto I\cap F$ is a bimonotonic bijection from B(E) onto B(F) with inverse $K\longmapsto K^{\perp\perp}$.

(Since the inclusions from an ideal of E into E as well as that of an Archimedean v.l. into its Dedekind completion are order continuous, by ii) $I \to I \cap F$ is a monotonic mapping from B(E) onto B(F). It remains to show $I=(I\cap F)^{\perp\perp}$ for any $I \in B(E)$. If F is an order dense ideal, we have (cf. [15] p. 106)

$$(I\cap F)^{\perp\perp} = I^{\perp\perp}\cap F^{\perp\perp} = I^{\perp\perp} = I.$$

In the other case, by the properties of the Dedekind completion, $(I\cap F)^{\perp}=I^{\perp}$, thus again $(I\cap F)^{\perp\perp} = I^{\perp\perp} = I$).

iv) Assume E to be complete. Then E is v.l. isomorphic with an order dense ideal $\tilde{E}$ in some complete v.l. G possessing a weak order unit. (For let U be any m.o.s. for E and take for G the direct product of the v. lattices $u^{\perp\perp}$, $u\in U$. It is clear that G is again a complete v.l. and that (writing the elements of G as functions $g:U\to E$, with $g(u)\in u^{\perp\perp}$ for each $u\in U$) the identity function is a weak order unit for G. Also, $x \longmapsto g$: $g(u) = x_u$ (the component of x in the projection band $u^{\perp\perp}$) is a v.l. isomorphism from E onto some sublattice $\tilde{E}\subset G$, since band projections are v.l. homomorphisms and U is a maximal orthogonal system. $\tilde{E}$ is an order ideal (and trivially order dense in G), because for $g\in G_+$ and $x\in E_+$ such that $g(u)\leqq x_u$ for each $u\in U$, the supremum $\bigvee_{u\in U} g(u)=:y$ exists in E by completeness, and $y_u = g(u)$ for all u by the order continuity of band projections).

v) For a complete v.l. G with weak order unit, the "continuous" representation cor.2 of section 3 gives a compact space Ω such that G is isomorphic to a v.l. $\hat{G}\subset\mathscr{C}^{\infty}(\Omega)$ containing $\mathscr{C}(\Omega)$ as an ideal. In fact, as pointed out following cor.2 above, Ω is the Stone space of the Boolean algebra of all band projections i.e. the Stone space of P(G), if band projections are identified with their ranges. Furthermore, for complete G we have P(G)=B(G) by the Riesz decomposition theorem (cf. [20] p.62). Finally, let us note that, since Ω is extremely disconnected (because $\mathscr{C}(\Omega)$, being an ideal of $\hat{G}$, is a complete v.l.), $\mathscr{C}^{\infty}(\Omega)$ itself is already a v.l. with respect to v.l. operations on sets of finiteness (cf. [15] p.323). Also $\hat{G}$ is even an ideal in $\mathscr{C}^{\infty}(\Omega)$. (If $0\leqq f\leqq g$ for $f\in\mathscr{C}^{\infty}(\Omega)$ and $g\in\hat{G}$, we have $f\wedge n1\in\hat{G}$ with $0\leqq f\wedge n1\leqq g$ for each n, so that $\bigvee_{n=1}^{\infty}(f\wedge n1)=:f_o$ exists in $\hat{G}$. Using Urysohn's lemma it is easily seen that $f_o=f$).

vi) For arbitrary Archimedean E with $\bar{E}$ its Dedekind completion, by iv) and v) we have the following chain of inclusions:

$$E \subset \overline{E} \simeq \overset{\cong}{E} \subset G \simeq \hat{G} \subset \mathcal{C}^{\infty}(\Omega) ;$$

here $\overset{\cong}{E}$ is a dense ideal in the complete v.l. G with weak order unit and Ω denotes the compact Stone space for B(G). Furthermore, by iii) there are bi-monotone bijections from B(E) onto $B(\overline{E})$ and from $B(\overline{E})$ onto B(G). This shows that also B(E) is a Boolean algebra and Ω its Stone space. Composing the two isomorphisms above we obtain a v.l. isomorphism of E onto some sublattice $\hat{E}$ of $\mathcal{C}^{\infty}(\Omega)$. If $E=\overline{E}$ is complete, $\hat{E}$ is an order dense ideal of $\hat{G}$ and hence of $\mathcal{C}^{\infty}(\Omega)$; otherwise $\hat{E}$ generates an order dense ideal in $\mathcal{C}^{\infty}(\Omega)$, since the ideal in $\overline{E}$ generated by E is all of $\overline{E}$. Summarizing, we have:

COROLLARY 3 (Ogasavara-Maeda): Every Archimedean v.l. E is isomorphic to a sublattice $\hat{E}$ of $\mathcal{C}^{\infty}(\Omega)$ with Ω the Stone space of the Boolean algebra B(E). If E is complete, $\hat{E}$ is an order dense ideal of $\mathcal{C}^{\infty}(\Omega)$; in general, the ideal generated by $\hat{E}$ is order dense in $\mathcal{C}^{\infty}(\Omega)$.

To close this section, let us briefly look at the points in the locally compact space Ω of the "continuous" representation; this seems to be worthwhile in view of the uniqueness result contained in cor.2. Denote by $x \longrightarrow \hat{x}$ the isomorphism of the representation $E \simeq \hat{E} \subset \mathcal{C}^{\infty}(\Omega)$.

Being precautious about the possible values $\pm\infty$, instead of point evaluations we rather consider, for each $t \in \Omega$, the mappings

$$\varphi_t : E \to \overline{\mathbb{R}}_+ \text{ with } \varphi_t(x) = |\hat{x}(t)| .$$

From the continuity of the functions $\hat{x} : \Omega \to \overline{\mathbb{R}}$ it is immediate that the φ_t are *valuations* of E (cf. [20] p. 161), i.e. on E_+ each φ_t is additive, positively homogeneous and lattice homomorph (with the usual operations in $\overline{\mathbb{R}}_+$), and $\varphi_t(x)=\varphi_t(|x|)$ for general $x \in E$. The kernels of valuations are, unlike the kernels of real lattice homomorphisms, no longer necessarily maximal ideals but rather *prime ideals*, defined only by the requirement that $x,y \in E$ and $x \wedge y = 0$ imply $x \in \ker\varphi_t$ or $y \in \ker\varphi_t$.

(For, this is obvious when both $\hat{x}(t)$ and $\hat{y}(t)$ are finite so that $0=(x\wedge y)\hat{}(t) = \min\{\hat{x}(t),\hat{y}(t)\}$. But if for example $\hat{x}(t)=\infty$, by continuity we must have $\hat{y}(t)=0$ in order to have $(x\wedge y)\hat{}(t)=0$).

Having identified the points of Ω with certain prime ideals for E, we also note that the locally compact topology on Ω, being the weakest topology for which the functions in $\mathcal{K}(\Omega)$ are continuous is the well-known "*hull-kernel topology*" (cf. [15], p.218) on sets of prime ideals: a subset $A \subset \Omega$ is closed iff

$$A=\{t \in \Omega: \hat{x}(t)=0 \text{ for each } x \in E \text{ for which } \hat{x} \text{ vanishes on } A\}.$$

Conversely, given any Archimedean v.l. E and topologizing appropriate sets Ω of prime ideals of E by the hull-kernel topology, one can obtain rather general representations of E by v. lattices $\hat{E}$ of continuous extended real valued functions (see e.g. the work of Johnson-Kist 1962, discussed in chapt. 7 of Luxemburg-Zaanen's book). However, these representations suffer from several pathologies concerning the description of $\hat{E}$ as well as the properties of the topology of Ω.

REFERENCES

[1] Alfsen, E.M.: Compact convex sets and boundary integrals (Springer, Berlin-Heidelberg-New York 1971)

[2] Alfsen, E.M. and Effros, E.G.: Structure in real Banach spaces I, II (Ann. of Math. 96 (1972) 98-173)

[3] Behrends, E. et al.: L^p-structure in real Banach spaces (to appear in Lecture Notes in Math. 1977)

[4] Davies, E.B.: The Choquet theory and the representation theory of ordered Banach spaces (Illinois J. Math. 13 (1969) 176-187)

[5] Gierz, G.: Darstellung von Banachverbänden durch Schnitte in Bündeln (to appear in: Mitteilungen des Mathematischen Seminars, Giessen 1977)

[6] Goullet de Rugy, A.: La structure idéale de M-espaces (J.Math. Pures Appl. IX Sér. 51 (1972) 331-373)

[7] Hackenbroch, W.: Zur Darstellungstheorie σ-vollständiger Vektorverbände (Math. Z. 128 (1972) 115-128)

[8] Hackenbroch, W.: Eindeutigkeit des Darstellungsraumes von Vektorverbänden (Math.Z. 135 (1974) 285-288)

[9] Hackenbroch, W.: Über den Freudenthalschen Spektralsatz (manuscripta math. 13 (1974) 83-99)

[10] Halmos, P.R.: Lectures on Boolean algebras (D. Van Nostrand, Princeton 1967)

[11] Kakutani, S.: Concrete representation of abstract L-spaces and the mean ergodic theorem (Ann. of Math. 42 (1941) 523-537)

[12] Lindenstrauss J. and Tzafriri, L.: Classical Banach spaces (Lecture Notes in Mathematics 338 (1973))

[13] Lotz, H.P.: Idealstruktur in Banachverbänden (Habilitationsschrift, Tübingen 1969)

[14] Lummer, G.: Algèbres de fonctions et espaces de Hardy (Lecture Notes in Mathematics 75 (1968))

[15] Luxemburg, W.A.J. and Zaanen, A.C.: Riesz spaces I (North Holland Publ. Comp., Amsterdam 1971)

[16] Meyer-Nieberg, P.: Charakterisierung einiger topologischer und ordnungstheoretischer Eigenschaften von Banachverbänden mit Hilfe disjunkter Folgen (Arch. Math. 24 (1973) 640-647)

[17] Pitcher, T.S.: A more general property than domination for sets of probability measures (Pacif. J. Math. 15 (1965) 597-611)

[18] Portenier, C.: Espaces de Riesz, espaces de fonctions et espaces de sections (Comm. Math. Helvetici 46 (1971) 289-313)

[19] Schaefer, H.H.: Topological vector spaces, 3rd printing (Springer, New York-Heidelberg-Berlin 1971)

[20] Schaefer, H.H.: Banach lattices and positive operators (Springer, Berlin-Heidelberg-New York 1974)

[21] Schaefer, H.H.: On the representation of Banach lattices by continuous numerical functions (Math.Z. 125 (1972) 215-232)

[22] Wright, J.D.M.: Measures with values in a partially ordered vector space (Proc. London Math. Soc. 25 (1972) 675-688)

K.-D. Bierstedt, B. Fuchssteiner (eds.)
Functional Analysis: Surveys and Recent Results

On the Theory of Approximation by Positive Operators in Vector Lattices

Manfred Wolff
FB Mathematik d. Universität
Auf d. Morgenstelle 10
D 7400 Tübingen 1

Summary: In this paper we give a survey and some new results in a field which can be described best by "theorems of Korovkin type in locally convex lattices". The survey is given in part I. More precisely in § 1 besides notational preliminaries we give a short historical review. In § 2 we report on recent results which concern the universal Korovkin closure of subsets and universal Korovkin systems in general. § 3 establishes some very elementary facts concerning universal Korovkin systems. For example most of the known Korovkin systems are universal. This paragraph contains also an important characterization of vector lattices with finite Korovkin systems.
Part II consists of new supplementary results with complete proofs. In § 4 we show to some extent the connection between Korovkin closures and general Choquet boundaries. The next paragraph is devoted to one main result stating that a sequentially complete locally convex lattice possesses a Korovkin system of 2 elements iff its dimension is at most 2. In the last paragraph we pose some open questions.

The list of references is by no means complete. However, most of the papers refered to contain extensive references on their own, so that it is not difficult to obtain a complete survey of what has been done in the field.

Acknowledgements: I wish to thank the organizers of this conference for their kind invitation and express my appreciation for the perfectly organized meeting.

In addition I would like to thank the "Gesellschaft der Freunde der Universität Dortmund" for their support of my participation in the conference on approximation theory, Jan. 1976, at Austin (Texas, USA). The valuable discussions at this Texas conference are reflected in this paper.

Part I

§ 1. Introduction

1.1) Preliminaries

In the theory of approximation by positive operators the elements to be approximated usually comprise a locally convex vector lattice (l c v l for short), i.e. a vector space over $\mathbb{R}$ bearing a lattice structure which is compatible with the algebraic operations and such that the locally convex topology is generated by a system of lattice seminorms; these are seminorms p satisfying the additional relation $p(x) \leq p(y)$ whenever $|x| = \sup(x,-x)$ is less or equal to $|y| = \sup(y,-y)$ [1]. Instead of going further into details we prefer to present some well-known and instructive examples.

1.1.1) Examples: 1) Let X be a compact Hausdorff-space. The space $\mathcal{C}(X)$ of all real valued continuous functions equipped with its usual structure and sup-norm is a Banach lattice.

2) If (X,Σ,μ) is a measure space then $L^p(X,\Sigma,\mu)(1 \leq p \leq \infty)$ are Banach lattices; more generally the so-called Banach function spaces on (X,Σ,μ) are Banach lattices.

3) If X is completely regular then $\mathcal{C}(X)$ equipped with the compact open topology is a l c v l, denoted by $\mathcal{C}_{co}(X)$.

4) Almost all Köthe sequence spaces are l c v l's.

5) Spaces of scalar-valued continuous functions with the strict topology are l c v l's.

1.1.2) Linear mappings: If E and F denote vector lattices and if S,T are linear mappings from E to F we write $S \leq T$ if $Sx \leq Tx$ holds for all $x \in E_+ = \{y \in E : 0 \leq y\}$. Thus a linear mapping T is isotonic or positive iff $0 \leq T$ holds. T is called a lattice homomorphism if it is linear and satisfies $T|x| = |Tx|$ for all x in E (equivalently: $T(\sup(x,y)) = \sup(Tx,Ty)$, and $T(\inf(x,y)) = \inf(Tx,Ty)$ for all $x,y \in E$). Important examples are the identity, embeddings (e.g. $\mathcal{C}(X) \to L^p(X,\mu)$, if μ is a Radon measure on the compact space X) and restrictions (e.g. $f \in \mathbb{R}^X \to f|_A$ for a fixed $A \subset X$) [2]

1.2) Some historical remarks

The starting point of the theory presented here is the following theorem of Korovkin [14,15] .

---

1) For details see [25].

2) $f|_A$ denotes the restriction of f to the subset A

1.2.1) Theorem: Let X denote the compact unit interval and let M be the subset of $E = \mathcal{C}(X)$ consisting of the three functions 1_X [3], $x \to x$, $x \to x^2$. A sequence (T_n) of positive linear operators [4] on E converges strongly to the identity on E already whenever $\lim_{n\to\infty} \|T_n g - g\| = 0$ holds for the three elements g in M.

The natural question as to which other spaces $\mathcal{C}(X)$ may possess such nice finite "Korovkin systems"[5] was answered completely by Šaškin [22] in the following way:

1.2.2) Theorem: $\mathcal{C}(X)$ (X compact) possesses a finite Korovkin system iff X is homoeomorphic to a subset of $\mathbb{R}^n$ for a suitable $n \in \mathbb{N}$.

The problem as to how to characterize Korovkin systems in a given space $E = \mathcal{C}(X)$ was attacked via the more general question: how to describe the greatest subspace $\mathcal{K}(M,I)$ on which convergence to the identity takes place whenever this is the case on M. After fruitful contributions by Šaškin [23] and Franchetti [11] (cf. [26], too) this problem was completely solved by Berens-Lorentz ([4], Thm. 2).

For a subset M let L_M denote the closed linear hull of M.

1.2.3) Theorem: Let $1_X \in M \subset E = \mathcal{C}(X)$. For an element f of E the following assertions are equivalent:

(a) $f \in \mathcal{K}(M,I)$.

(b) $f(x) = \inf\{g(x) : f \leq g \in L_M\} = \sup\{h(x) : f \geq h \in L_M\}$.

(c) If ε_x denotes the Dirac measure at the point x, if in addition μ is a positive Radon measure such that $\mu|_M = \varepsilon_x|_M$, then $\mu(f) = f(x)$ holds.

Obviously M is a Korovkin system iff $\mathcal{K}(M,I)$ equals E.

Thus a Korovkin system separates the point of X.

1.2.4) So far we considered merely spaces of type $\mathcal{C}(X)$.

There are a lot of generalizations to other function spaces; for example, the case $E = L^1([0,1],\mu)$ was treated by Dzjadyk [9], Wulbert [34] and Krasnosel'ski-Lifšič [16]. In these spaces $M = \{1,x,x^2\}$ is again a Korovkin system. A very nice theory for subspaces of $\mathcal{C}(X)$ (X locally compact) was initiated by Bauer [1,2,3], who made extensive use of the connection between Korovkin's theory and the Choquet-boundary. This theory was completely extended and

---

3) 1_A: indicator function of the set A.

4) By an operator we mean a mapping from E into itself.

5) More precisely a finite I-Korovkin system in the sense of def.2.1 below.

brought into a satisfactory final form by Donner [7,8] (cf. [28],too). The difference to our own approach is that Donner considers essentially all nets of positive operators where we restrict ourselves to equicontinuous nets. Though for $\mathcal{C}(X)$ and $1_X \in M$ both approaches agree, in general, this apparently slight difference leads to completely distinct theories (cf 3.6 below).
Further significant contributions to the general theory are mentioned in the subsequent sections.

1.2.5 In their paper [4] Berens and Lorentz discovered the following surprising universal property of the Korovkin closure $\mathcal{K}(M,I)$ in the case $E = \mathcal{C}(X)$ when 1_X is contained in M.

Theorem: Let F be an arbitrary Banach lattice, let $S:E \to F$ denote a lattice homomorphism, and let (T_n) be a sequence of positive linear mappings from E to F. If (T_n) converges to S pointwise on M, then it converges to S pointwise on $\mathcal{K}(M,I)$ too. In particular, any Korovkin system in E is a universal Korovkin system in an obvious sense.

§ 2 The universal Korovkin closure of subsets in vector lattices [6)]

It is now our aim to carry over the "universal Korovkin theorem" of Berens-Lorentz (see 1.2.5 above) to arbitrary locally convex lattices. Let E be a fixed locally convex vector lattice .
Let $\sigma = (F,S,(T_\alpha))$ be a triple consisting of another l c v l F, a continuous lattice homomorphism S from E to F and an equicontinuous net (T_α) of positive linear mappings from E to F. Then obviously the set of convergence $C(\sigma) = \{x \in E : \lim T_\alpha x = Sx\}$ is a closed linear subspace of E.

2.1) Definition: Let M denote a subset of E.

a) The universal Korovkin closure ("total shadow" in [4]) $\mathcal{K}_u(M)$ of M is the intersection of all those sets of convergence containing M, i.e. $\mathcal{K}_u(M) = \bigcap\{C(\sigma) : M \subset C(\sigma),\ \sigma \text{ arbitrary}\}$.
We call M a universal Korovkin system (uKS for short) if $\mathcal{K}_u(M)$ equals E.

b) Similarly the I-Korovkin closure $\mathcal{K}(M,I)$ (I denotes the identity on E) is defined by $\mathcal{K}(M,I) = \bigcap\{C(\sigma) : M \subset C(\sigma),\ \sigma = (E,I,(T_\alpha))\}$.
M is called an I-Korovkin system (I-KS for short) if $\mathcal{K}(M,I)$ equals E.

6) The results discussed here and in the next paragraph are taken from [33] .

Obviously $\mathcal{K}_u(M)$ is always contained in $\mathcal{K}(M,I)$ and any uKS is an I-KS. Moreover the closed linear hull L_M of M always lies in $\mathcal{K}_u(M)$. To obtain a full generalization of 1.2.3 in view of 1.2.5 we first remark that assertion 1.2.3c in general is meaningless (the point evaluations are lattice homomorphisms into $\mathbb{R}$ but $L^p([0,1],\lambda)$ (λ: Lebesgue measure) fail to have such lattice homomorphisms for $1 \leq p < \infty$)). There is however a substitute if we replace the codomain $\mathbb{R}$ by an arbitrary AL-space (see 2.3c below). Now we describe the counterpart to 1.2.3b): let $M^\wedge$ be the set $\{\inf A : \emptyset \neq A \subset L_M,\ A \text{ finite}\}$ and define $\mathcal{H}(M)$ to be the closed linear subspace $\overline{M^\wedge} \cap \overline{(-M^\wedge)}$.
Then we have

2.2) Theorem [31]: For any subset M the universal Korovkin closure $\mathcal{K}_u(M)$ always contains $\mathcal{H}(M)$ 7).

The counterpart of 1.2.3 reads as follows:

2.3) Theorem [33]: Let the set $\{x \in L_M : x \geq 0\}$ be total in L_M. For an x in E the following assertions are equivalent:

(a) $x \in \mathcal{K}_u(M)$.

(b) $x \in \mathcal{H}(M)$.

(c) For every space $F = L^1(X,\Sigma,\mu)$, for every continuous lattice homomorphism $S: E \to F$ with range dense in F and for every positive continuous linear mapping T from E to F the condition $T|_M = S|_M$ always implies $Tx = Sx$.

2.4) Corollary: Under the same assumptions of 2.3 on L_M the following assertions are equivalent:

(a) M is a universal Korovkin system.

(b) $\mathcal{H}(M) = E$.

(c) $\overline{M^\wedge} = E$.

2.5) Example: 2.4c can be used to give an easy proof to show that $M = \{\exp(-x^2),\ x\exp(-x^2),\ x^2\exp(-x^2)\}$ is an uKS in $E = \{f \in \mathcal{C}(\mathbb{R}) : \lim_{t\to\infty} f(t) = \lim_{t\to-\infty} f(t) = 0\}$ equipped with the sup-norm.

§ 3 More on universal Korovkin systems

Our first proposition is an easy consequence of the definition of an uKS.

3.1) Proposition: ([33]): Let M_E be a universal Korovkin system in the l c v l E, and in addition let S be a continuous lattice homomorphism from E onto a dense subspace of another l c v l F.

7) This theorem is true even in topological vector lattices; special cases of it were proved independently by Kitto-Wulbert [13] and Fakhoury [10].

<u>Then $S(M_E)$ is a universal Korovkin system in F.</u>

This rather simple statement together with 1.2.5 shows that many known I-Korovkin systems are, in fact, <u>universal</u> Korovkin systems.

<u>3.2) Examples:</u>

a) $F = \ell^p(\mathbb{N})(1 \leq p < \infty)$. $M = \{(n^{-2}),(n^{-3}),(n^{-4})\}$ is an uKS; for a proof consider $S: \mathcal{C}([0,1]) \to F$ given by $(Sf)(n) = n^{-2}f(\frac{1}{n})$ and $M_E = \{1,x,x^2\}$.

b) $\{\exp(-x), x\exp(-x), x^2\exp(-x)\}$ is an uKS in $\mathcal{C}_0(\mathbb{R}_+) = F$; to see this use $S: \mathcal{C}([0,1]) \to F$ given by $(Sf)(x) = (1+x)^2 \exp(-x).f(\frac{1}{1+x})$.

c) $\{1,x,x^2\}$ (X compact in $\mathbb{R}$) as well as $\{x,x^2,x^3\}$ are universal Korovkin systems in $L^p(X,\Sigma,\mu)(1 \leq p < \infty, \mu$ a Radon measure). The generalization to $X \subset \mathbb{R}^n$ is obvious.

d) By means of 2.4 (iii) we see $\{1,x,x^2\}$ to be an uKS in $\mathcal{C}_{co}(\mathbb{R}_+)$; hence $\{1,\frac{x}{1-x}, \frac{x^2}{(1-x)^2}\}$ is an uKS in $\mathcal{C}_{co}([0,1))$. This system may be used in connection with Bernstein - series (cf[19,6]).

Much deeper results depending on 3.1 and on an elaborated representation theory for Banach lattices are given in the theorems below (compare 1.2.2).

<u>3.3) Theorem</u> ([3o,27]): <u>Let E be a sequentially complete locally convex vector lattice. The following assertions are equivalent:</u>

(a) <u>E possesses a finite universal Korovkin system.</u>

(b) <u>E is finitely generated.</u>

<u>If in addition E is a Banach lattice then these statements are equivalent to</u>

(c) <u>E is representable as a generalized Banach function space on a compact set X of $\mathbb{R}^n$ for suitable $n \in \mathbb{N}$.</u>

The easiest proof of the equivalence of (a) and (b), based on 3.1, is given in [31] § 3. The equivalence itself was proved under a mild restriction in [3o] and in its full generality by Scheffold [27]. For details concerning (c) we refer to [29,3o]. The proof of the next somewhat surprising result uses [29].

Let us recall that among other Banach lattices the spaces $L^p(1 \leq p < \infty), c_0$ as well as all reflexive ones possess an order continuous norm, that means $\lim ||x_\alpha|| = 0$ holds for every decreaseing net (x_α) with $\inf x_\alpha = 0$.

3.4 Theorem [30,31]: Let E be a Banach lattice with order continuous norm. The following assertions are equivalent:

(a) E is separable.

(b) E possesses a universal Korovkin system with three elements.

(c) E is a Banach function space on a compact subset X of $\mathbb{R}$.

(d) E is lattice isomorphic (but not necessarily topologically) to a vector sublattice of $L^1([0,1],\Sigma,\lambda)$ (λ:Lebesgue measure).

It may at first be surprising, that $L^1([0,1]^n,\lambda^n)$ (λ^n:n-dimensional Lebesgue measure) possesses an uKS of three elements regardless of what n may be. The following simpler example sheds light on this fact.

3.5 Example: Let $\varphi:\mathbb{N}^2 \to [0,1]$ be an injection and consider $S: \mathcal{C}([0,1]) \to F = \ell^p(\mathbb{N}^2)$ $(1 \leq p < \infty)$, given by $(Sf)(k,\ell) = \exp(-(k+\ell))\, f(\varphi(k,\ell))$. By 3.1 $S(\{1,x,x^2\})$ is an uKS in F.

Let us finish this paragraph with an additional observation: Our theory developed so far depends heavily on our restriction to equicontinuous nets (see 2.1). If we would omit this additional assumption which in the classical cases (cf § 1) is superfluous we get a statement as in 3.3 only in special situations as the following theorem shows.

3.6) Theorem [32,33]: Let E be a sequentially complete barreled l c v l. Then the following assertions are equivalent:

(a) There exists a finite set M such that any net of positive continuous linear operators on E converges pointwise to the identity whenever this takes place on M.

(b) E is topologically and lattice isomorphic to (i.e. completely identifiable with) $\mathcal{C}(X)$ for a compact set X in $\mathbb{R}^n$ for an appropriate $n \in \mathbb{N}$.

Part II

Supplementary results

§ 4 On the generalized Choquet boundary

In this paragraph we try to attack the problem as to when one can replace the general L^1-space in condition (c) of 2.3 by $\mathbb{R}$.

To this end we denote by V(E) the set of all continuous real-valued lattice homomorphisms on the l c v l E (V(E) may only consist of O). Moreover for $M \subset E$ let $\Delta(M)$ be equal to $\{(x';y') \in E'_+ \times V(E) : x'(x) = y'(x)$ for all x in $M\}$, and set $\mathcal{E}_M = \{x \in E : x'(x) = y'(x)$ for all $(x';y') \in \Delta(M)\}$. Then, clearly, we have $\mathcal{H}_u(M) \subset \mathcal{E}_M$.

The <u>Choquet boundary</u> of M is $\partial M = \{y' \in V(E) : (x';y') \in \Delta(M) \Rightarrow x' = y'\}$. Obviously $\mathcal{E}_M$ is equal to E iff ∂M equals $V(E)$.
Call a l c v l E <u>dual atomic</u> if its dual space E' (which is in any case a vector lattice) is atomic, i.e. E' is the band generated by $V(E)$ (for definitions see [25]). Examples are all reflexive atomic vector lattices, e.g. $\ell^p(A)$ for an arbitrary index set A, $(1 \leq p < \infty)$, clearly $\ell^1(A)$, too; ℓ^∞ is atomic but not dual atomic; $\mathcal{C}(X)$ equipped with the topology of pointwise convergence is dual atomic but not atomic if X is connected. Recently in [17] were given nice examples of AM-spaces which are dual atomic but not atomic.
The following is another counterpart to 1.2.3
<u>4.1) Theorem:</u> <u>Let E be a dual atomic vector lattice, and let M be a subset of E such that $\{x \in L_M : x \geq o\}$ is total in L_M. Then we have $\mathcal{H}(M) = \mathcal{H}_u(M) = \mathcal{E}_M$</u>
Before giving the proof we add the following obvious
<u>4.2) Corollary:</u> <u>Under the assumption of 4.1 M is a universal Korovkin system iff ∂M equals $V(E)$</u>.
<u>Proof of 4.1</u> All we have to show is that $\mathcal{E}_M$ is contained in $\mathcal{H}(M)$ Thus let x_o be not in $\mathcal{H}(M)$. By 2.3 there exists a space $F = L^1(X,\mu)$, a continuous lattice homomorphism S from E onto a dense subset of F, and a positive linear continuous mapping T from E to F satisfying $T_{|M} = S_{|M}$ but with $Tx_o \neq Sx_o$.
The initial topology on E inherited from F by S^{-1} is coarser than the given one. Hence S' is easily seen to map F' onto a lattice ideal of E' (use [24],V 7.4). Since the latter is atomic, so is F'; this allows us to identify F with $\ell^1(A)$ for an appropriate index set A. Thus there exists an $a \in A$ with $(Tx_o)(a) \neq (Sx_o)(a)$ Set $x' : x \to (Tx)(a)$, $y' : x \to (Sx)(a)$. Then (x',y') is in $\Delta(M)$, and thus x_o is not in $\mathcal{E}_M$.

<u>Remark:</u> Let us point out that 4.1 does not imply 1.2.3; it is an open problem whether there exists a more general theorem including both statements as special cases.
We continue to carry over 1.2.3 to another class of l c v l's, the class of AM-spaces. For the sake of convenience we restrict our-

selves to Banach lattices (for the following cf. [12] and [25]). The generalizations to arbitrary AM-spaces are then quite clear. A Banach lattice is called AM-space if its norm satisfies $||\sup(|x|,|y|)|| = \sup(||x||,||y||)$. Typical examples are $\mathcal{C}(X)$ (X compact), $\mathcal{C}_o(X)$ (X locally compact), in particular c_o; for more complicated examples see [12,17,25] .

<u>4.3) Theorem: Let E be an AM-space, and let M be an arbitrary subset of E. Then the I-Korovkin closure $\mathcal{K}(M,I)$ is equal to $\mathcal{E}_M$.</u>

Before proving this theorem we again add an obvious corollary.

<u>4.4) Corollary: An arbitrary subset M of E is an I-Korovkin system iff the Choquet boundary ∂M equals V(E).</u>

To prove $\mathcal{E}_M \subset \mathcal{K}(M,I)$ we use the following extract of Korovkin's second proof of his theorem ([15]), which may be useful in other contexts, too.

<u>4.5) Lemma:</u> Let X be a topological space, let K and K' be compact subsets of X, and let (φ_α) be a net of continuous mappings from K to K'. For a subset M of $\mathcal{C}(X)$ set $x \sim y$ iff $f(x) = f(y)$ for all f in M. Let G be the set $\{g \in \mathcal{C}(X)\colon x \in K \text{ and } x\sim y \Rightarrow g(x) = g(y)\}$. If $\lim\limits_\alpha (\sup\{|f(\varphi_\alpha(x))-f(x)| \,:\, x \in K\}) = o$ holds for all f in M, then it holds for all f in G.

The proof is a slight modification of Korovkin's own proof and is done indirectly, using the compactness of K and K'.

<u>Proof of 4.3</u> (I) Set $X = E'$ (with ω^*-topology), and $K = U' \cap V(E)$[8]. By 1.3.1 in [12] we have $||y|| = \sup\{x'(|y|)\colon x' \in K\}$ for all y in E. For an equicontinuous net (T_α) of positive linear operators on E which converges to I pointwise on M we set $\varphi_\alpha = T'_{\alpha|K}$ and $K' = a\cdot U'_+$ where $\sup\limits_\alpha ||T_\alpha|| \leq a$. Because of the inclusions $M \subset \mathcal{E}_M \subset E \subset \mathcal{C}(X)$, 4.5 implies that $T_\alpha x$ converges to x uniformly on K for all x in $\mathcal{E}_M$. The above mentioned property of the norm proves $\mathcal{E}_M \subset \mathcal{K}(M,I)$.

(II) If x_o is not in $\mathcal{E}_M$, there exists (x',y') in $\Delta(M)$ with $x'(x_o) \neq y'(x_o)$. If $y' = o$, $T = I+|x_o|\otimes x'$ shows x_o not to be in $\mathcal{K}(M,I)$, since $T_{|M} = I_{|M}$ holds. If $y' \neq 0$, the set $J = \{x'' \in E'' : x''(y') = 0\}$ is a band of codimension 1 in E" (use [25], II, 4.4, and 5.5); hence there exists an a" in E" which is orthogonal to J with $a''(y') = 1$. $P = a'' \otimes y'$ then is the unique projection

---

8) U' denotes the dual unit ball, $U'_+ := U' \cap E'_+$.

onto the band orthogonal to J; this implies that $Q = I-P$ and hence $S = I+a'' \otimes (x'-y')$ are positive on E''. Now the bipolar theorem yields that the subset $U_\varepsilon^+ = U_\varepsilon \cap \mathcal{L}(E)_+$ of the ε-unit ball U_ε in $E' \otimes E$ is $\sigma(\mathcal{L}(E''), E \otimes E'')$-dense in $U_\varepsilon''^{+}$. This together with the metric approximation property of E'' implies the existence of an equicontinuous net (P_α) of positive operators of finite rank on E (viewed as a sublattice of E'') which converges to S with respect to $\sigma(\mathcal{L}(E''), E \otimes E')$ (for this argument compare [25], p.239). Thus (P_α) converges pointwise on M to the identity, but $\lim_\alpha P_\alpha x_o = x_o$ cannot hold because $Sx_o \neq x_o$, q.e.d.

§ 5 Lattices with Korovkin systems of two elements

The purpose of this paragraph is to show that those vector lattices are at most two-dimensional, i.e. sublattices of $\mathbb{R}^2$. Although the fact is not at all surprising, its proof is very hard and yields the assertion for sequentially complete lattices only. On the other hand, the following theorem due to H. Berens-G.G.Lorentz indicates that the problem above is not as dumb as it sounds. From a closer look at the definition of the various Korovkin systems given in 2.2 it is at once clear how to define the notion of the I-Korovkin system for nets of positive contractions.

Then we have

5.1) Theorem [4,5]: Let E be a Banach lattice with uniformly monotone norm (e.g. a uniformly convex lattice). Then the convergence set $C(\sigma)$ of $\sigma = (E,I,(T_\alpha))(\|T_\alpha\| \leq 1)$ is a closed sublattice.

Using the same representation methods as for the proof of 3.4 and keeping in mind the fact that a uniformly monotone norm is order continuous we get

Corollary: A space E as in 5.1 possesses an I-KS for positive contractions of at most 2 elements iff it is separable.

We now formulate our counterpart to this theorem.

5.2) Theorem: Let E be a sequentially complete locally convex vector lattice, and let M be an I-Korovkin system consisting of two elements. Then E is isomorphic to a sublattice of $\mathbb{R}^2$, i.e. E equals one of the three spaces $\{0\}$, $\mathbb{R}$, or $\mathbb{R}^2$.

5.3) The proof needs several lemmata and carries over for relative uniformly complete lattices, too (see [25], p.54, footnote).

Lemma 1: If $u > o$ and $M = \{u\}$ is an I-KS, then $\dim E = 1$, i.e. $E = \mathbb{R}$.

Proof: Let $u' \in E'_+$ satisfy $u'(u) = 1$. $T = u \otimes u'$ then must be the identity on E.

Lemma 2: Let E be a locally convex vector lattice, let $0 \leq P \leq I$ be a projection, and let M denote an I-KS in E. Then P(M) is an I-KS in P(E) = F

Proof: Note at first that the assumptions on P imply F to be a band in E ([25],II 2.9). Then any equicontinuous net (U_α) of positive operators on F can be extended to a similar net (T_α) on E by $T_\alpha x = U_\alpha Px+(I-P)x$; the lemma follows easily.

Lemma 3: The theorem is true for $E = \mathcal{C}(X)$(X compact) equipped with the sup-norm.

Proof: L_M contains a strictly positive function, which can be assumed to be 1_X([4],p. 11). If dim $L_M = 1$, Lemma 1 yields the assertion. Otherwise there is a function f in L_M which is linearly independent of 1_X and thus separates the points of X by 1.2.3.
If $f(x_j) = t_j$, $t_1 < t_2 < t_3$, then for $u' = (t_3-t_1)^{-1}((t_3-t_2)\varepsilon_{x_1}+ + (t_2-t_1)\varepsilon_{x_3})$ we have $u'|_M = \varepsilon_{x_2}|_M$; hence $u' = \varepsilon_{x_2}$ by 1.2.3, a contradiction. Thus X consists of 2 elements only.

Lemma 4: Let E be equal to $\mathcal{C}(X)$(X compact), now equipped with an arbitrary separated locally convex lattice topology $\mathfrak{T}$. Let f_1, f_2 be linearly independent satisfying $|f_1|+|f_2| = 1_X$. If $M = \{f_1,f_2\}$ is an I-KS in $(E,\mathfrak{T})$, then $E = \mathbb{R}^2$.

Proof: We remark first that by [24], V 7.3,$\mathfrak{T}$ is coarser than the sup-norm topology; hence by [24],V 7.4,$(E,\mathfrak{T})'$ is an ideal in $(E,||.||_\infty)' = M(X)$(space of all Radon measures on X) which obviously is $\sigma(M(X),E)$-dense (ω^*-dense) in M(X). We show that X must be finite. Then E is finite dimensional, and $\mathfrak{T}$ coincides with the sup-norm-topology; thus lemma 3 applies.

(I) Set $A = [f_1 f_2 \neq 0]$[9]; the open set $B = X\setminus\bar{A}$ is the disjoint union of the four sets $B \cap [f_j = 1]$, $B \cap [f_j = -1]$ $(j = 1,2)$ because of our assumption on f_1,f_2. These sets equal $B \cap [f_j > 0]$, $B \cap [f_j < 0]$ respectively; hence they are open. Suppose one of them, say $C = B \cap [f_1 = 1]$, is nonempty. Because of our first remark there exists a positive $\mathfrak{T}$-continuous Radon measure u' with support contained in C satisfying $u'(1_X) = 1$.
The set $H = \{h \in \mathcal{C}(X): 0 \leq h \leq 1_C\}$ is directed upwards. Define for each $h \in H$ an operator T_h by $T_h f = u'(f)h+(1_X-h)f$ for all f in $\mathcal{C}(X)$. Then $0 \leq T_h \leq 1_X \otimes u'+I$; thus (T_h) is an $\mathfrak{T}$-equicontinuous net of positive linear operators on E. Obviously $T_h f_j$ equals f_j, hence

9) $[g \neq 0]$ denotes the set $\{x:g(x)\neq 0\}$; similarly we use $[g>0]$,etc.

(T_h) converges strongly to I. Let $g \in E_+$ be arbitrary with support $S(g)$ contained in C. For all h in H with $1_{S(g)} \leq h \leq 1_C$ we have $T_h(g) = u'(g)\cdot h$; this implies $u'(g)h = g$ by the preceding remark. Since this equation holds for all such g, C cannot contain more than one point. Similar arguments hold for the other components of B; hence B is finite, moreover, $\bar{A}$ is clopen.

(II) Using lemma 2 for P given by $Pf = 1_{\bar{A}}f$ we can restrict ourselves to the case $\bar{A} = X$. A is the disjoint union of four open sets $[f_1^+f_2^+ > 0]$, $[f_1^-f_2^+ > 0]$, etc.[10]. We claim that $D = [f_1^+f_2^+ > 0]$ is finite. If both sets $f_j(D)$ consist of at most two points, D breaks up into at most four nonempty clopen sets D_k on each of which both functions are constant. Using lemma 2 with $Pf = 1_{D_k}f$ and lemma 1 we see that each of the D_k's consists of only one point; hence D is finite.

Thus [11] suppose w.l.o.g. that there are positive numbers $a_1 < a_2$ such that the open sets $D_1 = [f_1 < a_1] \cap D$, $D_3 = [a_1 < f_1 < a_2] \cap D$, and $D_2 = [f_1 > a_2] \cap D$ are nonempty. Because of our first remark there are $\mathcal{T}$-continuous Radon probability measures u_j' $(j = 1,2)$ having their support in D_j. If g is in L_M and $u_j'(g) \geq 0$, then a straightforward calculation shows $g|_{D_3} \geq 0$ since $g|_D = (a1_X + bf_1)|_D$ for suitable a,b in $\mathbb{R}$. Now choose $g_j \in L_M$ with $u_j'(g_k) = \delta_{jk}$ (Kronecker symbol). Then for $U = g_1 \otimes u_1' + g_2 \otimes u_2'$ we have $Uf|_{D_3} \geq 0$ for all $f \geq 0$.

The set $H = \{h \in E : 0 < h \leq 1_{D_3}\}$ is directed upwards. The net $(T_h)_{h \in H}$ given by $T_hf = hUf + (1_X - h)f$ consists of positive operators, bounded above by $I + |g_1| \otimes u_1' + |g_2| \otimes u_2'$; hence the net is equicontinuous. We have $T_h|_M = I|_M$. At least one of the functions g_1, g_2, say g_1, cannot vanish identically on D_3. Thus for $0 < g \leq 1_{D_1}$ with $u_1'(g) \neq 0$ there exists $h_0 \in H$ with $(T_{h_0}g)|_{D_3} \neq 0$. Choose an x' in $(E,\mathcal{T})_+'$ with support in D_3 and with $x'(T_{h_0}g) \neq 0$. It follows easily that (T_hg) does not converge to g, a contradiction. Thus D has to be finite.

(III) In an analogous manner the other three sets are shown to be finite (take $-f_1$ instead of f_1, etc.). Now the proof is complete.

Proof of 5.3: Let $M = \{x_1, x_2\}$ be an I-KS in E and set $u = |x_1| + |x_2|$; if $E_u = \bigcup_{n \in \mathbb{N}} \{x \in E : |x| \leq nu\}$ is not dense in E, there exists a positive u' in the polar of E_u ([25], p.155). $T = I + u \otimes u'$ coincides

---

10) $f^+ = \sup(f,0)$, $f^- = \sup(-f,0)$

11) The following idea goes back to Kitto and Wulbert [13] who proved 5.3 for L^p-spaces.

with I on M, hence M cannot be an I-KS. Thus E_u is dense. Since E is sequentially complete, the space E_u, normed by the gauge function of the set $\{x: |x| \leq u\}$, is an AM-space with unit ([25] ,II.7.2). Then there exists a compact space X and a lattice isomorphism U from E_u onto $\mathcal{C}(X)$, mapping u onto 1_X. Set $f_j = Ux_j$, and let $\mathcal{T}$ be the topology induced on $\mathcal{C}(X)$ by U. To apply lemma 4 we have to show $\{f_1, f_2\}$ to be an I-KS in $(\mathcal{C}(X), \mathcal{T})$.

Let (T_α) be an equicontinuous net of positive linear operators on $(\mathcal{C}(X), \mathcal{T})$ satisfying $\mathcal{T}\text{-}\lim_\alpha T_\alpha f_j = f_j$ $(j = 1,2)$. Then $U^{-1} T_\alpha$ can be extended to S_α on the whole of E, since E_u is dense (this is a crucial step!). The rest is now clear.

§ 6 Open problems and further supplements

6.1) Is $\mathcal{K}(M) = \mathcal{K}_u(M)$ even if $\{x \in L_M : x \geq 0\}$ is not total in L_M (cf 2.3)?

6.2) When does $\mathcal{K}_u(M) = \mathcal{K}(M,I)$ hold?

6.3) For which other lattices does the identity $\mathcal{E}_M = \mathcal{K}(M,I)$ hold (cf 4.3)?

6.4) If M is an I-KS in E, is it an I-KS in the completion of E?

6.5) (Trivial, if the answer to 6.4 is yes). Is there a normed but not complete lattice possessing an I-KS with 2 elements?

6.6) What about Korovkin-type theorems in ordered Banach spaces which are not lattices?(The most striking result is proved by A.G. Robertson [21] for C*-Algebras. His work yields even a new aspect to the classical Korovkin theorem)

6.7) Are there general Korovkin type theorems for the approximation of operators other than lattice homomorphism?(For some results in this direction see [20,18].)

References

1. H. Bauer: "Theorems of Korovkin type for adapted spaces", Ann.Inst. Fourier 23 (1973), 245-260
2. H. Bauer: "Convergence of monotone operators", Math.Z.136, (1974), 315-330
3. H. Bauer: "Approximationssatz und abstrakte Ränder", Mathem. Phys. Sem.-berichte 23 (1976), 141-173
4. H. Berens-G.G.Lorentz: "Theorems of Korovkin type for positive linear operators on Banach lattices", in: Approximation Theory, ed.by G.G.Lorentz, Acad.Press, New York, 1973, pp 1-30
5. H. Berens-G.G. Lorentz: "Korovkin theorems for sequences of contractions on L^p-spaces", in: Linear Operators and Approximation II,ed. by P.L. Butzer a. B. Sz.-Nagy, Birkhäuser Verlag

Basel u. Stuttgart, 1974, 367-375

6. E.W. Cheney, A. Sharma: "Bernstein power series", Canad.J. Math. 16 (1964), 241-252

7. K. Donner: "Korovkin theorems for positive linear operators", J.Approx.Theory 13 (1975), 443-45o

8. K. Donner: "Korovkin theorems and P-essential sets", Math. Z. (in print)

9. V.K. Dzjadyk: "Approximation of functions by positive linear operators and singular integrals", Math. Sborn.(N.S.) 7o (1966), 5o8-517

1o. H. Fakhoury: "Le théorème de Korovkin dans C(X) et $L^p(\mu)$", Sém. Choquet (Initiation à l'Analyse) 13e année 1973/74 no. 9

11. C. Franchetti: "Convergenza di operatori in sottospazi della spacio C(Q)", Boll.d.Un.Matem.Ital., Ser.IV,3 (1970),668-678

12. A. Goullet de Rugy: "La structure idéale des M-espaces", J.Math. pur. appl.IX. Sér., 51 (1972), 331-373

13. W. Kitto-D.E. Wulbert: "Korovkin approximations in L_p-spaces", Pac.J.Math. 63 (1976), 153-167

14. P.P.Korovkin: "Über die Konvergenz positiver linearer Operatoren im Raum der stetigen Funktionen", Dokl.Akad.Nauk. SSSR (N.S.) 9o (1953), 961-964

15. P.P.Korovkin: "Linear operators and approximation theory", Hindustan Publ.Corp., Delhi 196o

16. M.A.Krasnosel'ski -E.A.Lifšič: "A principle of convergence of sequences of positive linear operators", Studia Math. 31 (1968), 455-468

17. H.E. Lacey-P. Wojtaszczyk: "Nonatomic Banach Lattices can have l_1 as a dual space", Proc.AMS 57 (1976), 79-84

18. Le Baron O. Ferguson-M.D. Rusk: "Korovkin sets for an operator on a space of continuous functions", Pac.J. Math. 65 (1976), 337-345

19. W.Meyer-König-K.Zeller: "Bernsteinsche Potenzreihen", Stud. Math. 19 (196o), 89-94

2o. C.A. Micchelli: "Convergence of positive linear operators on C(X)", J.Approx. Th. 13 (1975), 3o5-315

21. A.G. Robertson: "A Korovkin theorem for Schwarz maps on C^*-algebras", Math. Z. (in print)

22. Yu.A. Šaškin: "Korovkin systems in spaces of continuous functions", Amer. Math. Soc. Transl., Ser. 2, 54 (1966), 125-144 (= Izv. Akad. Nauk. SSSR, Ser. Mat., 26 (1962), 495-512)

23. Yu.A. Šaškin: "The Milman-Choquet boundary and approximation theory", Funct. Anal. Appl. 1 (1967), 17o-171

24. H.H. Schaefer: "Topological vector spaces" 3rd pr., Springer-Verlag, Berlin-Heidelberg-New York 1971

25. H.H. Schaefer: "Banach lattices and positive operators", Springer-Verlag, Berlin-Heidelberg-New York 1974

26. E. Scheffold: "Über die punktweise Konvergenz von Operatoren in C(X)", Public. de la Rev. de la Acad. d.C. Zaragoza 28 (1973), 5-12

27. E. Scheffold: "Ein allgemeiner Korovkin-Satz für lokalkonvexe Vektorverbände", Math. Z. 132 (1973), 2o9-214

28. R.K. Vasil'ev: "The conditions for the convergence of isotonic operators in partially ordered sets with convergence classes", Matem. Zametki 12 (1972), 337-348

29. M. Wolff: "Darstellung von Banach-Verbänden und Sätze vom Korovkin-Typ", Math. Ann. 2oo (1973), 47-67

3o. M. Wolff: "Über Korovkin-Sätze in lokalkonvexen Vektorverbänden", Math.Ann. 2o4 (1973), 49-56

31. M. Wolff: "Über die Korovkinhülle von Teilmengen in lokalkonvexen Vektorverbänden", Math. Ann. 213 (1975), 97-1o8

32. M. Wolff: "Über die Charakterisierung von C(X) durch einen optimalen Satz vom Korovkintyp", Jber. Deutsch. Math.-Verein 78 (1976), 78-8o

33. M. Wolff: "On the universal Korovkin closure of subsets in vector lattices", J.approx. theory (in print)

34. D.E. Wulbert: "Convergence of operators and Korovkin's theorem", J.approx. theory 1 (1968), 381-39o

K.-D. Bierstedt, B. Fuchssteiner (eds.)
Functional Analysis: Surveys and Recent Results

SPACES OF CONTINUOUS FUNCTIONS

J. Schmets
Institut de Mathématique
Université de Liège
15, avenue des Tilleuls
B-4000 LIEGE (Belgium)

Let X be a completely regular and Hausdorff space and E be a locally convex topological Hausdorff vector space. We denote by $C(X)$ the space of the continuous functions on X and by $C(X;E)$ the space of the continuous functions on X with values in E.

We consider on $C(X)$ locally convex topologies corresponding to uniform convergence on subsets of the repletion υX of X and we introduce on $C(X;E)$ locally convex topologies obtained by use of subsets of υX and of the elements of a system P' of semi-norms on E finer than the one of E. Thus we get the spaces $C_{\mathcal{P}}(X)$ and $C_{P',\mathcal{P}}(X)$.

Our aim is to characterize the ultrabornological, bornological, barreled and evaluable spaces associated to $C_{\mathcal{P}}(X)$ and $C_{P',\mathcal{P}}(X)$.

For the study of $C(X)$ endowed with topologies becoming more and more general, we refer to [1, 2, 3, 4, 6, 9, 11, 12, 14 and 15]; a systematic study is available in [12]. As $C(X;E)$ is concerned, some material can already be found in [7, 12 and 13].

1. THE SPACE $C_{\mathcal{P}}(X)$

We denote once for all by X a completely regular and Hausdorff space and by $C(X)$ the linear space of the continuous scalar (i.e. with values in $\mathbb{R}$ or in $\mathbb{C}$) functions on X.

Up to a few years ago, only two locally convex topologies were usually considered on $C(X)$: the compact-open topology, and the simple or pointwise topology giving respectively the spaces $C_c(X)$ and $C_s(X)$. But this situation has proved to be unsatisfactory at least in (a) the research of the locally convex properties of $C_c(X)$, and (b) the theory of measures and of spaces of measures on X.

A natural way to introduce new locally convex topologies on $C(X)$ is the use of semi-norms of uniform convergence on suitable subsets of X. This has proved to be not the best way. A much better technique needs unhappily a large roundabout, passing through the characters of $C(X)$ and the repletion of X.

DEFINITION 1.1. A <u>character</u> of $C(X)$ is a linear functional τ on $C(X)$ which is different from 0 and multiplicative [i.e. such that $\tau(fg) = \tau(f)\tau(g)$ for every $f,g \in C(X)$].

Let us recall one of their basic properties.

THEOREM 1.2. *If τ is a character of $C(X)$, then, for every sequence $f_n \in C(X)$, there is $x \in X$ such that $\tau(f_n) = f_n(x)$ for every* n. #

In particular, if τ is a character of $C(X)$, one has $\tau(1) = 1$ and $\tau(f)$ belongs to $f(X)$ for every $f \in C(X)$.

The repletion (or realcompactification) of X can be described easily by use of the characters of $C(X)$.

DEFINITION 1.3. The *repletion* υX *of* X is the uniform subspace

$$\{\tau : \tau \text{ is a character of } C(X)\}$$

of the weak* algebraic dual of $C(X)$.

Let us recall a few facts.

a) For every $x \in X$, the Dirac measure ε_x is a character of $C(X)$. Up to the isomorphism ε, X appears as a dense uniform subspace of υX; X is *replete* if υX equals X. Every Lindelöf space is replete.

b) Every continuous application τ from X into a completely regular and Hausdorff space Y has a unique continuous extension τ^{υ} from υX into υY. In particular, for every $f \in C(X)$, f^{υ} is given at every $\tau \in \upsilon X$ by $f^{\upsilon}(\tau) = \tau(f)$.

c) Up to an isomorphism leaving X invariant, υX is the largest uniform space containing X as a dense subspace and where every $f \in C(X)$ has a continuous extension.

So the linear algebras $C(X)$ and $C(\upsilon X)$ are isomorphic and υX appears as the largest space wherefrom it is possible to derive semi-norms of uniform convergence.

DEFINITION 1.4. A subset B of υX is *bounding* if every $f \in C(X)$ is bounded on B. As υX is a complete uniform subspace of the weak* algebraic dual of $C(X)$, the bounding subsets of υX are exactly the relatively compact subsets of υX.

Of course $B \subset \upsilon X$ is bounding if and only if

$$||\cdot||_B = \sup_{x \in B} |\cdot(x)|$$

is a semi-norm on $C(X)$. Moreover $\{||\cdot||_B : B \in P\}$ is a system of semi-norms on $C(X)$ if and only if

a) every $B \in P$ is bounding and $\cup\{B : B \in P\}$ is dense in υX,

b) for every $B', B'' \in P$, there is $B \in P$ such that $B' \cup B'' \subset \overline{B}^{\upsilon X}$.

DEFINITION 1.5. A space $C_P(X)$ is $C(X)$ endowed with such a system of semi-norms $\{||\cdot||_B : B \in P\}$. But, for convenience, we impose that P contains the closure as well as the subsets of its elements, which of course makes no difference as far as the locally convex topological vector space $C_P(X)$ is concerned.

2. ASSOCIATED SPACES

A very natural question is to ask whether $C_c(X)$ is ultrabornological, bornological, barreled or evaluable. The answer is given by criteria which relate these locally convex properties of $C_c(X)$ to topological properties of X.

A second very natural question consists then in the characterization of the "closest space to $C_c(X)$ which satisfies the desired property" in the sense of Komura [8]. Let us explain this.

Once for all let E denote a locally convex topological Hausdorff vector space. Moreover let R be a locally convex property
(a) stable for Hausdorff inductive limits,
(b) satisfied by every linear space when endowed with its finest locally convex topology.
Examples for R are "the space is ultrabornological, bornological, barreled or evaluable".

DEFINITION 2.1. The R-*space associated to* E is the linear space E endowed with the coarsest locally convex topology τ_R which (α) is finer than the one of E, (β) makes E satisfy R.
This definition makes sense : in fact, τ_R is the inductive limit of all the locally convex topologies on E satisfying (α) and (β).

A good way to get properties of these R-spaces associated to E is to search for special constructions of them. For instance,
a) the bornological space associated to E is the inductive limit E_b of the normed space E_B where B runs through the family of the bounded disks of E.
b) [2] the ultrabornological space associated to E is the inductive limit E_{ub} of the Banach spaces E_B where B runs through the family of the bounded Banach disks of E.

As the properties are concerned, let us mention that every continuous linear map from an ultrabornological (resp. bornological; barreled) F into E is also continuous from F into the ultrabornological (resp. bornological; barreled) space associated to E. A consequence of this is the following result.

THEOREM 2.2. [12] *If* (E,τ) *is ultrabornological and webbed* ([5]), *then, for every locally convex topology* τ' *on* E *coarser than* τ, (E,τ) *is the ultrabornological space associated to* (E,τ').

Proof. Since the identity map from the ultrabornological space (E,τ) into (E,τ') is continuous, it is also continuous from (E,τ) into $(E,\tau')_{ub}$. Therefore the identity map from the ultrabornological space $(E,\tau')_{ub}$ into the webbed space (E,τ) has closed graph and therefore is continuous. Hence the conclusion.#

3. THE ULTRABORNOLOGICAL SPACE ASSOCIATED TO $C_{P}(X)$

It turns out that the associated spaces to $C_c(X)$ are of the kind $C_{P}(X)$ and moreover that the associated spaces to $C_{P}(X)$ are spaces of the same kind. This gives some more interest to the family of the $C_{P}(X)$ spaces.

In what follows we shall not try to give the characterization of all the associated spaces to $C_{P}(X)$ - for this we refer to [12]. However we shall develop a very interesting example : the one of the ultrabornological case.

If Y is a dense subset of υX, one can consider the space $C_{\mathcal{A}(Y)}(X)$ where $\mathcal{A}(Y)$ is the family of all finite subsets of Y. Then $\upsilon_Y X$ is the set of the elements of υX which are bounded on the bounded Banach disks of $C_{\mathcal{A}(Y)}(X)$. In other words, the elements of $\upsilon_Y X$ are those characters of $\mathcal{C}(X)$ which are continuous on the ultrabornological space associated to $C_{\mathcal{A}(Y)}(X)$.

It can be shown that $\upsilon_Y X$ is equal to υX in the following cases :

a) when Y contains X.

b) when every element of X is the limit in υX of a sequence of elements of Y (this is always the case if X is metrizable and if Y is a subset of X).

c) when every element of X has a bounding neighborhood in X (this is always the case if X is locally compact or pseudo-compact).

d) when X is a P-space.

a) is a direct consequence of theorem 1.2. For b), c) and d), we refer to [12].

Now we can state the following general result.

THEOREM 3.1. [12]. <u>If Y is a dense subspace of υX such that $\upsilon_Y X$ equals υX, then $C_c(\upsilon X)$ is the ultrabornological space associated to $[\mathcal{C}(X),P]$ for any system of semi-norms P on $\mathcal{C}(X)$ in between the ones of $C_{\mathcal{A}(Y)}(X)$ and $C_c(\upsilon X)$.</u>

In fact it is sufficient to prove that the ultrabornological space associated to $C_{\mathcal{A}(Y)}(X)$ is $C_c(\upsilon X)$. But this needs some developments.

The first step of the proof is the following fundamental result of L. Nachbin, which gives a hint to get all the associated spaces to $C_{P}(X)$.

In its statement we shall use the following notations :

a) Δ denotes the set $\{f \in \mathcal{C}(X) : |f(x)| \leq 1, \forall x \in X\}$. More generally, for every $f \in \mathcal{C}(X)$, $\Delta(f)$ is equal to $\{g \in \mathcal{C}(X) : |g| \leq |f|\}$, so that $\Delta = \Delta(1)$.

b) if f belongs to $\mathcal{C}(X)$, $\tilde{f}$ denotes its unique continuous extension from βX (the Stone-Čech compactification of X) into $\beta\mathbb{C}$. Moreover if A is a subset of βX, the notation $||\tilde{f}||_A < r$ where $r \in]0,+\infty[$ means that $\tilde{f}(x)$ belongs to $\{z : |z| < r\}$ for every $x \in A$.

THEOREM 3.2. [9].

a) <u>If T is an absolutely convex subset of $\mathcal{C}(X)$, there exists a smallest compact subset K(T) of βX such that $f \in \mathcal{C}(X)$ belongs to T if $\tilde{f}$ is equal to 0 on a neigh-</u>

borhood of $K(T)$ _in_ βX.

b) _If_ T _is an absolutely convex subset of_ $C(X)$ _containing_ Δ, _then_ $K(T)$ _is the smallest compact subset of_ βX _such that_ $f \in C(X)$ _belongs to_ T _if_ $\tilde{f}$ _is equal to_ 0 _on_ $K(T)$. _Moreover one has then_

$$f \in C(X) \ \& \ \|\tilde{f}\|_{K(T)} < 1 \Rightarrow f \in T.$$

Proof. a) Of course βX is a compact subset of βX such that $f \in C(X)$ belongs to T if $\tilde{f}$ is equal to 0 on a neighborhood of βX in βX, i.e. on βX. Therefore, to conclude, it is sufficient to prove that any intersection of compact subsets of βX such that $f \in C(X)$ belongs to T if $\tilde{f}$ is equal to 0 on a neighborhood of one of them, verifies the same property. But, since that property depends on the neighborhoods of the intersection, it is easy to see that the relation is true if it holds for the intersection of two such compact subsets K_1 and K_2 of βX.

Let $f \in C(X)$ be such that $\tilde{f}$ vanishes on a neighborhood V of $K_1 \cap K_2$ in βX. There is then $g \in C(X)$ with values in $[0,1]$ such that $\tilde{g} = g^{\beta}$ is equal to 1 on a neighborhood V_1 of K_1 in βX and to 0 on a neighborhood V_2 of $K_2 \setminus V$ in βX. In this way, $2fg$ belongs to $C(X)$ and $(2fg)^{\sim}$ vanishes on $V \cup V_2$, i.e. on a neighborhood of K_2 in βX, hence $2fg$ belongs to T. In the same way $2(1-g)f$ belongs to $C(X)$ and $[2(1-g)f]^{\sim}$ vanishes on V_1, hence $2(1-g)f$ belongs to T. The conclusion then follows from the fact that T is absolutely convex since $f = \frac{1}{2}[2fg+2(1-g)f]$.

b) By a), it is enough to prove that if T contains Δ and if K is a compact subset of βX such that $f \in C(X)$ belongs to T if $\tilde{f}$ is equal to 0 on a neighborhood of K in βX, then $g \in C(X)$ belongs to T if $\tilde{g}$ is equal to 0 on K. For every $r > 0$, the function θ_r defined on $\mathbb{C}$ by $\theta_r(z) = z$ if $|z| \leq r$ and $\theta_r(z) = r\frac{z}{|z|}$ if $|z| > r$ is continuous and bounded by r as one can easily check. Therefore $h = \theta_{1/2} \circ g$ belongs to $\frac{1}{2}\Delta$ and $2h$ to T. Moreover $2(g-h)$ belongs to $C(X)$ and $[2(g-h)]^{\sim}$ is equal to 0 on a neighborhood of K in βX, hence $2(g-h)$ belongs to T. Finally

$$g = \frac{1}{2}[2h+2(g-h)]$$

belongs to T since T is absolutely convex.

To prove the last part of b), it is sufficient to note that, if $f \in C(X)$ is such that $\|\tilde{f}\|_{K(T)} < 1-\varepsilon$ with $\varepsilon > 0$, then

$$f = \theta_{1-\frac{\varepsilon}{2}} \circ f + [f - \theta_{1-\frac{\varepsilon}{2}} \circ f] \in (1-\frac{\varepsilon}{2})T + \frac{\varepsilon}{2}T. \#$$

DEFINITION 3.3. The _hold_ of an absolutely convex subset T of $C(X)$ is the compact subset $K(T)$ defined in theorem 3.2.

REMARK 3.4. The impact of theorem 3.2 is tremendous. By its part b), the research of the associated spaces $C_P(X)$ becomes almost equivalent to the one of characteristic properties of the hold of T when moreover T satisfies another property such as : it absorbs the bounded Banach disks (to get the ultrabornological associated space), it absorbs the bounded disks (to get the bornological associated space), it is a barrel (to get the barreled associated space), ...

A first improvement of theorem 3.2 is given by the following result where $\mathcal{H}$ denotes the family of the absolutely convex, equicontinuous, closed and bounded subsets of $C_s(X)$.

PROPOSITION 3.5. [2] *If T is an absolutely convex and absorbing subset of $C(X)$ containing Δ, then the following properties are equivalent :*

(a) *the hold $K(T)$ of T is contained in υX.*

(b) *T is a neighborhood of 0 in $C_c(\upsilon X)$.*

(c) *T is bornivorous in $C_c(\upsilon X)$.*

(d) *for every $f \in C(X)$, T absorbs $\Delta(f)$.*

(e) *for every $H \in \mathcal{H}$, T absorbs H.*

Proof. (a) $\Rightarrow$ (b) is an immediate consequence of part b) of theorem 3.2.

(b) $\Rightarrow$ (c) and (c) $\Rightarrow$ (d) are trivial.

(d) $\Rightarrow$ (e) is immediate since the function f_H defined on X by

$$f_H(x) = \sup_{f \in H} |f(x)|, \ \forall x \in X,$$

is continuous on X and such that $H \subset \Delta(f_H)$.

(e) $\Rightarrow$ (a). If τ belongs to $\beta X \setminus \upsilon X$, we know that there is a bounded, continuous and strictly positive function f on X such that $\tau(f) = 0$. Let us then consider the sets

$$G_n = \{x \in \beta X : f^{\beta}(x) = \tilde{f}(x) \notin [0,1/n]\}, \ (n \in \mathbb{N});$$

they are open and increasing in βX, and cover X. Since τ does not belong to $\overline{G}_n^{\beta X}$ whatever be $n \in \mathbb{N}$, it is sufficient to prove that there exists $n \in \mathbb{N}$ such that $K(T) \subset \overline{G}_n^{\beta X}$. If this were not the case, there would exist a sequence $f_n \in C(X)$ such that $f_n \in T$ and $\tilde{f}_n(\overline{G}_n^{\beta X}) = \{0\}$. It is easy to check that such a sequence would be locally finite on X. Therefore the closed absolutely convex hull in $C_s(X)$ of the sequence $n f_n$ would belong to $\mathcal{H}$ and could not be absorbed by T, which is contradictory.#

The next step consists then in the following criterion for $C_c(X)$ to be ultrabornological or bornological.

THEOREM 3.6. *The following are equivalent :*

(a) [6] *the space $C_c(X)$ is ultrabornological.*

(b) [9], [14] *the space $C_c(X)$ is bornological.*

(c) [2] *the space $C_c(X)$ is semi-bornological* [i.e. every bounded linear functional on $C_c(X)$ is continuous].

(d) *the space X is replete.*

Moreover under these assumptions, the space $C_c(X)$ is the inductive limit of the Banach spaces $C(X)_{\Delta(f)}$ for $f \in C(X)$ [6] *and of the Banach spaces $C(X)_H$ for $H \in \mathcal{H}$* [2].

Proof. (a) $\Rightarrow$ (b) and (b) $\Rightarrow$ (c) are trivial.

(c) $\Rightarrow$ (d). By theorem 1.2, we know that every $\tau \in \upsilon X$ is a bounded linear functional on $C_s(X)$, hence on $C_c(X)$. So, if $C_c(X)$ is semi-bornological, every $\tau \in \upsilon X$ is a continuous linear functional on $C_c(X)$: there are a compact subset K of X and a constant $C > 0$ such that

$$|\tau(f)| \leq C||f||_K, \ \forall f \in \mathcal{C}(X). \qquad (*)$$

This implies that τ belongs to K for otherwise there would exist $f \in \mathcal{C}(X)$ such that $f(K) = \{0\}$ and $\tau(f) = 1$, which is contradictory to $(*)$.

(d) $\Rightarrow$ (a) and the characterizations of $C_c(X)$ as inductive limits. First of all, it is easy to check that, for every $f \in \mathcal{C}(X)$ and every $H \in \mathcal{H}$, the spaces $\mathcal{C}(X)_{\Delta(f)}$ and $\mathcal{C}(X)_H$ are Banach spaces. Let us call then E the inductive limit of the spaces $\mathcal{C}(X)_{\Delta(f)}$ for $f \in \mathcal{C}(X)$ and E' the one of the spaces $\mathcal{C}(X)_H$ for $H \in \mathcal{H}$. Of course, the spaces E and E' are ultrabornological, the space E has a finer topology than $C_c(X)$ and the space E' has a finer topology than E since, for every $H \in \mathcal{H}$, H is contained in $\Delta(f_H)$ where f_H is defined as in the proof of (d) $\Rightarrow$ (e) in proposition 3.5. To conclude, it is then sufficient to show that the topology of $C_c(X)$ is finer than the one of E'. To do so, let T be an absolutely convex neighborhood of 0 in E'. Then T absorbs Δ for otherwise there would exist a sequence $f_n \in \Delta$ such that $f_n \notin n^2 T$ for every $n \in \mathbb{N}$ and therefore the closed absolutely convex hull of the sequence f_n/n would belong to $\mathcal{H}$ and would not be absorbed by T, which is contradictory. The conclusion then follows immediately from proposition 3.5.#

To prove theorem 3.1, we need one more result. It concerns the linear space $\mathcal{C}^b(X)$ of the bounded continuous scalar functions on X. When it is endowed with the norm of the uniform convergence on X, it is the well known Banach space $C^b(X)$.

THEOREM 3.7. [12]. <u>For any locally convex topology</u> τ <u>on</u> $\mathcal{C}^b(X)$ <u>coarser than the one of</u> $C^b(X)$, $C^b(X)$ <u>is the ultrabornological space associated to</u> $[\mathcal{C}^b(X),\tau]$.

<u>Proof</u>. This is an immediate consequence of theorem 2.2 since every Banach space is ultrabornological and webbed.#

Now we are in order to prove theorem 3.1.

<u>Proof of theorem</u> 3.1. By theorem 3.6, we know that the space $C_c(\upsilon X)$ is ultrabornological.

So, to conclude, it is sufficient to show that $C_c(\upsilon X)$ is the ultrabornological space associated to $C_{\mathcal{A}(Y)}(X)$ or, what amounts to the same, that the spaces $C_{\mathcal{A}(Y)}(X)$ and $C_c(\upsilon X)$ have the same bounded Banach disks. Since $C_c(\upsilon X)$ has a finer topology than $C_{\mathcal{A}(Y)}(X)$, the conclusion will then follow if we prove that every bounded Banach disk T of $C_{\mathcal{A}(Y)}(X)$ is bounded in $C_c(\upsilon X)$.

To see this, let K be a compact subset of υX and denote by $T|_K$ the set of the restrictions to K of the elements of T. Of course $T|_K$ is a bounded Banach disk of the space $C_s(K)$, hence of $C^b(K)$ by use of theorem 3.7. Hence the conclusion.#

Two comments are in order.

1) As a consequence of theorem 3.1, we get the fact that, if Y is a dense subset of υX such that $\upsilon_Y X = \upsilon X$, then every continuous linear map from an ultra-bornological space E into $C_{A(Y)}(X)$ is also continuous from E into $C_c(\upsilon X)$. When Y contains X, we can better this result as follows.

PROPOSITION 3.8. Every continuous linear map from a bornological and barreled space E into $C_s(X)$ is also continuous from E into $C_c(\upsilon X)$.

Proof. If x belongs to υX, denote by Q_x the functional defined on E by

$$Q_x(e) = (Te)(x), \forall e \in E.$$

Of course Q_x is a linear functional on E. Moreover it is a bounded functional on E as one can check easily. Therefore Q_x is continuous. Then it is immediate to verify that the map $x \to Q_x$ is continuous from υX into the weak* dual of E. In this way, for every compact subset K of υX, the set $\{Q_x : x \in K\}$ is compact in the weak* dual of E, hence is equicontinuous on E. The conclusion follows then from the equality

$$||Te||_K = \sup\{|Q_x(e)| : x \in K\}, \forall e \in E.\#$$

Another way to prove proposition 3.8 is to establish first that the weak* dual of a semi-bornological space is replete (cf. [4] or [12], pp. 68-69 for details).

2) Theorem 3.1 allows to give a very simple example of a non ultrabornological space which contains a linear subspace of codimension 1 which is ultrabornological.

EXAMPLE 3.9. If x_o is a non isolated point of a replete P-space X, then $L = \{f \in C(X) : f(x_o) = 0\}$ is an ultrabornological subspace of codimension 1 of the non ultrabornological space $C_{A(X/\{x_o\})}(X)$. Of course the codimension of L in $C(X)$ is equal to 1. Moreover the space $C^o_{A(X/\{x_o\})}(X)$ is not ultrabornological : by theorem 3.1 and the remark d) which precedes it, the ultrabornological space associated to it is $C_c(X)$, hence $C_s(X)$ since every bounding subset of a P-space is finite. Finally L is ultrabornological because it is a closed subspace of codimension 1 of $C_s(X)$ which is ultrabornological.

4. THE SPACES $C_{P',\mathcal{P}}(X;E)$

Let as before X be a completely regular and Hausdorff space and E be a locally convex topological vector space. We denote the elements of X by x or y and the ones of $C(X)$ by f or g. As E is concerned, we shall use the following notations :

e for the elements of E,

P for the system of the semi-norms of E,

P' for a system of semi-norms on E which is finer than P.

Then the linear space of the continuous functions on X with values in E is denoted by $C(X;E)$ and its elements are represented by ϕ.

If ϕ belongs to $C(X;E)$, f to $C(X)$ and e to E, let us mention that the functions $f\phi$ and $f\cdot e$ belong to $C(X;E)$ and that the inclusion $C[X;(E,P')] \subset C(X;E)$ holds.

Now we must face the problem of endowing the space $C(X;E)$ with suitable systems of semi-norms.

A very natural way is the following. For every $p \in P$ and every $\phi \in C(X;E)$, $p[\phi(.)]$ is of course a continuous and positive function on X. So, if K is a relatively compact subset of υX and if $[p(\phi)]^{\sim}$ designates the unique continuous extension of $p(\phi)$ from υX into the replete space $\mathbb{R}$, the expression

$$||\phi||_{p,K} = \sup_{x \in K}[p(\phi)]^{\sim}(x)$$

is defined for every $\phi \in C(X;E)$. It is moreover easy to check that it is a semi-norm on $C(X;E)$. Then one can ask for the most general way to endow $C(X;E)$ with a system of semi-norms of the type $||.||_{p,K}$. The answer is easy to obtain : it is the space $C(X;E)$ equipped with $\{||.||_{p,K} : p \in P, K \in \mathcal{P}\}$ where $\mathcal{P}$ is such that the space $C_{\mathcal{P}}(X)$ can be considered; this is what we call a space $C_{P,\mathcal{P}}(X;E)$.

Let us remark that this procedure is possible because the function $p[\phi(.)]$ is continuous on X for every $p \in P$. If one also wants to replace P by a finer system of semi-norms P' on E, then he may not proceed as generally as previously. However there is one case when the procedure is valid : when every element of $\mathcal{P}$ is finite and contained in X. This gives the following space. If Y is a dense subspace of X, the space $C_{P',\mathcal{A}(Y)}(X;E)$ is $C(X;E)$ endowed with $\{||.||_{p',A} : p' \in P', A \in \mathcal{A}(Y)\}$, where $||.||_{p',A}$ is obviously defined by

$$||\phi||_{p',A} = \sup_{x \in A} p'[\phi(x)], \ \forall \phi \in C(X;E);$$

this set is of course a system of semi-norms on $C(X;E)$. If Y equals X, we write $C_{P',s}(X;E)$ instead of $C_{P',\mathcal{A}(X)}(X;E)$.

A little bit more generally, if $\mathcal{P}$ is a family of subsets of a dense subspace Y of X such that

a) the space $C_{\mathcal{P}}(X)$ can be considered,

b) for every $p' \in P'$ and every $K \in \mathcal{P}$,

$$||\phi||_{p',K} = \sup_{x \in K} p'[\phi(x)] < +\infty, \ \forall \phi \in C(X;E),$$

then one can consider the space $C_{P',\mathcal{P}}(X)$, i.e. the space $C(X;E)$ endowed with the system of semi-norms

$$\{||.||_{p',K} : p' \in P', K \in \mathcal{P}\}.$$

When P' is equivalent to P, these two general spaces $C_{P,\mathcal{P}}(X;E)$ and $C_{P',\mathcal{P}}(X;E)$ are isomorphic and deserve the same notation, which is the case. However it is essential to keep in mind that when P' is strictly finer than P, the notation $C_{P',\mathcal{P}}(X;E)$ implies that $Y = \cup\{B : B \in \mathcal{P}\}$ is a dense subset of X.

5. THE HOLD OF AN ABSOLUTELY CONVEX SUBSET OF $C(X;E)$

For every $\phi \in C(X;E)$, let us denote by $\tilde{\phi}$ the unique continuous extension of ϕ from βX into βE, and, for every $p \in P$, by $[p(\phi)]^{\sim}$ the unique continuous extension of $p(\phi)$ from βX into $\beta[0,+\infty[$. Then the following generalization of theorem 3.2 is at hand.

THEOREM 5.1. [12] and [13]. a) *If T is an absolutely convex subset of $C(X;E)$, there exists a smallest compact subset K(T) of βX such that $\phi \in C(X;E)$ belongs to T if $\tilde{\phi}$ is equal to 0 on a neighborhood of K(T) in βX.*
b) *If T is an absolutely convex subset of $C(X;E)$ for which there exist $p \in P$ and $r > 0$ such that*

$$T \supset \{\phi \in C(X;E) : p[\phi(x)] \leq r, \forall x \in X\},$$

then K(T) is the smallest compact subset of βX such that $\phi \in C(X;E)$ belongs to T if $\tilde{\phi}$ is equal to 0 on K(T). Moreover one has

$$\phi \in C(X;E) \ \& \ \{[p(\phi)]^{\sim}(x) : x \in K(T)\} \subset [0,r[\Rightarrow \phi \in T.$$

Proof. The proof of part a) goes on as in theorem 3.2. For part b), it is sufficient to replace $g = \theta_{1/2} \circ f$ by

$$\Psi_{p,r/2}(x) = \begin{cases} \phi(x) & \text{if } p[\phi(x)] \leq \frac{r}{2} \\ \frac{r}{2} \frac{\phi(x)}{p[\phi(x)]} & \text{if } p[\phi(x)] > \frac{r}{2} \end{cases} .$$

This function ψ belongs to $C(X;E)$ (but we need the condition $p \in P$: $p \in P'$ would not fit). To establish the last part of b), note that K(T) being compact there is $r' \in]0,r[$ such that

$$\sup_{x \in K(T)} [p(\phi)]^{\sim}(x) = r' .$$

So, for $r'' \in]r',r[$, one has

$$\phi = \Psi_{p,r''} + (\phi - \Psi_{p,r''}) \in \frac{r''}{r} T + (1 - \frac{r''}{r})T = T. \#$$

DEFINITION 5.2. The *hold* of an absolutely convex subset T of $C(X;E)$ is the compact subset K(T) defined in theorem 5.1.

Now we try to give some more information on the hold of T when we impose some property on T.

A first way to do so is given by use of the following criterion which is a direct consequence of theorem 5.1.

CRITERION 5.3. [12]. *If T is an absolutely convex subset of $C(X;E)$, then an element x of βX belongs to K(T) if and only if, for every neighborhood V of x in βX, there is an element ϕ of $C(X;E)$ such that $\phi \notin T$ and $\tilde{\phi}(\beta X \setminus V) = \{0\}$.*#

In fact, by use of this criterion, we get the following result.

PROPOSITION 5.4. [13]. *The hold of an absolutely convex and absorbing subset T of $C(X;E)$ is finite if and only if T absorbs every sequence $\phi_n \in C(X;E)$ such that the closures in βX of the sets $\{x \in X : \phi_n(x) \neq 0\}$ are disjoint.*

Proof. The condition is necessary. If $\phi \in C(X;E)$ does not belong to T, $\{x \in \beta X : \tilde{\phi}(x) = 0\}$ is not a neighborhood of K(T) and therefore the closure in βX of $\{x \in X : \phi(x) \neq 0\}$ is not disjoint from K(T). Hence, if K(T) is finite, only a finite number of the ϕ_n 's do not belong to T. The conclusion is then immediate since T is absorbing.

The condition is sufficient. Suppose that K(T) is not finite. There is then a sequence G_n of open subsets of βX which closures in βX are disjoint and such that $G_n \cap K(T) \neq \emptyset$ for every n (use lemma II.11.6 of [12], for instance). By the previous criterion, there exists then a sequence $\phi_n \in C(X;E)$ such that $\phi_n \notin n\,T$ and $\tilde{\phi}_n(\beta X \setminus G_n) = \{0\}$ for every n. Hence a contradiction.#

Another way is given by the following localization lemma.

LOCALIZATION LEMMA 5.5. [12]. *If T is an absolutely convex subset of $C(X;E)$ which absorbs every bounded subset of $C_{P',P}(X;E)$ which is equicontinuous on $Y = \cup\{B : B \in P\}$, then, for every increasing sequence G_n of open subsets of βX which cover Y, there is an integer n_0 such that $K(T) \subset \overline{G_{n_0}}^{\beta X}$.*

Proof. It is enough to show that there exists an integer n such that $\phi \in C(X;E)$ belongs to T if $\tilde{\phi}$ is equal to 0 on a neighborhood in βX of $\overline{G_n}^{\beta X}$.

If this is not the case, there is a sequence $\phi_n \in C(X;E) \setminus T$ such that $\tilde{\phi}_n$ is equal to 0 on a neighborhood in βX of $\overline{G_n}^{\beta X}$. But then the sequence $n\phi_n$ is bounded in $C_{P',P}(X;E)$ and equicontinuous on Y, and cannot be absorbed by T. Hence a contradiction.#

PROPOSITION 5.6. [13]. *If T is an absolutely convex subset of $C(X;E)$ which absorbs the bounded Banach disks of $C_{P',P}(X;E)$, then K(T) is a subset of υX.*

Proof. Suppose that $\beta X \setminus \upsilon X$ contains at least one element x of K(T). Then we know that there is a function $f \in C(X)$ with values in $]0,1]$ which unique continuous extension f^{β} on βX vanishes at x. Therefore the sets

$$G_n = \{y \in \beta X : f^{\beta}(y) \notin [0,1/n]\}, \quad (n \in \mathbb{N}),$$

constitute an open and increasing cover of υX in βX.

Consider now a sequence $\phi_n \in C(X;E)$ such that $\tilde{\phi}_n(G_n) = \{0\}$ and $\phi_n \notin n\,D$ for every $n \in \mathbb{N}$. Of course the sequence $n\phi_n$ tends to 0 in $C_{P',P}(X;E)$. Moreover for every ℓ_1-sequence c_n, the series $\sum_{n=1}^{\infty} c_n n\phi_n$ converges in $C_{P',P}(X;E)$ since the series is in fact locally finite on υX. In this way,

$$B = \{\sum_{n=1}^{\infty} c_n n\phi_n : \sum_{n=1}^{\infty} |c_n| \leq 1\}$$

is a bounded Banach disk of $C_{P',P}(X;E)$ and cannot be absorbed by T, which is contradictory.#

COROLLARY 5.7. *If T is an absolutely convex and bornivorous subset of $C_{P',A(Y)}(X;E)$, then K(T) is a finite subset of υX.*

Proof. This follows easily from propositions 5.4 and 5.6.#

Now let us turn to part b) of theorem 5.1.

First of all, let us give conditions which imply the existence of $p \in P$ and $r > 0$ such that

$$T \supset \{\phi \in C(X;E) : p[\phi(x)] \leq r, \forall x \in X\}.$$

PROPOSITION 5.8. [13]. _If_ T _is an absolutely convex subset of_ $C(X;E)$ _and if one of the following two conditions is satisfied_

(a) T _is bornivorous in_ $C_{P',\mathcal{P}}(X;E)$ _and_ (E,P') _is metrizable_,

(b) T _absorbs the bounded Banach disks of_ $C_{P',\mathcal{P}}(X;E)$ _and_ (E,P') _is a Fréchet space_,

then there are $p' \in P'$ _and_ $r > 0$ _such that_

$$T \supset \{\phi \in C(X;E) : p'[\phi(x)] \leq r, \forall x \in X\}.$$

Proof. Let $\{p_n : n \in \mathbb{N}\}$ be a countable system of semi-norms on E equivalent to P'.

If the result is false, there is a sequence $\phi_n \in C(X;E)$ such that

$$\sup_{x \in X} p_n[\phi_n(x)] \leq n^{-4} \text{ and } \phi_n \notin T, \forall n \in \mathbb{N}.$$

(a) The sequence $n^2\phi_n$ is of course bounded in $C_{P',\mathcal{P}}(X;E)$ which is contradictory with the fact that it is not absorbed by T.

(b) The sequence $n^2\phi_n$ converges obviously to 0 in $C_{P',\mathcal{P}}(X;E)$. Moreover, for every ℓ^1-sequence c_n, the series $\Sigma_{n=1}^{\infty} c_n n^2\phi_n$ converges in that space as one can verify easily. Therefore

$$B = \{\Sigma_{n=1}^{\infty} c_n n^2 \phi_n : \Sigma_{n=1}^{\infty}|c_n| \leq 1\}$$

is a bounded Banach disk in $C_{P',\mathcal{P}}(X;E)$ and cannot be absorbed by T, which is contradictory.#

Finally we give some conditions which assure that $\phi \in C(X;E)$ belongs to T if $\tilde{\phi}$ takes the value 0 at every element of K(T).

PROPOSITION 5.9. [13]. _If_ T _is an absolutely convex and bornivorous subset of_ $C_{P',s}(X;E)$ _and if_ X _is replete and satisfies the first axiom of countability, then an element_ ϕ _of_ $C(X;E)$ _belongs to_ T _if_ ϕ _vanishes on_ K(T).

Proof. By corollary 5.7, we know already that K(T) is a finite subset of X. Therefore, since X satisfies the first axiom of countability, there is a sequence of functions $f_n \in C(X)$ with values in [0,1], equal to 1 on neighborhoods of K(T) in X and to 0 outside decreasing neighborhoods of K(T) in X which intersection coincides with K(T).

Now let $\phi \in C(X;E)$ vanish on K(T). As the sequence $n f_n \phi$ is bounded in $C_{P',s}(X;E)$, there is $C > 0$ such that $n f_n \phi \in CT$ for every $n \in \mathbb{N}$, hence $f_n \phi$ belongs to T for n large enough. Moreover, for every $n \in \mathbb{N}$, $[(1-f_n)\phi]^{\sim}$ vanishes on a neighborhood in βX of K(T), so $(1-f_n)\phi$ belongs to T. Therefore

$$\phi = \frac{1}{2}[(1-f_n)\phi + f_n \phi]$$

belongs to T for n large enough.#

PROPOSITION 5.10. [13]. *If T is an absolutely convex subset of $C(X;E)$ which absorbs the bounded Banach disks of $C_{P,\mathcal{P}}(X;E)$ and such that $K(T)$ is contained in X, and if (E,P) is metrizable, then an element ϕ of $C(X;E)$ belongs to T if ϕ vanishes on $K(T)$.*

Proof. Suppose that $\phi \in C(X;E)$ vanishes on $K(T)$ and let $\{p_n : n \in \mathbb{N}\}$ be a countable system of semi-norms on E, equivalent to P. Then the sets

$$G_n = \{x \in X : p_n[\phi(x)] < n^{-4}\}, \ (n \in \mathbb{N}),$$

are decreasing open neighborhoods of $K(T)$ in X. Therefore there exists a sequence of functions $f_n \in C(X)$ with values in $[0,1]$, which are equal to 1 on a neighborhood of $K(T)$ and to 0 outside G_n.

Of course the sequence $n^2 f_n \phi$ converges to 0 in $C_{P,\mathcal{P}}(X;E)$. Moreover, for any ℓ^1-sequence c_n, let us prove that the series $\Sigma_{n=1}^{\infty} c_n n^2 f_n \phi$ converges in $C_{P,\mathcal{P}}(X;E)$. Every $x \in \upsilon X \setminus \{y \in \upsilon X : \tilde{\phi}(y) = 0\}$ belongs to the complement of some $\overline{G_n}^{\upsilon X}$, hence the series converges locally uniformly on $\upsilon X \setminus \{y \in \upsilon X : \tilde{\phi}(y) = 0\}$. To conclude, it is then sufficient to note that, for every $x \in G_n$, one has

$$p_n[\Sigma_{m=1}^{\infty} c_m m^2 f_m(x)\phi(x)]$$

$$\leq \Sigma_{m=1}^{n-1}|c_m|m^2 p_n[\phi(x)] + \Sigma_{m=n}^{\infty}|c_m|m^2 p_m[f_m(x)\phi(x)]$$

$$\leq n^{-4} \Sigma_{m=1}^{n-1}|c_m|m^2 + n^{-2} \Sigma_{m=n}^{\infty}|c_m|$$

and the last member of these inequalities tends to 0 if $n \to \infty$.

Therefore

$$B = \{\Sigma_{n=1}^{\infty} c_n n^2 f_n \phi : \Sigma_{n=1}^{\infty}|c_n| \leq 1\}$$

is a bounded Banach disk of $C_{P,\mathcal{P}}(X;E)$. Hence there exists $C > 0$ such that $f_n \phi$ belongs to $(C/n)T$ for every $n \in \mathbb{N}$.

Moreover $(1-f_n)\phi$ belongs to εT for every $n \in \mathbb{N}$ and every $\varepsilon > 0$ since $[(1-f_n)\phi]^{\sim}$ is equal to 0 on a neighborhood of $K(T)$ in βX.

Finally we have

$$\phi = f_n \phi + (1-f_n)\phi, \ \forall n \in \mathbb{N},$$

and therefore ϕ belongs to T. Hence the conclusion.#

6. SOME BORNOLOGICAL AND ULTRABORNOLOGICAL $C_{P,\mathcal{P}}(X;E)$ SPACES

PROPOSITION 6.1. [13].

a) *If X is replete and if (E,P) is metrizable, then the space $C_{P,c}(X;E)$ is bornological.*

b) *If X is replete and if (E,P) is a Fréchet space, then the space $C_{P,c}(X;E)$ is ultrabornological.*

Proof. We just need to prove that every absolutely convex subset T of $C(X;E)$ which absorbs the bounded disks (resp. the bounded Banach disks) of $C_{P,c}(X;E)$ is a neighborhood of 0.

By proposition 5.6, the hold $K(T)$ of T is a compact subset of $\upsilon X = X$. By proposition 5.8, there are $p \in P$ and $r > 0$ such that

$$T \supset \{\phi \in C(X;E) : p[\phi(x)] \leq r, \forall x \in X\}.$$

The conclusion follows then immediately from the last part of theorem 5.1. b) which says that we have then

$$T \supset \{\phi \in C(X;E) : \sup_{x \in K(T)} p[\phi(x)] < r\}.\#$$

Part b) of the previous proposition allows to get the following result.

PROPOSITION 6.2. _If_ X _is replete, if_ $Y \subset X$ _is such that_ $\upsilon_Y X = X$ _and if_ (E,P) _is a Fréchet space, then_ $C_{P,c}(X;E)$ _is the ultrabornological space associated to_ $[C(X;E);\tau]$ _where_ τ _is any locally convex topologies in between the ones of_ $C_{P,A(Y)}(X;E)$ _and of_ $C_{P,c}(X;E)$.

Proof. By part b) of proposition 6.1, we know already that $C_{P,c}(X;E)$ is ultrabornological. So to conclude, it is sufficient to prove that $C_{P,c}(X;E)$ is the ultrabornological space associated to $C_{P,A(Y)}(X;E)$ or, what amounts to the same, that every bounded Banach disk B of $C_{P,A(Y)}(X;E)$ is bounded in $C_{P,c}(X;E)$.

Let K be a compact subset of X and denote by $B|_K$ the set of the restrictions on K of the elements of B. Since we have $\upsilon_Y K = K$, it is easy to check that $B|_K$ is a bounded Banach disk of $C_{P,s}(K;E)$, hence of the Fréchet space $C_{P,c}(K;E)$ by use of theorem 2.2. Hence the conclusion.#

7. COMMENTS AND REMARKS

In [12], one can find the characterization of the barreled, (d-barreled, σ-barreled) and evaluable (d-evaluable, σ-evaluable) spaces associated to $C_{P',s}(X;E)$. Adapting the same techniques to the space $C_{P',A(Y)}(X;E)$, the results extend and provide in particular that

a) the space $C_{P',A(Y)}(X;E)$ is a Mackey space if and only if (E,P') is a Mackey space.

b) the space $C_{P',A(Y)}(X;E)$ is evaluable if and only if (E,P') is evaluable.

More generally the evaluable space associated to $C_{P',A(Y)}(X;E)$ is the space $C_{P'_e,A(Y)}(X;E)$ where P'_e denotes the system of semi-norms of the evaluable space associated to (E,P').

c) the space $C_{P',A(Y)}(X;E)$ is barreled if and only if the spaces $C_{A(Y)}(X)$ and (E,P') are barreled.

More generally, if $C_{A(Y)}(X)$ is barreled, then the barreled space associated to $C_{P',A(Y)}(X;E)$ is the space $C_{P'_t,A(Y)}(X;E)$ where P'_t denotes the system of semi-norms of the barreled space associated to (E,P').

In [13], one can find some further examples of ultrabornological and bornological $C_{P',s}(X;E)$ spaces such as

a) if $C_s(X)$ is bornological and if (E,P) is metrizable, then $C_{P,s}(X;E)$ is bornological.

b) if $C_s(X)$ is ultrabornological and if (E,P) is a Fréchet space, then $C_{P,s}(X;E)$ is ultrabornological.

c) if $C_s(X)$ is bornological and if X satisfies the first axiom of countability, then the bornological space associated to $C_{P',s}(X;E)$ is the space $C_{P'_b,s}(X;E)$ where P'_b is the system of semi-norms of the bornological space associated to (E,P').

d) if $C_s(X)$ is ultrabornological, if (E,P) is metrizable and ultrabornological and if X satisfies the first axiom of countability, then $C_{P,s}(X;E)$ is ultrabornological.

In this way we have gathered the results known so far in this direction and therefore many questions are still open. Let us state just one, asked by K.-D. BIERSTEDT: if X is compact and if (E,P) is a bornological DF-space, is $C_{P,c}(X;E)$ a bornological space ? *)

REFERENCES

[1] Buchwalter, H. (1969-1970). Parties bornées d'un espace topologique complètement régulier, Sém. Choquet 9, n° 14, 15 pages.

[2] Buchwalter, H. (1972). Sur le théorème de Nachbin-Shirota, J. Math. Pures et Appl. 51, 399-418.

[3] Buchwalter, H., et Noureddine, K. (1972). Topologies localement convexes sur les espaces de fonctions continues, C.R. Acad. Sc. Paris 274A, 1931-1934.

[4] Buchwalter, H., et Schmets, J. (1973). Sur quelques propriétés de l'espace $C_s(T)$, J. Math. Pures et Appl. 52, 337-352.

[5] De Wilde, M. Closed graph theorem and webbed spaces, (to appear).

[6] De Wilde, M., et Schmets, J. (1971). Caractérisation des espaces C(X) ultrabornologiques, Bull. Soc. Roy. Sc. Liège 40, 119-121.

[7] Katsaras, A. (1976). On the space C(X,E) with the topology of simple convergence, Math. Ann. 223, 105-117.

[8] Komura, Y. (1962). On linear topological spaces, Kumamoto Math. J. of Sc. 5A, 148-157.

[9] Nachbin, L. (1954). Topological vector spaces of continuous functions, Proc. Nat. Acad. USA 40, 471-474.

[10] Noureddine, K. (1973). L'espace infratonnelé associé à un espace localement convexe, C.R. Acad. Sc. Paris 174A, 116-124.

[11] Noureddine, K., et Schmets, J. (1973). Espaces associés à un espace localement convexe et espaces de fonctions continues, Bull. Soc. Roy. Sc. Liège 42, 116-124.

[12] Schmets, J. (1976). Espaces de fonctions continues, Lecture Notes in Mathematics 519, Springer Verlag, Berlin.

[13] Schmets, J. (1977). Bornological and ultrabornological C(X;E) spaces, Manuscripta Math., (to appear).

[14] Shirota, T. (1954). On locally convex vector spaces of continuous functions, Proc. Japan Acad. 30, 294-298.

[15] Warner, S. (1958). The topology of compact convergence on continuous function spaces, Duke Math. J. 25, 265-282.

*) Editors' remark: A partial affirmative answer to this question was recently given by J. Mujica, Representation of analytic functionals by vector measures, preprint (1977), Campinas, Brazil .

K.-D. Bierstedt, B. Fuchssteiner (eds.)
Functional Analysis: Surveys and Recent Results

THE STRICT TOPOLOGY AND (DF) SPACES

W. Ruess
Institut für Angewandte Mathematik
Universität Bonn
Bonn, Federal Republic of Germany

INTRODUCTION

This paper is devoted to a study of the locally convex structure of several analogues of the strict topology β on the space $C(S)$ of bounded continuous functions on a locally compact Hausdorff space S. This original strict topology has been introduced by R.C. Buck in two papers in the years 1952 and 1958 and, later on, has been extended to the completely regular setting by various other authors. The respective topologies first of all turned out to be of particular importance for topological measure theory: for T completely regular Hausdorff and ζ any of the strict topologies on $C(T)$, the dual of $(C(T),\zeta)$ represents the space of certain measures on T. This fact made it possible to establish various relations between topological properties of T, the locally convex structure of $(C(T),\zeta)$ and properties of sets of special measures on T.

In recent years, various generalizations and extensions of these (locally convex) strict topologies have been considered, both in the commutative (continuous function space) and the non-commutative (general Banach algebra) setting. Accordingly, today's strict topologies have a broad spectrum of applications in probability and topological measure theory, approximation theory, the spectral analysis of bounded continuous functions, analytic function theory and certain aspects of C^*-algebra theory.

This positive aspect of broad applicability, however, is confronted with a very unpleasant counterpart: from the point of view of locally convex structure theory, spaces with strict topologies are rather bad objects. In general, they do not belong to any of the classical classes of locally convex spaces; they are neither (F) nor (DF) spaces, neither barrelled nor of any other of the usual barrelledness types. Therefore, from the very beginning, in dealing with strict topologies, the methods of general locally convex structure theory rarely could be applied and, very often, troublesome ad-hoc-methods were needed.

It is the object of this paper to give a survey of our work on trying to proceed with resolving this dilemma, and to give a systematic approach to the significant locally convex structure properties of strict topologies.

Before going into details, let us first specify the three basic types of strict topologies we shall be concerned with in this paper:
(1) Strict topologies on the space $C(T)$, T completely regular Hausdorff [13], [39],
(2) the strict topology β_B on a Banach module X over a Banach algebra B with bounded approximate identity [40] and
(3) the strict topology β_A on the double centralizer algebra $M(A)$ of a C^*-algebra A [5], [44].

Our investigations are based on one very general property of these topologies: If (X,β) denotes a strict space of any of the above types (1), (2) or (3), then it is known (consult [7]) that β is the finest locally convex topology on X agreeing with itself on the unit ball of the respective (Banach space) norm on X. It follows from this localization property of β, that (X,β) is an (L)-space (and, a fortiori, an (Lb)-space) in the sense of [32] with a fundamental sequence of bounded sets.

Starting from this observation and from several concrete problems of strict topologies, we carry on our systematic study in two directions:
(i) Closed graph theorems for (L)-spaces;
(ii) Locally convex structure properties of (Lb)-spaces with a fundamental sequence of bounded sets.

The particular interest in the class of locally convex spaces specified in (ii) is due to the fact that, besides all strict spaces, it also contains the classical (DF) spaces of Grothendieck [15]. Accordingly, the primary object is to determine which of the properties of (DF) spaces also hold true for this wider class.

Our results on the above two points will in fact provide us with the desired systematic insight in the locally convex structure of strict topologies. One of the main steps in this course will be furnished by showing that, although not being countably evaluable, (Lb)-spaces with a fundamental sequence of bounded sets indeed share almost all of the other nice properties of (DF) spaces. (It is for this reason that we shall call such spaces generalized (DF) spaces.)

Our investigations will essentially be centered around the following concrete problems of strict topologies:

1. Characterization of semi-reflexive and of semi-Montel linear subspaces of spaces with strict topologies.

2. The Mackey problem for the space $(H^\infty(G),\beta)$ of bounded holomorphic functions on a plane region G with the strict topology.

3. Non-nuclearity of strict topologies.

4. (Weakly) compact linear operators on strict spaces.

We shall also touch on several other concrete problems considered in the literature.

In closing this introductory part, we want to refer the interested reader to three survey papers on strict topologies and their applications: the one by R.A. Hirschfeld [19] mainly concentrates on the measure theoretic aspects, that by L.A. Rubel [28] deals with the function theoretic aspect, and the most recent one by H.S. Collins [7] covers the whole spectrum of today's strict topologies and indicates most of the interesting applications. Finally, for the measure theoretic aspects, we should also like to call the reader's attention to the approach of the Buchwalter school (consult [2]) via compactologies, which is somehow dual to the one of [13] and [39] via strict topologies.

The present exposition is mainly based on our papers [34], [35] and [36].

TERMINOLOGY AND NOTATION

Throughout this paper, T (resp. S) will denote a completely regular (resp. locally compact) Hausdorff topological space. The space of bounded continuous scalar-valued functions on T is denoted by C(T). $C_o(S)$ denotes the space of all continuous scalar-valued functions on S that vanish at infinity.

Terminology and notation concerning locally convex spaces is generally taken from Horváth [21]. In particular, if (X,τ) is a locally convex space, then $\tau(X,X')$ and $\beta(X,X')$ denote the Mackey and strong topology respectively on X. All locally convex spaces considered in this paper are assumed to be Hausdorff. The notion "locally convex space" is abbreviated by "lcs". A convex circled subset of a linear space is called a disk.

If (X,τ) is an lcs, then we denote by $\mathcal{U}_\tau$ the filter of τ-neighbourhoods of zero. An absorbing disk U in X is called $\aleph_o$-barrel, if there exists a sequence $(U_n)_{n\in\mathbb{N}}$ of closed disks $U_n\in\mathcal{U}_\tau$ such that $U = \bigcap\{U_n \mid n\in\mathbb{N}\}$. An lcs (X,τ) is called $\aleph_o$-barrelled (resp. $\aleph_o$-evaluable) if every $\aleph_o$-barrel (resp. bornivorous $\aleph_o$-barrel) is a τ-neighbourhood of zero in X. Recall that a (DF) space [15] is an $\aleph_o$-evaluable lcs with a fundamental sequence of bounded sets.

The cone of τ-continuous seminorms on an lcs (X,τ) is denoted by C_τ. The (real) linear space generated by C_τ is $X_\tau = C_\tau - C_\tau$. The (locally convex) topology on X_τ of uniform convergence on the class of all bounded subsets of (X,τ) is denoted by β_X. (For a definition of this topology through polarity for the pair (X,X_τ), the reader is referred to [32].)

1. STRICT TOPOLOGIES

In this section we give a brief survey of the spectrum of strict topologies and of some of their applications. We restrict ourselves to those aspects and problems which are of special importance for our direction of study. For further details we again refer to the above mentioned papers [7], [19] and [28].

1.1 STRICT TOPOLOGIES ON SPACES OF CONTINUOUS FUNCTIONS

The original strict topology β has been introduced by R.C. Buck [3],[4] for a locally compact Hausdorff space S to be the locally convex topology on C(S) generated by the seminorms p_ρ, $\rho \epsilon C_o(S)$, where, for $x \epsilon C(S)$, $p_\rho(x) = \sup\{|x(s)\rho(s)| \mid s \epsilon S\}$. Immediate from the definition is β's relation to two other canonical topologies on C(S): β is finer than the compact-open and coarser than the sup-norm topology on C(S), and it coincides with these in case S is compact.

The interest in the strict topology stems from the fact that β turned out to be the natural topology to replace the sup-norm topology on the space of bounded continuous functions in passing from a compact base space K to a locally compact base space S: as a rule of thumb, β does for C(S) what the sup-norm does for C(K). In his fundamental paper [4] on β, Buck already showed that two of the important results on the sup-norm algebra C(K), the Riesz representation theorem and the Stone-Weierstrass theorem, directly extend to C(S) if this space is endowed with the strict topology β. We quote these and two other results illustrating the above rule:

(1.1) Riesz representation theorem for β [4]: *The dual of $(C(S),\beta)$ is the space M(S) of bounded regular Borel measures on S (via the usual integral representation).*

(1.2) Stone-Weierstrass theorem for β [4], [14]: *If A is a closed self-adjoint subalgebra of $(C(S),\beta)$ which separates the points of S, then A = C(S).*

(1.3) *If, for two locally compact spaces S_1 and S_2, $(C(S_1),\beta)$ and $(C(S_2),\beta)$ are isomorphic as locally convex algebras, then S_1 and S_2 are homeomorphic* ([4]).

(1.4) *There exist integral representations for β-(weakly) compact operators between C(S) spaces, S locally compact, analogous to the classical ones for the case of compact base spaces* ([37,38]).

Starting from the year 1967, several authors considered the measure theoretic aspect of strict topologies in the more general setting of completely regular base spaces. In particular, in 1972, Sentilles [39] and, independently, Fremlin, Garling and Haydon [13] introduced for a completely regular space T three types of strict topologies on C(T), the substrict (β_o), strict (β) and the superstrict (β_1) topology, the respective duals of which turned out to be the spaces of tight, τ-additive and σ-additive measures on T (for the spaces of measures cf. [45]). In this way, strict topologies became a strong new tool in the study of topological measure theory. For a survey of the extensive literature on the relationship between topological properties of T, the locally convex structure of $(C(T),\zeta)$ and the corresponding spaces $M_\zeta(T)$ of measures on T, and on open problems in this area, we refer again to the papers [7], [19].

1.2 THE STRICT TOPOLOGY ON SPACES OF HOLOMORPHIC FUNCTIONS

The object of study here is the space $H^\infty(G)$ of bounded holomorphic functions on a plane region G. The work of Rubel and Shields [30] on this linear subspace of C(G), endowed with the induced strict topology β, indicated for the first time that β not only is an appropriate topology for investigating spaces of bounded *continuous* functions but also a strong tool for the analysis of bounded *holomorphic* functions. To illustrate this fact, we quote some of the relevant results together with several other general facts which will be needed later on:

(1.5) *The space $(H^\infty(G),\beta)$ is a semi-Montel locally convex space (bounded sets are relatively compact); its strong dual is the space $M_o(G) = M(G)/(H^\infty(G))^\perp$ with the (Banach space) norm of total variation. Contrary to $M(G)$ itself, $M_o(G)$ is a separable Banach space.*

A sequence $(f_n)_{n\in\mathbb{N}}\subset H^\infty(G)$ is β-convergent to $f\in H^\infty(G)$ if and only if it is boundedly convergent to f, i.e. (i) $|f_n(z)|\leq M<\infty$ for all $z\in G$ and all $n\in\mathbb{N}$ and (ii) $f_n(z)\longrightarrow f(z)$ for each $z\in G$ (or, equivalently, uniformly on compact subsets of G) ([30]).

(1.6) *There exists a countable dominating subset $(z_n)_{n\in\mathbb{N}}$ of G (for every $f\in H^\infty(G)$: $\sup\{|f(z)| \mid z\in G\} = \sup\{|f(z_n)| \mid n\in\mathbb{N}\}$) that has no limit point in G* ([30]).

In case G = D = open unit disk in $\mathbb{C}$ we have:

(1.7) *(i) The principal ideal (f) = $fH^\infty(D)$ generated by $f\in H^\infty(D)$ is dense in (the locally convex algebra) $(H^\infty(D),\beta)$ if and only if f is an outer function.*
(ii) The principal ideal (f) is β-closed if and only if the outer factor of f is a unit ([30]).

(For the factorization of H^p-functions into the product of an inner and an outer function, see [20] or [31,Thm.17.17].)

(1.8) The β-ideal theorem [30]: *Every closed ideal in $(H^\infty(G),\beta)$ is the principal ideal generated by an inner function.*

(1.9) The β-corona conjecture [28,pp.20-21]: *The point evaluations are dense in the space of distinguished (i.e. β-continuous) complex homomorphisms.*

For further information we refer again to the paper [30] of Rubel/Shields and to Rubel's survey [28].

1.3 STRICT TOPOLOGIES IN THE NON-COMMUTATIVE SETTING

Starting from the year 1968, the strict topology β has been extended from the commutative case of C(T) spaces to the not necessarily commutative settings of general Banach modules and C^*-algebras.

R.C. Busby [5] introduced a "two-sided" strict topology β_A on the double centralizer algebra M(A) of a C^*-algebra A: A is viewed as a closed two-sided ideal in M(A), and β_A is generated by the seminorms $(\lambda_a,\rho_a)_{a\in A}$, where, for $x\in M(A)$, $\lambda_a(x)=\|ax\|$ and $\rho_a(x)=\|xa\|$. Two motivating examples for the double centralizer algebra concept are the algebra $C_o(S)$, whose double centralizer algebra is C(S) [4], and the algebra K(H) of compact operators on a Hilbert space H, whose double centralizer algebra is the space L(H) of all bounded linear operators on H [5]. In case A = $C_o(S)$, we have $\beta_A = \beta$ [4].

Busby used β_A in his study of extensions of C^*-algebras and derived some of the basic results about the (complete locally convex) topological algebra $(M(A),\beta_A)$. For further information on strict topologies in the C^*-algebra setting see also the papers [12] and [44].

In the general setting of a left Banach B-module $(X,\|\ \|)$, where B is a Banach algebra with bounded approximate identity, a "left" strict topology β_B has been introduced by F.D. Sentilles and D.C. Taylor [40]: β_B is that locally convex topology on X generated by the seminorms $(\lambda_b)_{b\in B}$, where, for $x\in X$, $\lambda_b(x) = \|bx\|$. Sentilles and Taylor established most of Buck's basic results on β also for β_B. Further results on the space (X,β_B) can be found in the papers [42], [43] and [35], the latter including an application of the specific properties of (X,β_B) to results in Rieffel's paper [27].

1.4 SOME PROBLEMS OF STRICT TOPOLOGIES

We specify now those four problems that will give the orientation for our general study of the locally convex structure of strict topologies.

Problems 1-3 all come from the space $(H^\infty(G),\beta)$:

Problem 1: Does β coincide with the equicontinuous weak*-topology on $H^\infty(G)$ as the dual of the (separable) Banach space $M_o(G)$?

Problem 2: Is β the Mackey topology of the dual pair $(H^\infty(G),M_o(G))$?

Problem 3: Is $(H^\infty(G),\beta)$ a nuclear locally convex space?

Problem 1 is motivated by the following two facts:

(i) (P. Hessler, unpublished): A subset of $(H^\infty(G),\beta)$ is closed as soon as it is sequentially closed.

(ii) In the dual X^* of a separable Banach space X, a subset is closed in the equicontinuous weak*-topology if and only if it is sequentially closed. (Recall that the equicontinuous weak*-topology on the dual X' of a locally convex space X is defined to be the finest topology on X' agreeing with $\sigma(X',X)$ on the equicontinuous subsets of X'.)

Problem 1 has been answered affirmatively by Rubel and Ryff [29], and Shapiro [41], [42] has placed this result in the general setting of semi-Montel linear subspaces of a (left) Banach B-module X with the left strict topology β_B.

Problem 2 also is concerned with the position of β amongst other canonical topologies on $H^\infty(G)$. Moreover, it is closely related to the corresponding Mackey problem for the whole space C(G). The Mackey problem for $(C(S),\beta)$, S locally compact, is of particular importance and has extensively been studied (consult [7]), because the β-equicontinuous subsets of M(S) are exactly the norm-bounded uniformly tight ones of Le Cam [24], so that the coincidence of β with the Mackey topology of the dual pair (C(S),M(S)) immediately gives a characterization of these sets as subsets of the weak*-compact convex subsets of M(S). It is known that when S is locally compact and paracompact, then $(C(S),\beta)$ is a Mackey space [8],[10] (and even strong Mackey, see the remark following the proof of Theorem 2.10 in section 2), but that $(H^\infty(D),\beta)$, D the open unit disk in $\mathbb{C}$, is not [9]. The Mackey problem for $(H^\infty(G),\beta)$ for general G still seems to be open, cf. [28].

Problem 3 is motivated by the positive answer to problem 1, for the dual of a Banach space endowed with the equicontinuous weak*-topology is a Schwartz space.

Problem 4 arises from the following result of Sentilles [38], proved by using the kernel's techniques he developped in [37]: Given S_1, S_2 locally compact Hausdorff, for a linear operator u from $(C(S_1),\beta)$ into $(C(S_2),\beta)$ to be (weakly) compact, it is necessary and sufficient that u be β-$\|\ \|$-continuous and transforms bounded subsets of $(C(S_1),\beta)$ into (weakly) relatively compact subsets of $(C(S_2),\beta)$. Because of the similarity of this result with Grothendieck's classical result on (weakly) compact operators from quasinormable spaces into Banach spaces, it is natural to ask:

1. Are spaces with strict topology quasinormable? and, more generally,
2. Which are the general structure properties of strict spaces these characterizations of (weakly) compact operators are based on?

2. THE LOCALLY CONVEX STRUCTURE OF STRICT TOPOLOGIES

Throughout this section, by a "strict space" (X,β) we shall always mean a space of any of the above types (1) $(C(T),\zeta)$, $\zeta\in\{\beta_o,\beta,\beta_1\}$, (2) (X,β_B) (for a Banach B-module $(X,\|\ \|)$), or (3) $(M(A),\beta_A)$ (for a C^*-algebra A). $(X,\|\ \|)$ will denote the respective Banach space, i.e. (1) (C(T),sup-norm), (2) the Banach module $(X,\|\ \|)$ or (3) the C^*-algebra $(M(A),\|\ \|)$.

2.1 BASIC RESULTS

As pointed out in the introduction, our investigations are based on the fact that strict spaces are (L)-spaces with a fundamental sequence of bounded sets.

Before stating this central result (Theorem 2.2 below), we recall the definition of (L)-spaces:

2.1 Definition [32,34]: *(i) An lcs (X,τ) is said to be of type (L) (resp. (Lb)) if, for every absorbing (resp. bornivorous) disk T in (X,τ), the finest locally convex topology on X agreeing with τ on T is equal to τ.*

(ii) An lcs (X,τ) is said to be of type (LC) (resp. (LCb)) if, for every absorbing (resp. bornivorous) disk T in (X,τ), the finest locally convex topology on X agreeing with τ on T is compatible with the dual pair (X,X'), i.e. coarser than the Mackey topology $\tau(X,X')$.

For the relation of these localization properties to barrelledness properties, we note here the following implications (see also [32-34]):

$$\begin{array}{ccccccc} \text{barrelled} & \Longrightarrow & \aleph_0\text{-barrelled} & \Longrightarrow & (L) & \Longrightarrow & (LC) \\ \Downarrow & & \Downarrow & & \Downarrow & & \Downarrow \\ \text{evaluable} & \Longrightarrow & \aleph_0\text{-evaluable} & \Longrightarrow & (Lb) & \Longrightarrow & (LCb) \end{array}$$

None of the reverse implications holds.

2.2 Theorem [35]: *Let (X,β) be an lcs with strict topology and let $(X,\| \ \|)$ be the respective Banach space.*
(1) (X,β) is of type (L). In general, it is not barrelled nor even $\aleph_0$-evaluable or σ-evaluable.
(2) Every β-barrel in X is β-bornivorous. In particular, the β-bounded subsets of X are exactly those absorbed by the β-closure of the unit ball of $(X,\| \ \|)$.
(3) The strong dual $(X'_\beta,\beta(X'_\beta,X))$ of (X,β) is a Banach space; $\beta(X'_\beta,X)$ is given exactly by the restriction onto X'_β of the dual (Banach space) norm on $(X,\| \ \|)'$.

2.3 Corollary:
(1) The topologies β_o,β,β_1 and the sup-norm topology on $C(T)$ have the same bounded sets.
(2) If (X,β_B) is an lcs with strict topology of type (2), we denote its dual by X'_B, by $\| \ \|_B$ the norm $\|x\|_B=\sup\{\|bx\| \mid b\in B, \|b\|\le 1\}$, $x\in X$, and by $\| \ \|'$ and $\| \ \|'_B$ the respective dual norms of the norms $\| \ \|$ and $\| \ \|_B$. With these notations we have:
(2a) $\beta_B \subset \| \ \|_B$-topology $\subset \| \ \|$-topology.
(2b) β_B and the $\| \ \|_B$-topology have the same bounded sets.
(2c) $\beta(X'_B,X) = \| \ \|' | X'_B$-topology $= \| \ \|'_B | X'_B$-topology = Banach space topology.

Remarks: 1. Assertion (1) of Corollary 2.3 is Theorem 4.7 of [39]. Assertion (2.c) answers the question of Sentilles and Taylor [40,p.149] whether $(X'_B,\beta(X'_B,X))$ is a Banach space, and shows that the equivalent conditions of Theorem 4.3 in [40] are always fulfilled.
2. The proof of Theorem 2.2 is based on the fact that β is the finest locally convex topology on X agreeing with β on the norm balls of $(X,\| \ \|)$ (consult [7]). First, this implies that (X,β) is an (L)-space, for every β-barrel is $\| \ \|$-bornivorous ($(X,\| \ \|)$ is a Banach space and $\beta \subset \| \ \|$-topology!). The rest now follows from the facts that in an (LC)-space every barrel is bornivorous ([33]) and that the strong dual of an (LCb)-space is Mackey complete ([33, Final Remarks]). (For the details of proof see [35].)

One further weakened barrelledness property has been considered in connection with strict topologies: Sentilles and Taylor [40] showed that in the dual of a strict space (X,β_B) of type (2) every strong nullsequence is equicontinuous. Spaces with this property have been introduced by J.H. Webb [46]: he defined an lcs (X,τ) to be sequentially barrelled (resp. sequentially evaluable) if every nullsequence in $(X',\sigma(X',X))$ (resp. $(X',\beta(X',X))$) is τ-equicontinuous. Sentilles' and Taylor's result now is a special case of the following general results:

2.4 Proposition [35]:
(1) If (X,τ) is an lcs with a fundamental sequence of bounded sets, then, of the following conditions, (i) implies (ii) and (ii) implies (iii):

(i) (X,τ) is an (Lb)-space.
(ii) Every β_X-nullsequence of seminorms in C_τ is equicontinuous.
(iii) (X,τ) is sequentially evaluable.

(2) If (X,τ) is a Mackey space of type (LC) (resp. (LCb)), then (X,τ) is sequentially barrelled (resp. sequentially evaluable).

(3) If (X,τ) is a sequentially barrelled (resp. sequentially evaluable) lcs, then it is of type (LC) (resp. (LCb)).

According to Theorem 2.2, spaces with strict topologies always belong to the following two new classes of locally convex spaces:

I. The class of (L)-spaces.
II. The class of (Lb)-spaces with a fundamental sequence of bounded sets.

Thus, to get a systematic insight into the structure of strict topologies, we investigate the specific properties of these two classes. As for the special directions of study, we start from the problems of section 1.4:

Class I: Problems 1 and 2 are concerned with comparing two locally convex topologies on a linear space. This is a special case of the closed graph theorem, i.e. the question, under which conditions a closed graph linear map between two locally convex spaces (in this special case the identity mapping on the linear space) is continuous. We are thus led to study closed graph theorems for (L)-spaces.

Class II: On the basis of the known results about (DF) spaces, Problems 3 and 4 could easily be answered if spaces with strict topologies would belong to this class of spaces. Thus, noting that the class of (Lb)-spaces with a fundamental sequence of bounded sets contains the class of (DF) spaces, we are led to analyse which of the (DF) space properties also hold true for the spaces of class II.

2.2 CLOSED GRAPH THEOREMS FOR (L)-SPACES

In this section we specify classes of locally convex spaces (Y,ρ) with the property that every closed graph linear map from an (L)-space (X,τ) into (Y,ρ) is continuous, and use our results for an investigation of problems 1 and 2 of section 1.4.

Notation: Let C be any class of locally convex spaces. By $B_r(C)$ (resp. $B(C)$) we denote the class of locally convex spaces (Y,ρ) with the property that for every lcs (X,τ) in C every linear map $u:(X,\tau)\rightarrow(Y,\rho)$ (resp. every surjective linear map $u:(Y,\rho)\rightarrow(X,\tau)$) with closed graph is continuous (resp. open).

In case C is an inductive class of lcs, i.e. it contains all finite-dimensional spaces and is stable under locally convex final topologies, we have the following general properties of $B_r(C)$ and $B(C)$:

1. An lcs (Y,ρ) belongs to $B(C)$ if and only if every separated quotient of (Y,ρ) belongs to $B_r(C)$. In particular, we have: $B(C)\subset B_r(C)$.
2. If (Y,ρ) is in $B_r(C)$ or in $B(C)$, then, for every Hausdorff locally convex topology ρ_1 on Y coarser than ρ, (Y,ρ_1) also belongs to $B_r(C)$ or $B(C)$ respectively.

We shall concentrate on the inductive classes $C=(L)$=spaces of type (L) and $C=(LM)$=Mackey spaces of type (L).

Terminology: An increasing sequence $(A_n)_{n\in\mathbb{N}}$ of disks in an lcs (X,τ) is called absorbent (resp. bornivorous) if every $x\in X$ (resp. every τ-bounded subset of X) is absorbed by some A_n.

2.5 Theorem [34]:
(1) Every lcs with an absorbent sequence of compact disks belongs to B(L). In particular, we have: Every semi-Montel lcs with a fundamental sequence of bounded sets belongs to B(L).

(2) Every lcs with an absorbent sequence of weakly compact disks belongs to B(LM). In particular, we have: Every semi-reflexive lcs with a fundamental sequence of bounded sets belongs to B(LM).

(We want to point out that these results are just special cases of much more general closed graph theorems for locally convex spaces whose topologies have certain specific localization properties, see [34] and [36].)

2.6 Theorem [34]:
(1) Every B_r-complete (resp. B-complete) Schwartz space belongs to $B_r(L)$ (resp. $B(L)$).
(2) Every B_r-complete (resp. B-complete) semi-reflexive and quasinormable lcs belongs to $B_r(LM)$ (resp. $B(LM)$).

For the sake of completeness, we also cite a result on the necessity of the conditions imposed in the above theorems:

2.7 Theorem [34]:
(1) Every sequentially barrelled lcs belonging to $B_r(L)$ (resp. $B(L)$) is B_r-complete (resp. B-complete) and semi-Montel.
(2) Every sequentially barrelled lcs belonging to $B_r(LM)$ (resp. $B(LM)$) is B_r-complete (resp. B-complete) and semi-reflexive.

For details and further results in the context of the above three theorems, the reader is referred to [34] and [36].
We note here only the idea of a direct proof of Theorem 2.5, for we shall use this theorem in attacking problems 1 and 2:
Given a closed graph linear map u from an (L)-space (X,τ) into an lcs (Y,ρ) with an absorbent sequence $(A_n)_{n\in\mathbb{N}}$ of ρ-compact disks, there exists a locally convex Hausdorff topology ρ_1 on Y, coarser than ρ, such that $u:(X,\tau)\longrightarrow(Y,\rho_1)$ is continuous, e.g. $\rho_1=\{u(U)+V \mid U\in\mathcal{U}_\tau,\ V\in\mathcal{U}_\rho\}$. Since (X,τ) is an (L)-space and $(B_n) = (\overset{(-1)}{u}(A_n))$ is an absorbent sequence in X, the map u is τ-ρ-continuous if and only if all restrictions $u|B_n$ are $\tau|B_n$-ρ-continuous [32]. However, the continuity of these restricted maps follows from the τ-ρ_1-continuity of u and the fact that $\rho|A_n = \rho_1|A_n$ (the A_n's are ρ-compact!). Thus we have proved that (Y,ρ) belongs to $B_r(L)$. The rest follows from result 1 on $B(C)$ and the fact that the property of possessing an absorbent sequence of compact disks is preserved by passage to separated quotients.

We now turn to problems 1 and 2 of section 1.4.

PROBLEM 1: The solution of this problem and of some related ones is included in the following general results:

2.8 Theorem [35]: *Let (X,τ) be an lcs with a fundamental sequence of bounded sets and A a linear subspace of X.*
(1) If (X,τ) is of type (LCb) and $(A,\tau|A)$ is semi-reflexive, then $(A,\tau|A)$ is of type (LC) and we have: $\beta(A',A) = \beta(X',X)/A^\perp$.
(2) If (X,τ) is sequentially evaluable and $(A,\tau|A)$ is semi-reflexive, then, in addition to (1), $(A,\tau|A)$ is sequentially evaluable and B-complete and $\tau|A$ is finer than the equicontinuous weak-topology $\lambda(A,A')$ on A as the dual of (the (F) space) $(A',\beta(A',A))$.*
(3) If (X,τ) is sequentially evaluable and $(A,\tau|A)$ is semi-Montel, then, in addition to (2), $\tau|A$ is equal to $\lambda(A,A')$ and thus is B-complete, of type (L) and belongs to $B(L)$.

For the special case of a strict space of type (2) we get (for the notation see Corollary 2.3(2)):

2.9 Corollary [35]: *Let (X,β_B) be a strict space of type (2).*
(1) If $(A,\beta_B|A)$ is a semi-reflexive linear subspace of (X,β_B), then we have: 1. $\|\ \|'/A^\perp = \|\ \|_B'/A^\perp = \beta(A',A)$. 2. $(A,\beta_B|A)$ is B-complete. 3. $(A,\beta_B|A)$ is sequentially evaluable, and thus every relatively compact subset of $(A',\beta(A',A))$ is equicontinuous.
(2) If $(A,\beta_B|A)$ is semi-Montel, then, in addition to (1), $\beta_B|A$ is equal to the equicontinuous weak-topology on A as the dual of the Banach space $(A',\|\ \|'/A^\perp)$.*

Theorem 2.8 and Corollary 2.9 place the main results of [41] and [42] on semi-reflexive and semi-Montel linear subspaces of $(C(S),\beta)$ and (X,β_B), and, in particular, our problem 1 on $(H^\infty(G),\beta)$, in the general setting of sequentially evaluable spaces with a fundamental sequence of bounded sets. The closed graph theorem 2.5(1) enters the proof of Theorem 2.8(3) in the following way: by assertion (2) of this theorem, $\lambda(A,A')$ is coarser than $\tau|A$. $\lambda(A,A')$ being of type (L) and $\tau|A$ having an absorbent sequence of compact disks, the identity mapping on A is $\lambda(A,A')$-$\tau|A$-continuous.

We prove now assertions (1) and (2) of Theorem 2.8: (1) The coincidence of the two topologies on A' is a consequence of Ptak's closed graph theorem, for $(X',\beta(X',X))$ is an (F) space and $(X'/A^\perp,\beta(A',A))$ is barrelled. By [33,Final Remarks], the completeness of $(A',\beta(A',A))$ implies that $(A,\tau|A)$ is an (LCb)-space. Since $(A,\tau|A)$ is quasi complete, it is thus also of type (LC) [33,Lemma 1.4].
(2) Let $(a_n')_{n\in\mathbb{N}}\subset A'$ be a $\beta(A',A)$-nullsequence in A'. By assertion (1), we know that $\beta(A',A)$ is the quotient (with respect to $A^\perp$) of the (F) space topology $\beta(X',X)$. Hence, the Lemma on p. 274 of [21] applies: there exists a compact subset C of $(X',\beta(X',X))$ such that $(a_n')_{n\in\mathbb{N}}\subset\pi_A(C)$. By the Banach-Dieudonné Theorem, C is contained in the $\beta(X',X)$-closed convex circled hull of a $\beta(X',X)$-nullsequence in X', and thus, by assumption on (X,τ), is equicontinuous. Consequently, also the sequence $(a_n')_{n\in\mathbb{N}}$ is equicontinuous. This shows that $(A,\tau|A)$ is sequentially evaluable and that we have: $\lambda(A,A')\subset\tau|A\subset\tau(A,A')$. Thus, $\tau|A$ is also B-complete.

PROBLEM 2: By result (1.6) on the space $H^\infty(G)$, there exists for any plane region G a countable dominating set $(z_n)_{n\in\mathbb{N}}\subset G$ without limit point in G. Collins [6,Lemma 3.4] showed that the corresponding isometric embedding $u:(H^\infty(G),\|\ \|)\longrightarrow(l^\infty,\|\ \|)$, $f\longmapsto(f(z_n))_{n\in\mathbb{N}}$, is continuous for the respective strict topologies on $H^\infty(G)$ and l^∞, the strict continuity of its inverse being doubtful. Theorem 2.5, however, also yields the β-continuity of u^{-1}:

2.10 Theorem [35]:

(1) The mapping $u : H^\infty(G)\longrightarrow l^\infty$, $u(f) = (f(z_n))_{n\in\mathbb{N}}$, is a linear norm isometry and an isomorphism of $(H^\infty(G),\beta)$ onto a closed linear subspace H_G of (l^∞,β).

(2) $(H^\infty(G),\beta)$ is Mackey if and only if Schur's property holds for $l^1/H_G^\perp$, i.e. weak and norm convergence for sequences are equivalent.

Rubel [28, p.17] noted that "the evidence is now in that $\beta\neq m$ (m=Mackey topology on $H^\infty(G)$) as soon as $H^\infty(G)$ is not trivial, but to prove it is another matter". Theorem 2.10 relates the Mackey problem for $(H^\infty(G),\beta)$ with the Schur problem in Banach space theory. For this latter problem, though, there is, up to now, not much hope for a satisfactory general solution (compare [18]), so that also from this point of view Rubel's scepticism is again substantiated.

Proof of Theorem 2.10: (1) $u:H^\infty(G)\longrightarrow l^\infty$ being a β-β-continuous norm isometry, the norm unit ball in $H_G=u(H^\infty(G))$ is β-compact, so that $(H_G,\beta|H_G)$ is a semi-Montel linear subspace of (l^∞,β). By Theorem 2.8(3), it is of type (L). According to the closed graph theorem 2.5, $(H^\infty(G),\beta)$ belongs to B(L). Hence u^{-1} is strictly continuous.
(2) If $(H^\infty(G),\beta)$ is Mackey, then, by Proposition 2.4(2), it is sequentially barrelled. Thus, every weak nullsequence in $l^1/H_G^\perp$ is β-equicontinuous. According to [35, Proposition 2.10], the norm topology on $l^1/H_G^\perp$ agrees with the weak topology on β-equicontinuous sets. This shows that $l^1/H_G^\perp$ has the Schur property.
For the converse, let B be any weakly relatively countably compact subset of $l^1/H_G^\perp$. By [23,§24,3(8)], B is weakly relatively sequentially compact. Thus, if $l^1/H_G^\perp$ has the Schur property, B is norm relatively compact. Assertion (3) of Theorem 2.8 now completes the proof that B is β-equicontinuous, for we have $\beta|H_G = \lambda(H_G,l^1/H_G^\perp) =$ = topology of uniform convergence on the norm precompact subsets of $l^1/H_G^\perp$.

Note that we proved that, in case $l^1/H_G^\perp$ has the Schur property, even every weak*-relatively countably compact subset of the dual of $(H^\infty(G),\beta)$ is equicontinuous. Locally convex spaces with this latter property are said to be strong Mackey. At the bottom of our above more specific result on the strong Mackey property of $(H^\infty(G),\beta)$ is the fact [35,Proposition 2.12] that a semi-reflexive linear subspace of an (LCb)-

space with a fundamental sequence of bounded sets is Mackey if and only if it is strong Mackey.

2.3 GENERALIZED (DF) SPACES

According to our considerations at the end of section 2.1, we investigate in this section which of the properties of (DF) spaces are shared by (Lb)-spaces with a fundamental sequence of bounded sets. Our object of study thus is the following class of locally convex spaces:

2.11 Definition: An (Lb)-space with a fundamental sequence of bounded sets is called a *generalized (DF) space ((gDF) space)*.

According to the diagram in section 2.1, every (DF) space is a (gDF) space. It can be shown that the (gDF) spaces are exactly the D_b-spaces of Noureddine [25,26], and, if semi-Montel, the (dF) spaces of Brauner [1]. We intentionally chose to call them generalized (DF) spaces, for we shall be able to show that, except of being $\aleph_0$-evaluable, they have almost all other essential properties of Grothendieck's classical (DF) spaces.

We first list those properties that extend more or less directly from (DF) spaces to (gDF) spaces (note that, by Proposition 2.4(1), (gDF) spaces are sequentially evaluable):

(2.1) *The strong dual of a (gDF) space (X,τ) is an (F) space; the cone C_τ of τ-continuous seminorms is β_X-complete (as a subset of X_τ).*

(2.2) *For a metrizable lcs (X,τ) both the strong dual $(X',\beta(X',X))$ (which is a (DF) space) and the λ-dual $(X',\lambda(X',X))$ are (gDF) spaces (here $\lambda(X',X)$ denotes the topology of uniform convergence on the precompact subsets of (X,τ)).*

(2.3) *If (X,τ) is an lcs with a fundamental sequence of bounded sets and with the property that every β_X-nullsequence in C_τ is equicontinuous (in particular, if (X,τ) is a (gDF) space), then we have:*
(*) For every sequence $(U_n)_{n\in\mathbb{N}}$ of τ-zero-neighbourhoods in X there exists another such, U say, which is absorbed by any of the U_n's.

(2.4) *Let (X,τ) be an lcs and A a linear subspace of X.*
(i) If $(A,\tau|A)$ is sequentially evaluable and has a fundamental sequence of bounded sets, then we have: $\beta(A',A) = \beta(X',X)/A^\perp$.
(ii) If A is closed and (X,τ) is sequentially evaluable and has a fundamental sequence of bounded sets, then $(X/A,\tau/A)$ has the same properties and we have: $\beta(A^\perp,X/A) = \beta(X',X)|A^\perp$. In particular, any separated quotient of a (gDF) space is a (gDF) space as well.

(2.5) *(i) If (X,τ) is an lcs and $(A,\tau|A)$ a sequentially evaluable linear subspace of (X,τ) with a fundamental sequence of bounded sets, then every bounded subset of $\bar{A}$ is contained in the closure of a bounded subset of A.*
(ii) The completion of a sequentially evaluable lcs with a fundamental sequence of bounded sets is again of this type. In particular, the completion of a (gDF) space is a (gDF) space as well.
(iii) A sequentially evaluable lcs with a fundamental sequence of bounded sets is complete if and only if it is quasi complete.

For the above results (2.4) and (2.5), see [35, Proposition 2.3 and Corollary 2.4] and [25, section 3] . This latter paper also contains the following permanence property known to hold for (DF) spaces:

(2.6) *Every (separated) inductive limit $(X,\tau) = \bigcup_{n\in\mathbb{N}} \iota_n(X_n)$ of a sequence of (gDF) spaces (X_n,τ_n) is a (gDF) space as well. Every τ-bounded subset of X is contained in the closed convex circled hull of a finite union of sets $\iota_n(B_n)$ for suitable bounded subsets B_n of (X_n,τ_n).*

We turn now to the problems of quasinormability and of nuclearity of (gDF) spaces. For the subclass of (DF) spaces, it is known that

1. every (DF) space is quasinormable [22] and

2. a (DF) space is nuclear if and only if its strong dual is nuclear [16].

These two properties also hold for (gDF) spaces. More generally, we have:

2.12 Theorem [35]: *Every lcs with a fundamental sequence of bounded sets and the property that every β_X-nullsequence of continuous seminorms in C_τ is equicontinuous (in particular, every (gDF) space) is quasinormable.*

Recall that an lcs (X,τ) is quasinormable [15], if for every equicontinuous subset H of X' there exists a neighbourhood U of zero in (X,τ) such that on H the topology induced by $\beta(X',X)$ coincides with the topology of uniform convergence on U, and that a Schwartz space is a quasinormable space whose bounded sets are precompact.

2.13 Theorem [35]: *A sequentially evaluable lcs with a fundamental sequence of bounded sets (in particular, a (gDF) space) is nuclear if and only if its strong dual is nuclear.*

Proof of Theorem 2.13: If (X,τ) is a nuclear sequentially evaluable lcs with metrizable strong dual, then, by result (2.5), its completion $(\tilde{X},\tilde{\tau})$ also has all of these properties and, moreover, the respective strong topologies $\beta(X',X)$ and $\beta(X',\tilde{X})$ on X' coincide. Thus, $\tilde{X}$ is the dual of the (F) space $(X',\beta(X',X))$, and the topology $\lambda(\tilde{X},X')$ is an (L)-topology (Theorem of Banach-Dieudonné) coarser than $\tilde{\tau}$ ($(\tilde{X},\tilde{\tau})$ being sequentially evaluable). By the closed graph theorem 2.5, we conclude that $\tilde{\tau}$ is equal to $\lambda(\tilde{X},X')$. Nuclearity of $(X',\beta(X',X))$ now follows from the following result of Brauner [1 ,Proposition 1.11]: If (X,τ) is an (F) space such that $(X',\lambda(X',X))$ is nuclear, then so is (X,τ) itself.
For the converse, let (X,τ) be a sequentially evaluable lcs with nuclear and metrizable strong dual. According to Proposition 2.4(3) and [34,Proposition 2.3], the space $(X',\beta(X',X))$ is even a nuclear (F) space. This yields that (X,τ) is evaluable: if T is a bornivorous barrel in (X,τ), then, by the Theorem of Banach-Dieudonné,the ($\beta(X',X)$-compact) set T^o is contained in the $\beta(X',X)$-closed convex circled hull of the range of a $\beta(X',X)$-nullsequence in X' and thus, by assumption on (X,τ), is equicontinuous. Since (X,τ) is evaluable, it can be viewed as a linear subspace of $(X'',\beta(X'',X'))$ which, by the classical result of Grothendieck [16], is nuclear. This completes the proof.

Among other things, the importance of the class of (DF) spaces is founded on the nice properties of both the cartesian product and the π-tensor product of two such spaces. The most remarkable of these can also be extended to (gDF) spaces [36]:

(1) *A set of bilinear maps of the product of two (gDF) spaces X and Y into an lcs Z is equicontinuous if and only if it is equihypocontinuous. (One of the spaces X or Y can also be any quasinormable lcs with the property (*) of result (2.3).)*

(2) *The π-tensorproduct $X\otimes_\pi Y$ of two (gDF) spaces X and Y is a (gDF) space as well; the strong topology on its dual, the space B(X,Y) of continuous bilinear maps on X×Y, is equal to the topology of bi-bounded convergence.*

For the details and related more general results, the reader is referred to [36]. Result (2) extends both Grothendieck's classical result [16,§1,3 Proposition 5] (X and Y (DF) spaces) and Noureddine's variant [26,Théorème 2] (X and Y semi-Montel (gDF) spaces), and thus gives another answer to Grothendieck's "Problème des topologies" [16,p.33] and solves Problème 2 of [26].

We turn now to the application of our general results to problems 3 and 4 of section 1.4.

PROBLEM 3: To our knowledge, there are by now three results on the non-nuclearity of strict topologies: Collins [6] showed that $(C(S),\beta)$ is nuclear only if it is finite-dimensional, Shapiro [43] notes the non-nuclearity of infinite-dimensional linear subspaces of $(C(S),\beta)$,and in [12] Fontenot proves that $(M(A),\beta_A)$ is nuclear only if A is finite-dimensional, A a C^*-algebra. By Theorem 2.13, no (infinite-dimensional) sequentially evaluable lcs with normable strong dual can be nuclear (this strong dual being a Banach space, see [34, Proposition 2.3]!). Thus we have the following general result on the non-nuclearity of strict topologies, including all the above ones and, in particular, the case of the space $(H^\infty(G),\beta)$:

2.14 Theorem [35]: *Neither any of the spaces $(C(T),\zeta)$, $\zeta\in\{\beta_o,\beta,\beta_1\}$, (X,β_B) and $(M(A),\beta_A)$ nor any other infinite-dimensional linear subspace of these spaces which is of type (LCb) is nuclear.*

Note that, according to Theorem 2.8(3) and Theorem 2.12, every semi-Montel linear subspace of the above strict spaces is a Schwartz space.

The proof of Theorem 2.14 is a consequence of Theorem 2.13 and the fact that,for a sequentially evaluable lcs with normable strong dual, a linear subspace is sequentially evaluable if and only if it is of type (LCb).

PROBLEM 4: First of all, Theorem 2.12 gives a direct answer to the question of quasinormability of strict topologies.

2.15 Theorem [35]: *Every lcs (X,τ) with any of the strict topologies considered in this paper is quasinormable. Moreover, if $(A,\tau|A)$ is a semi-Montel linear subspace of (X,τ), then every linear subspace of $(A,\tau|A)$ is a Schwartz space which (if infinite-dimensional) is not nuclear. This last assertion in particular applies to the spaces $(H^\infty(G),\beta)$ and (l^∞,β).*

As for the characterization of (weakly) compact linear operators on strict spaces, we have the following general result for quasinormable lcs:

2.16 Theorem [35]: *Let (X,τ) be a quasinormable lcs and (Y,ρ) a quasi complete lcs on which there exists a norm topology ρ_1 finer than ρ.*

(1) Every linear operator from X into Y which is τ-ρ_1-continuous and transforms τ-bounded subsets of X into (weakly) relatively compact subsets of (Y,ρ) is (weakly) compact from (X,τ) into (Y,ρ).

(2) If, in addition, (X,τ) has property (*) of result (2.3) (in particular, if (X,τ) is a (gDF) space), then the same conclusions hold also in case ρ_1 is only supposed to be metrizable.

2.17 Theorem [35]: *Let (X_i,β_{B_i}) be two strict spaces of type (2) with respective (Banach space) norms $\|\ \|_i$, $i\in\{1,2\}$, and such that (X_2,β_{B_2}) is complete. Then a linear operator $u : (X_1,\beta_{B_1})\longrightarrow(X_2,\beta_{B_2})$ is (weakly) compact if and only if it is β_{B_1}-$\|\ \|_2$-continuous and transforms β_{B_1}-bounded subsets of X_1 into (weakly) relatively compact subsets of (X_2,β_{B_2}).*

Note that this special case of Theorem 2.16 in particular includes Sentilles' results [38] for $(C(S),\beta)$ spaces. Theorem 2.16 itself extends both Grothendieck's classical result [17,Ch.IV,§4,3] (where $\rho=\rho_1$=Banach space norm) and van Dulst's variant [11] (where $\rho=\rho_1$=(F) space topology). Extensions of Theorem 2.16 can be found in [36,Ch.IV].

We close this paper with a proof of assertion (2) of Theorem 2.16. The proof is a slight refinement of Grothendieck's proof [17,Ch.IV,§4,3]: Let $(V_n)_{n\in\mathbb{N}}$ be a ρ_1-neighbourhood base at zero consisting of ρ_1-barrels V_n such that $V_{n+1}\subset V_n$. By assumption, there exists a sequence $(\alpha_n)_{n\in\mathbb{N}}$ of positive real numbers such that $U=\cap\{\alpha_n u^{(-1)}(V_n)\mid n\in\mathbb{N}\}$ is a τ-neighbourhood of zero in X. Also, there exists an equicontinuous $\sigma(X',X)$-closed disk H in X' such that $q_H|U^o=\beta(X',X)|U^o$, where q_H is the Minkowski functional of H in X'_H = span of H in X'. Since ρ is coarser than ρ_1, one can show that for every ρ-zero neighbourhood V there exists $n\in\mathbb{N}$ such that $u'(V^o)\subset\alpha_n U^o$. According to Théorème 12 and 13 of [17,Ch.II,18], u' transforms the

ρ-equicontinuous subsets of Y' into $\beta(X',X)$-precompact (resp. $\sigma(X',X'')$-relatively compact) subsets of X'. Altogether, we conclude that $u'(V^o)$ is compact (resp. weakly compact) as a subset of the (Banach) space (X'_H, q_H), V any ρ-zero neighbourhood in Y. Since, according to the choice of H, u can be viewed as a continuous linear map of the space X endowed with the seminorm topology generated by the Minkowski functional q_{H^o} of H^o into (Y,ρ), and since X'_H is the dual of (X, q_{H^o}), it follows again by the above theorems of [17,Ch.II,18] that $u(H^o)$ is a (weakly) relatively compact subset of (Y,ρ). This completes the proof.

REFERENCES

1. BRAUNER,K.: Duals of Fréchet spaces and a generalization of the Banach-Dieudonné Theorem. *Duke Math.J.* 40,845-855(1974)
2. BUCHWALTER,H.: Fonctions continues et mesures sur un espace complètement régulier. *Proc.Summer School Top.Vector Spaces (Bruxelles 1972)*, Lecture Notes in Math. 331,183-202(1973)
3. BUCK,R.C.: Operator algebras and dual spaces.*Proc.Amer.Math.Soc.*3,681-687(1952)
4. BUCK,R.C.: Bounded continuous functions on a locally compact space.*Michigan Math. J.*5,95-104(1958)
5. BUSBY,R.C.: Double centralizers and extensions of C*-algebras. *Trans.Amer.Math. Soc.*132,79-99(1968)
6. COLLINS,H.S.: On the space $l^\infty(S)$, with the strict topology.*Math.Z.*106,361-373 (1968)
7. COLLINS,H.S.: Strict, weighted, and mixed topologies and applications.*Advances Math.*19,207-237(1976)
8. CONWAY,J.B.: The strict topology and compactness in the space of measures. *Bull. Amer.Math.Soc.*72,75-78(1966)
9. CONWAY,J.B.: Subspaces of $C(S)_\beta$, the space (l^∞,β) and (H^∞,β).*Bull.Amer.Math.Soc.* 72,79-81(1966)
10. CONWAY,J.B.: The strict topology and compactness in the space of measures II. *Trans.Amer.Math.Soc.*126,474-486(1967)
11. VAN DULST,D.: (Weakly) compact mappings into (F) spaces. *Math.Ann.*224,111-115 (1976)
12. FONTENOT,R.A.: The double centralizer algebra as a linear space.*Proc.Amer.Math. Soc.*53,99-103(1975)
13. FREMLIN,D.H.,GARLING,D.J.H.,HAYDON,R.G.: Bounded measures on topological spaces. *Proc.London Math.Soc.*III.Ser.25,115-136(1972)
14. GLICKSBERG,I.: Bishop's generalized Stone-Weierstrass theorem for the strict topology. *Proc.Amer.Math.Soc.*14,329-333(1963)
15. GROTHENDIECK,A.: Sur les espaces (F) et (DF). *Summa Brasil.Math.*3,57-122(1954)
16. GROTHENDIECK,A.: Produits tensoriels topologiques et espaces nucléaires. *Mem. Amer.Math.Soc.*16(1955)
17. GROTHENDIECK,A.: Espaces vectoriels topologiques. *Sociedade Matemática S. Paulo*, Sao Paulo 1964
18. HAGLER,J.: A counterexample to several questions about Banach spaces. Preprint 1975
19. HIRSCHFELD,R.A.: Riots. *Nieuw Archief voor Wiskunde* (3)22,1-43(1974)
20. HOFFMAN,K.: Banach spaces of analytic functions. *Prentice-Hall,Inc.* Englewood Cliffs, N.J.1962
21. HORVATH,J.: Topological vector spaces and distributions I. *Addison-Wesley*, Reading 1966
22. KATS,M.P.: Every (DF) space is quasinormable. *Functional Analysis Appl.*7, 157-158(1973)
23. KÖTHE,G.: Topologische lineare Räume I. *Springer* Berlin-Heidelberg-New York 1966
24. LE CAM,L.: Convergence in distribution of stochastic processes. *Statistica* 2, 207-236(1957)
25. NOUREDDINE,K.: Nouvelles classes d'espaces localement convexes. *Publ.Dépt.Math. Lyon* 10 (1973)
26. NOUREDDINE,K.: Note sur les espaces D_b. *Math.Ann.*219,97-103(1976)

27. RIEFFEL,M.A.: Induced Banach representations of Banach algebras and locally compact groups. *J.Functional Analysis* 1,443-491(1967)
28. RUBEL,L.A.: Bounded convergence of analytic functions. *Bull.Amer.Math.Soc.*77, 13-24(1971)
29. RUBEL,L.A.,RYFF,J.V.: The bounded weak*-topology and the bounded analytic functions. *J. Functional Analysis* 5,167-183(1970)
30. RUBEL,L.A.,SHIELDS,A.L.: The space of bounded analytic functions on a region. *Ann.Inst.Fourier* 16,235-277(1966)
31. RUDIN,W.: Real and complex analysis. *McGraw-Hill* 1970
32. RUESS,W.: A Grothendieck representation for the completion of cones of continuous seminorms. *Math.Ann.*208,71-90(1974)
33. RUESS,W.: Generalized inductive limit topologies and barrelledness properties. *Pacific J. Math.*63,499-516(1976)
34. RUESS,W.: Closed graph theorems for generalized inductive limit topologies. To appear in *Math.Proc. Cambridge Phil.Soc.*
35. RUESS,W.: On the locally convex structure of strict topologies. *Math.Z.*153, 179-192(1977)
36. RUESS,W.: Halbnorm-Dualität und induktive Limestopologien in der Theorie lokalkonvexer Räume. Habilitationsschrift Bonn 1976.
37. SENTILLES,F.D.: Compactness and convergence in the space of measures. *Illinois J. Math.*13,761-768(1969)
38. SENTILLES,F.D.: Compact and weakly compact operators on $C(S)_\beta$. *Illinois J. Math.* 13,769-776(1969)
39. SENTILLES,F.D.: Bounded continuous functions on a completely regular space. *Trans.Amer.Math.Soc.*168,311-336(1972)
40. SENTILLES,F.D.,TAYLOR,D.C.: Factorization in Banach algebras and the general strict topology. *Trans.Amer.Math.Soc.*142,141-152(1969)
41. SHAPIRO,J.H.: Weak topologies on subspaces of C(S). *Trans.Amer.Math.Soc.*157, 471-479(1971)
42. SHAPIRO,J.H.: The bounded weak star topology and the general strict topology. *J. Functional Analysis* 8,275-286(1971)
43. SHAPIRO,J.H.: Noncoincidence of the strict and strong operator topologies. *Proc. Amer.Math.Soc.*35,81-87(1972)
44. TAYLOR,D.C.: The strict topology for double centralizer algebras. *Trans.Amer. Math.Soc.*150,633-643(1970)
45. VARADARAJAN,V.S.: Measures on topological spaces. *Amer.Math.Soc.Translations* II.Ser.48,161-228(1965)
46. WEBB,J.H.: Sequential convergence in locally convex spaces. *Proc. Cambridge Phil. Soc.*64,341-364(1968)

K.-D. Bierstedt, B. Fuchssteiner (eds.)
Functional Analysis: Surveys and Recent Results

ESPACES DE MESURES ET PARTITIONS CONTINUES DE L'UNITE

Henri Buchwalter
Département de Mathématiques
Université Claude-Bernard - LYON I
43, bd du Onze Novembre 1918
69621 Villeurbanne, France

On construit, avec l'aide des partitions continues de l'unité, toute une gamme de topologies complètes sur les espaces $M_\sigma(T)$ et $M_\beta(T)$ donnant comme dual l'espace $C^\infty(T)$. La caractérisation des parties relativement compactes pour ces topologies fournit un nouveau critère de compacité étroite dans l'espace $M_\sigma(T)$.

INTRODUCTION

Dans tout l'article T désigne un espace complètement régulier séparé quelconque, auquel on associe son compactifié de Stone-Čech βT et son replété υT (ou "real-compactification" de Hewitt), [6], [3]. On introduit aussi l'algèbre de Banach $C^\infty(T)$ des fonctions réelles continues et bornées sur T, ainsi que l'espace $M_\beta(T) = M(\beta T) = C^\infty(T)'$ des mesures de Radon sur βT, muni de sa norme d'espace dual. L'espace $M_\beta(T)$, qui dépend en réalité de βT plus que de T, est en général trop grand pour être qualifié d'espace de "mesures" sur T. Ce qui importe c'est essentiellement de l'utiliser comme cadre suffisamment vaste pour contenir les espaces de mesures intéressants. Ces espaces intéressants sont principalement les espaces $M_\sigma(T)$, $M^\infty(T)$, $M_\tau(T)$ et $M_t(T)$, [1], [9], mais en fait nous allons nous limiter ici à l'étude de quelques aspects de la théorie de l'espace $M_\sigma(T)$ des mesures σ-régulières sur T, en nous attachant à l'examen d'un problème précis.

Pour énoncer le problème posé rappelons qu'on topologise traditionnellement les espaces $M_\sigma(T)$ et $M_\beta(T)$ de deux façons différentes : soit avec la norme du dual $C^\infty(T)'$, soit avec la topologie faible $\sigma(M_\beta, C^\infty)$, dite topologie étroite. Les deux manières ont leurs avantages et leurs inconvénients ; par exemple la topologie de la norme est évidemment complète mais le dual correspondant est un espace assez mal connu ; au contraire la topologie étroite fournit comme dual l'espace $C^\infty(T)$, mais elle n'est pas complète en général. On peut alors poser le problème :

PROBLEME

Existe-t-il sur chacun des espaces $M_\sigma(T)$ *et* $M_\beta(T)$ *des topologies localement convexes assez simples, qui soient complètes et qui donnent en même temps pour dual l'espace* $C^\infty(T)$ *?*

Ce problème est résolu positivement pour l'espace $M_\sigma(T)$ d'une façon élégante par Berruyer-Ivol [2]. A notre connaissance il ne l'est pas pour l'espace $M_\beta(T)$ et nous offrons ici par une méthode tout à fait différente un choix d'autres solutions.

Avant d'entrer dans le vif du sujet signalons encore que l'essentiel des résultats figurent dans [4] ; aussi nous permettrons nous de fréquentes références à cet article.

Pour résoudre le problème posé, il existe un outil particulièrement bien adapté, et d'ailleurs utilisé par Berruyer-Ivol, [2] : c'est celui des compactologies, [3]. Il suffit en effet de construire sur l'espace $C^\infty(T)$ des compactologies convenables $\mathcal{K}$ donnant pour dual soit l'espace $M_\sigma(T)$, soit l'espace $M_\beta(T)$, pour construire, du même coup, par dualité, des topologies complètes sur $M_\sigma(T)$ ou $M_\beta(T)$ donnant pour dual $C^\infty(T)$. Pour obtenir de telles compactologies sur $C^\infty(T)$, nous utilisons la notion de partition continue de l'unité sur T, déjà exploitée d'ailleurs de différentes manières par De Marco-Wilson [5], Rome [7], Sentilles-Wheeler [8].

Pour bien préciser les choses rappelons que nous appelons partition continue de l'unité sur T toute suite $\varphi = (\varphi_n)$ de $C^\infty(T)$ telle que $\varphi_n \geqslant 0$ et $\Sigma\, \varphi_n = 1$. Nous abandonnons expressément la condition de finitude locale sur T de la famille des supports (supp φ_n), condition trop restrictive pour résoudre le problème posé. Pour simplifier nous dirons que $\varphi = (\varphi_n)$ est une pcu et nous désignerons par Φ l'ensemble de toutes ces pcu.

1. LES COMPACTOLOGIES $\mathscr{H}_q$ ET LES TOPOLOGIES $\mathscr{T}_p$

L'abandon de la condition de finitude locale sur T de la famille des supports (supp φ_n) n'est en réalité pas gênant. L'important est que subsiste le résultat suivant :

(1.1) LEMME ([4], [8])

On fixe une pcu $\varphi = (\varphi_n) \in \Phi$.
a) Pour toute suite $\xi = (\xi_n) \in \ell^\infty$, *la fonction* $\Sigma\, \xi_n \varphi_n$ *est continue et bornée sur* T.
b) L'ensemble $H_{\varphi,\infty} = \{\Sigma\, \xi_n \varphi_n \;;\; \|\xi\|_\infty \leqslant 1\}$ *est équicontinu et uniformément borné sur* T.

Fixons maintenant un couple (p,q) d'exposants conjugués, $1 \leqslant p \leqslant +\infty$, $1 \leqslant q \leqslant +\infty$, $\frac{1}{p} + \frac{1}{q} = 1$. Alors la boule unité A_q de l'espace ℓ^q est compacte et métrisable pour la topologie faible σ_q égale à $\sigma(\ell^q, \ell^p)$ si $q > 1$ et à $\sigma(\ell^1, c_\circ)$ si q=1. Il suit facilement de là que, pour toute $\varphi = (\varphi_n) \in \Phi$, l'application

$$U_\varphi \;:\; \xi \longrightarrow \Sigma\xi_n \varphi_n$$

de ℓ^q dans $C^\infty(T)$, est linéaire et continue pour la topologie σ_q et la topologie sur $C^\infty(T)$ de la convergence simple sur T. On peut alors introduire les disques équicontinus simplement compacts

$$H_{\varphi,q} = \{\Sigma\, \xi_n \varphi_n \;;\; \|\xi\|_q \leqslant 1\}$$

de $C^\infty(T)$, qui sont les images $U_\varphi(A_q)$, toutes contenues quand q varie, dans le disque équicontinu $H_{\varphi,\infty}$.

Pour q=1, on reconnaît en $H_{\varphi,1}$ l'enveloppe disquée simplement fermée de la suite (φ_n) dans l'espace $C^\infty(T)$. Pour q quelconque il est facile de voir que la famille $(H_{\varphi,q})_{\varphi\in\Phi}$ est filtrante croissante, car pour $\varphi = (\varphi_n) \in \Phi$ et $\psi = (\psi_n) \in \Phi$, on peut poser $\zeta = (\zeta_n) \in \Phi$ avec $\zeta_{2n} = \frac{1}{2}\varphi_n$ et $\zeta_{2n+1} = \frac{1}{2}\psi_n$, et constater que $H_{\zeta,q}$ contient $\frac{1}{2}(H_{\varphi,q} \cup H_{\psi,q})$; d'où il **résulte** que les disques équicontinus simplement compacts $H_{\varphi,q}$, $\varphi \in \Phi$, constituent la base d'une compactologie $\mathscr{H}_q$ sur l'algèbre $C^\infty(T)$.

Le lien avec les espaces $M_\sigma(T)$ et $M_\beta(T)$ se fait avec le lemme suivant :

(1.2) LEMME (Lemme de commutation)

On fixe une pcu $\varphi = (\varphi_n) \in \Phi$. *On a la formule de commutation*

$$\mu(\Sigma\, \xi_n\, \varphi_n) = \Sigma\, \xi_n\, \mu(\varphi_n)$$

dans chacun des deux cas :

a) $\mu \in M_\beta(T)$ *et* $\xi \in c_o$
b) $\mu \in M_\sigma(T)$ *et* $\xi \in \ell^\infty$.

PREUVE

Il suffit de voir que la suite $f_n = \sum_{n+1}^{\infty} \xi_k\, \varphi_k$ converge uniformément vers zéro si $\xi \in c_o$ pour obtenir a). Pour obtenir b) on se ramène à supposer $\xi_n \geq 0$, ce qui assure la condition $f_n \downarrow 0$. □

Il importe maintenant de déterminer les duals des espaces compactologiques $(C^\infty(T), \mathscr{H}_q)$, c'est-à-dire en suivant [3], les espaces de formes linéaires sur $C^\infty(T)$ dont les restrictions aux parties $H_{\varphi,q}$, $\varphi \in \Phi$, sont continues quand on munit chacune de ces parties de la topologie de la convergence simple sur T. Notant $(C^\infty(T), \mathscr{H}_q)^*$ ces duals, on a le résultat clé :

(1.3) THEOREME ([4])

a) *Pour* $q < +\infty$ *on a* $(C^\infty(T), \mathscr{H}_q)^* = M_\beta(T)$
b) *Pour* $q = +\infty$ *on a* $(C^\infty(T), \mathscr{H}_\infty)^* = M_\sigma(T)$.

PREUVE

a) Fixons $\mu \in M_\beta(T)$ et $\varphi \in \Phi$. D'après le lemme de commutation (1.2.a)), et le fait évident que la suite $(\mu(\varphi_n))$ est élément de ℓ^1, on voit que la fonction composée $\mu \circ U_\varphi$ est une forme linéaire sur ℓ^q, dont la restriction à la boule unité A_q de ℓ^q est σ_q-continue, ce qui implique facilement, par compacité de A_q, que μ est continue sur le disque $H_{\varphi,q} = U_\varphi(A_q)$. Réciproquement soit μ une forme linéaire sur $C^\infty(T)$, telle que $\mu \notin M_\beta(T)$; il existe donc une suite (f_n) dans $C^\infty(T)$, telle que $0 \leq f_n \leq 1$ et $|\mu(f_n)| \geq 2^n$. On pose alors $\varphi_n = 2^{-n} f_n$ pour $n \geq 1$ et $\varphi_o = 1 - \sum_1^\infty \varphi_n$, ce qui fournit une pcu $\varphi = (\varphi_n) \in \Phi$ telle que $\varphi_n \to 0$ dans $H_{\varphi,q}$ tandis que $|\mu(\varphi_n)| \geq 1$, ce qui prouve que μ n'est pas élément du dual $(C^\infty(T), \mathscr{H}_q)^*$.

b) On démontre de la même façon que toute $\mu \in M_\sigma(T)$ est continue sur $H_{\varphi,\infty}$ en remplaçant (1.2.a)) par (1.2.b)). Réciproquement soit $\mu \in (C^\infty(T), \mathscr{H}_\infty)^*$. On a déjà $\mu \in M_\beta(T)$ d'après la preuve de a). Pour voir que μ est élément de $M_\sigma(T)$, il suffit, ce qui est classique, de prouver que μ est simplement continue sur chaque partie $H \subset C^\infty(T)$, équicontinue, métrisable (pour la topologie de la convergence simple) et contenue dans la boule unité de $C^\infty(T)$. Or d'après le théorème (3.5) de Rome, [7], adapté au cas particulier d'une partie H métrisable ([7], p. 57-58), il existe, pour tout $\varepsilon > 0$, une pcu $\varphi = (\varphi_n)$ et une suite (t_n) de points de T telles que

$$\| f - \Sigma\, f(t_n)\, \varphi_n \| \leq \varepsilon$$

pour toute $f \in H$. On voit donc que toute $f \in H$ est égale, à ε près en norme, à une fonction $g \in H_{\varphi,\infty}$, l'application $f \to g$ étant même linéaire et simplement continue de H dans $H_{\varphi,\infty}$. Il en résulte aisément la continuité simple de μ sur H. □

REMARQUE

Le théorème, qui est bien entendu fondamental pour résoudre le problème posé dans l'introduction, devient faux si l'on utilise, au lieu de Φ, l'ensemble des pcu

$\varphi = (\varphi_n)$ localement finies sur T. En effet, si T est par exemple compact, ces pcu localement finies sont nécessairement finies, de sorte que les duals compactologiques associés sont tous égaux au dual algébrique de l'espace de Banach $C^\infty(T)$.

Introduisons maintenant, pour $1 \leqslant p \leqslant +\infty$, les topologies $\mathcal{T}_p$ sur les espaces $M_\sigma(T)$ et $M_\beta(T)$. On a déjà remarqué que la suite $(\mu(\varphi_n))$ est élément de ℓ^1 pour toute $\mu \in M_\beta(T)$ et toute $\varphi \in \Phi$. On peut donc définir les applications

$$V_\varphi \; : \mu \longrightarrow (\mu(\varphi_n))$$

de $M_\beta(T)$ dans ℓ^p et considérer sur $M_\beta(T)$ le système des semi-normes

$$\|\mu\|_{\varphi,p} = \|V_\varphi(\mu)\|_p \quad ; \quad \varphi \in \Phi$$

On construit ainsi une topologie localement convexe $\mathcal{T}_p$ sur $M_\beta(T)$ et sur son sous-espace $M_\sigma(T)$, ce qui fournit, avec des notations évidentes, les elc $M_{\beta,p}(T)$ et $M_{\sigma,p}(T)$, dont les topologies sont les topologies initiales associées au jeu des applications $V_\varphi : M_\beta(T) \to \ell^p$.

Pour étudier la topologie $\mathcal{T}_p$, revenons aux exposants conjugués (p,q) et distinguons les deux cas $q < +\infty$, $p > 1$ et $q = +\infty$, $p = 1$. Dans le premier cas il est clair que la formule de commutation (1.2.a)) garantit que les applications $U_\varphi : \ell^q \to C^\infty(T)$ et $V_\varphi : M_\beta(T) \to \ell^p$ sont transposées l'une de l'autre et que la valeur $\|\mu\|_{\varphi,p} = \|V_\varphi(\mu)\|_p$ n'est autre que la borne supérieure de $|\mu(f)|$ lorsque f décrit la partie $H_{\varphi,q}$. Il en résulte que l'espace $M_{\beta,p}(T)$ coïncide exactement, compte tenu de (1.3.a)), avec l'elc complet $(C^\infty(T), \mathcal{H}_q)^*$ tel qu'il est défini dans [3]. Ainsi $M_{\beta,p}(T)$ est dual (topologisé comme elc complet) d'un espace compactologique et les résultats généraux de [3], que l'on peut aisément retrouver avec le théorème de Banach-Grothendieck, fournissent une première solution au problème posé dans l'introduction sous la forme :

(1.4) THEOREME

Soit $1 < p \leqslant +\infty$, *d'exposant conjugué* q. *L'espace* $M_{\beta,p}(T)$ *est un elc complet, dont le dual est l'espace* $C^\infty(T)$. *Ses bornés sont les parties de* $M_\beta(T)$ *bornées en norme, de sorte que* $C^\infty(T)$ *est le dual fort de* $M_{\beta,p}(T)$. *L'espace* $M_\sigma(T)$ *est dense dans* $M_{\beta,p}(T)$. *Plus précisément* T *est total dans* $M_{\beta,p}(T)$.

PREUVE

Rajoutons à ce qui a été dit que les bornés de $M_{\beta,p}(T)$ se déterminent avec le théorème de Banach-Steinhaus puisque la topologie $\mathcal{T}_p$ est intermédiaire entre la topologie de la norme et la topologie faible $\sigma(M_\beta, C^\infty)$. Le fait que T soit total dans $M_{\beta,p}(T)$ provient du théorème de Hahn-Banach, une fois connu le dual $M_{\beta,p}(T)' = C^\infty(T)$. □

REMARQUE

Sur l'espace $M_\beta(T)$, qui ne dépend que de βT, la topologie $\mathcal{T}_p$ dépend essentiellement de T (ou plus précisément de υT). En effet une pcu $\varphi = (\varphi_n)$ peut se prolonger à βT selon $\varphi^\beta = (\varphi_n^\beta)$ mais on obtient ainsi une suite dont la somme $\Sigma\, \varphi_n^\beta$ n'est pas continue sur βT et qui vaut 1 sur υT. Par ailleurs le fait que T soit total dans $M_{\beta,p}(T)$, ou le théorème de Hahn-Banach, assure que βT est contenu dans l'enveloppe disquée fermée $\overline{\Gamma}(T)$ dans $M_{\beta,p}(T)$, qui n'est autre que la boule unité de $M_\beta(T)$. On peut montrer toutefois ([4]), que l'adhérence $\overline{T}$ de T dans $M_{\beta,p}(T)$ est toujours réduite au replété υT.

Le second cas $q = +\infty$, $p = 1$ se traite de la même façon mais la formule de commutation utilisée est (1.2.b)). Il en résulte que l'application transposée de $U_\varphi : \ell^\infty \to C^\infty(T)$ est l'application $V_\varphi : M_\sigma(T) \to \ell^1$. Alors, avec (1.3.b)) on a :

(1.5) THEOREME

L'espace $M_{\sigma,1}(T)$ est un elc complet dont les bornés sont les parties bornées en norme, et dont le dual fort est l'espace $C^\infty(T)$. De plus T est total dans $M_{\sigma,1}(T)$.

CAS DE L'ESPACE $M_{\beta,1}(T)$.

Il est déjà immédiat que $M_{\beta,1}(T)$ possède une base de voisinages de zéro formée des disques $W_\varphi(\varepsilon) = \{\mu ; \Sigma|\mu(\varphi_n)| \leqslant \varepsilon\}$ qui sont fermés pour la topologie étroite, et a fortiori fermés dans l'espace $M_{\beta,\infty}(T)$. Comme ce dernier espace est complet et que sa topologie est moins fine que celle de $M_{\beta,1}(T)$, on en déduit que $M_{\beta,1}(T)$ est complet. Cependant l'espace $M_{\beta,1}(T)$ ne répond pas au problème posé dans l'introduction puisque son dual n'est pas, en général, l'espace $C^\infty(T)$. On va voir toutefois que ce dual reste un espace de fonctions sur βT. Pour cela désignons par $\widetilde{U}_\varphi : \ell^\infty \to M_{\beta,1}(T)'$ la transposée de l'application $V_\varphi : M_{\beta,1}(T) \to \ell^1$. Pour $\xi \in \ell^\infty$ et $\mu \in M_\beta(T)$ on a

$$< \mu, \widetilde{U}_\varphi(\xi) > = < V_\varphi(\mu), \xi > = \Sigma\, \xi_n\, \mu(\varphi_n).$$

L'image $\widetilde{U}_\varphi(A_\infty)$ de la boule unité de ℓ^∞ est un disque faiblement compact du dual $M_{\beta,1}(T)'$ et la formule de dualité précédente montre que le polaire $[\widetilde{U}_\varphi(A_\infty)]°$ n'est autre que le voisinage de zéro $W_\varphi = W_\varphi(1)$ de $M_{\beta,1}(T)$. Par le théorème du bipolaire on obtient $W° = \widetilde{U}_\varphi(A_\infty)$, ce qui suffit pour voir que tout élément $L \in M_{\beta,1}(T)'$ est déterminé par la donnée d'une pcu $\varphi = (\varphi_n)$ et d'une suite $\xi = (\xi_n) \in \ell^\infty$ selon la formule $L(\mu) = \Sigma\, \xi_n\, \mu(\varphi_n)$.

Si l'on traite maintenant chaque $\mu \in M_\beta(T)$ comme une mesure de Radon sur βT, on est amené à introduire la fonction $g = \Sigma\, \xi_n\, \varphi_n^\beta$ sur βT, où φ_n^β est la prolongée continue canonique de φ_n. On remarquera que g est une fonction bornée de la première classe de Baire sur βT et que sa restriction $g_{|T} = \Sigma\, \xi_n\, \varphi_n$ est continue sur T. De plus on a

$$L(\mu) = \Sigma\, \xi_n\, \mu(\varphi_n) = \Sigma\, \xi_n \int \varphi_n^\beta d\mu = \int g d\mu$$

ce qui fournit une représentation fonctionnelle de L par l'intermédiaire de g. On peut aller plus loin en considérant l'espace $\Phi(\beta T)$ de toutes les fonctions de Baire sur βT qui admettent une représentation (non unique) de la forme $g = \Sigma\, \xi_n\, \varphi_n^\beta$, avec $\xi \in \ell^\infty$ et $\varphi \in \Phi$. Chaque $g \in \Phi(\beta T)$ détermine un élément $L_g \in M_{\beta,1}(T)'$ selon la formule

$$L_g(\mu) = \int g d\mu = \Sigma\, \xi_n\, \mu(\varphi_n)$$

d'où l'existence d'une application $g \to L_g$ de $\Phi(\beta T)$ dans $M_{\beta,1}(T)'$. Cette application est surjective d'après ce qu'on a vu plus haut ; elle est aussi injective, car si $L_g=0$ alors $L_g(u) = 0$ pour tout caractère $u \in \beta T$, donc g=0. En résumé

(1.6) THEOREME

L'espace $M_{\beta,1}(T)$ est un elc complet dont les bornés sont les parties bornées en

norme et dont le dual fort s'identifie à l'espace $\Phi(\beta T)$ *de toutes les fonctions de Baire sur* βT *de la forme* $g = \Sigma\, \xi_n \varphi_n^\beta$, *avec* $\xi = (\xi_n) \in \ell^\infty$ *et* $\varphi = (\varphi_n) \in \Phi$, *espace que l'on munit de la norme uniforme sur* βT. *De plus* βT *est total dans* $M_{\beta,1}(T)$.

La dernière affirmation provient du fait que toute $L = L_g \in M_{\beta,1}(T)'$, qui est nulle sur βT, est nécessairement nulle puisque $g = L_g|_{\beta T}$. Le rapport qui existe entre le dual $\Phi(\beta T)$ de $M_{\beta,1}(T)$ et le dual $C^\infty(T) = C(\beta T)$ de son sous-espace fermé $M_{\beta,1}(T)$ est facile à expliciter. Si l'on désigne par $\Phi_\circ(\beta T)$ le sous-espace formé des fonctions $g \in \Phi(\beta T)$ telles que $g|_T = 0$, on obtient facilement, en introduisant la fonction

$$f = (\Sigma\, \xi_n \varphi_n)^\beta = (g|_T)^\beta$$

la représentation en somme directe topologique (pour la norme de $\Phi(\beta T)$)

$$\Phi(\beta T) = C(\beta T) \oplus \Phi_\circ(\beta T).$$

Comme pour toute $\mu \in M_\sigma(T)$, on a $L_g(\mu) = \mu(f)$ comme conséquence de la formule de commutation (1.2.a), on voit encore que $\Phi_\circ(\beta T)$ est exactement l'orthogonal, dans $M_{\beta,1}(T)'$, du sous-espace $M_\sigma(T)$ de $M_{\beta,1}(T)$.

Pour terminer ce paragraphe 1, on peut voir que les espaces $M_{\sigma,p}(T)$ et $M_{\beta,p}(T)$ n'ont pas des topologies très classiques. En particulier ces topologies ne sont pratiquement jamais infratonnelées. En effet :

(1.7) PROPOSITION

Les assertions suivantes sont équivalentes :

a) T *est fini ;*
b) L'un des espaces $M_{\beta,p}(T)$ *est infratonnelé pour* $1 \leqslant p \leqslant +\infty$ *;*
c) L'un des espaces $M_{\sigma,p}(T)$ *est infratonnelé pour* $1 < p \leqslant +\infty$.

PREUVE

b $\Longrightarrow$ a : Fixons $p \in [1,+\infty]$. D'après l'hypothèse et la forme du dual de $M_{\beta,p}(T)$ pour le cas p=1, on voit que la boule unité Δ de $C^\infty(T)$ est une partie équicontinue du dual $M_{\beta,p}(T)'$. A fortiori c'est une partie équicontinue sur T, d'où l'on déduit déjà que T est un espace discret. De plus il existe une $\varphi = (\varphi_n) \in \Phi$ et une constante $M > 0$ telles que $\Delta \subset MH_{\varphi,q}$, ce qui prouve l'inégalité

$$\|\mu\| \leqslant M\|\mu\|_{\varphi,p}$$

qui permet d'affirmer la continuité de la norme $\|.\|$ sur l'espace $M_{\beta,p}(T)$. Mais alors $M_{\beta,p}(T)$ est topologiquement égal à l'espace de Banach $M_\beta(T) = \ell^\infty(T)'$. Si l'on considère maintenant l'application $V_\varphi : M_{\beta,p}(T) \to \ell^p$ (ou $c_\circ$ si $p = +\infty$), on voit que c'est une injection isométrique pour la norme $\|.\|_{\varphi,p}$, ce qui implique la séparabilité de l'espace de Banach $\ell^\infty(T)'$, donc la finitude de l'espace T.

c $\Longrightarrow$ a : La preuve est la même jusqu'au point où l'on montre que $M_{\sigma,p}(T)$ est topologiquement égal à l'espace de Banach $M_\sigma(T) = \ell^1(T)$. De ce fait $M_{\sigma,p}(T)$ est complet et, puisque l'on a ici $p > 1$, égal à $M_{\beta,p}(T)$ ce qui ramène au cas précédent. □

Pour le cas de l'espace $M_{\sigma,1}(T)$ on a :

(1.8) PROPOSITION

Les assertions suivantes sont équivalentes :

a) T est un espace discret fini ou dénombrable ;
b) L'espace $M_{\sigma,1}(T)$ est infratonnelé.

PREUVE

b $\Longrightarrow$a : Comme dans la preuve de (1.7) on aboutit au fait que T est discret et que $M_\sigma(T) = \ell^1(T)$ est séparable, d'où la condition card T $\leqslant \chi_0$.

a $\Longrightarrow$b : Lorsque T=$\mathbb{N}$, on peut choisir pour $\varphi = (\varphi_n)$ la base canonique faible de $C^\infty(T) = \ell^\infty$, ce qui assure la continuité de la norme de $M_\sigma(T) = \ell^1$ pour la topologie de $M_{\sigma,1}(T)$. Mais cela signifie que $M_{\sigma,1}(T)$ est exactement l'espace de Banach ℓ^1, d'où la condition b). □

Comme conséquence on a :

(1.9) COROLLAIRE

Soit I un ensemble infini non dénombrable. Alors la topologie sur l'espace $\ell^1(I)$, définie par l'espace $M_{\sigma,1}(I)$, est complète, non infratonnelée, de dual $\ell^\infty(I)$ et telle qu'elle induit sur chaque sous-espace $\ell^1(J)$, J partie dénombrable de I, la topologie de la norme.

2. COMPACITE DANS LES ESPACES $M_{\beta,p}(T)$ ET $M_{\sigma,p}(T)$

Pour déterminer les compacts des espaces $M_{\beta,p}(T)$, on va tenir compte simultanément de la complétude de ces espaces, du fait que leurs topologies sont les topologies initiales associées aux applications $V_\varphi : M_{\beta,p}(T) \to \ell^p$, et aussi du fait que, pour $p = +\infty$, l'image de V_φ est contenue dans l'espace c_0.

(2.1) THEOREME ([4])

Les espaces $M_{\beta,p}(T)$, $1 \leqslant p \leqslant +\infty$, ont tous les mêmes parties relativement compactes, qui sont les parties bornées (en norme) K telles que l'on ait, pour chaque $\varphi = (\varphi_n) \in \Phi$

$$\sup_{\mu \in K} |\mu(\varphi_n)| \to 0$$

PREUVE

Les parties relativement compactes de $M_{\beta,1}(T)$ [resp. de $M_{\beta,\infty}(T)$] sont les parties K dont l'image par chacune des applications V_φ, $\varphi \in \Phi$, est relativement compacte dans ℓ^1 [resp. dans c_0]. Ce sont donc les parties K bornées en norme qui vérifient respectivement les conditions.

(C_1) Pour toute $\varphi \in \Phi$ et tout $\varepsilon > 0$, il existe un entier N tel que, pour toute partie finie $J \subset [N,\infty)$ et toute $\mu \in K$, on ait $\left|\sum_{n \in J} \mu(\varphi_n)\right| \leqslant \varepsilon$.

(C_∞) Pour toute $\varphi \in \Phi$ et tout $\varepsilon > 0$, il existe un entier N tel que, pour tout $n \geqslant N$ et toute $\mu \in K$, on ait $|\mu(\varphi_n)| \leqslant \varepsilon$.

en sorte qu'il s'agit de prouver l'équivalence de (C_1) et (C_∞), autrement dit l'implication (C_∞) $\Longrightarrow$(C_1).

Or supposons (C_1) non vérifiée. Il existe alors $\varphi \in \Phi$ et $\varepsilon > 0$ tels que l'on puisse trouver une suite $(J_k)_{k \geqslant 1}$ de parties finies disjointes de $\mathbb{N}$ et une suite (μ_k) dans K telles que

$$\left| \sum_{n \in J_k} \mu_k(\varphi_n) \right| > \varepsilon .$$

Posons $L = \cup J_k$ et $\psi_k = \sum_{n \in J_k} \varphi_n$ pour chaque k.

La fonction $\sum_{k \geqslant 1} \psi_k = \sum_{n \in L} \varphi_n$ est continue sur T, positive et majorée par 1.
En rajoutant $\psi_o = 1 - \sum_{k \geqslant 1} \psi_k$, on construit ainsi une suite $\psi = (\psi_k)_{k \geqslant o}$, élément de Φ, pour laquelle on a $|\mu_k(\psi_k)| > \varepsilon$ pour tout $k \geqslant 1$, ce qui contredit (C_∞). □

REMARQUE

Le critère de compacité exprimé par le théorème est remarquablement simple et maniable, en particulier pour l'espace $M_{\beta,1}(T)$. On en déduit aisément (voir par exemple [4]) que l'enveloppe solide s(K) de toute partie relativement compacte des $M_{\beta,p}(T)$ est encore relativement compacte.

Pour l'espace $M_{\beta,1}(T)$ le lemme de Schur fournit encore des précisions intéressantes.

(2.2) PROPOSITION

L'espace $M_{\beta,1}(T)$ *est faiblement séquentiellement complet et ses parties faiblement compactes sont relativement compactes.*

PREUVE

Si K est faiblement relativement compacte dans $M_{\beta,1}(T)$, elle est déjà bornée et son image $V_\varphi(K)$ par toute application V_φ est faiblement relativement compacte dans l'espace ℓ^1. Elle y est donc relativement compacte d'après le lemme de Schur, ce qui ramène à la condition (C_1). On démontre de la même manière que toute suite de Cauchy faible dans $M_{\beta,1}(T)$ est en réalité relativement compacte, donc convergente. □

Le théorème (2.1) permet aisément la description des parties relativement compactes des espaces $M_{\sigma,p}(T)$, $1 \leqslant p \leqslant +\infty$. Rappelons déjà que le dual fort de chaque espace $M_{\sigma,p}(T)$ est l'espace de Banach $C^\infty(T)$ comme il résulte de (1.5) pour p=1 et de (1.4) pour $p > 1$.

(2.3) THEOREME ([4])

Les espaces $M_{\sigma,p}(T)$, $1 \leqslant p \leqslant +\infty$, *ont tous les mêmes parties relativement compactes, qui sont aussi les parties faiblement (ou étroitement) relativement compactes, qui sont aussi les parties précompactes. Ces parties communes sont exactement les traces sur* $M_\sigma(T)$ *des parties relativement compactes communes des espaces* $M_{\beta,p}(T)$.

PREUVE

Grâce aux diagrammes d'applications continues

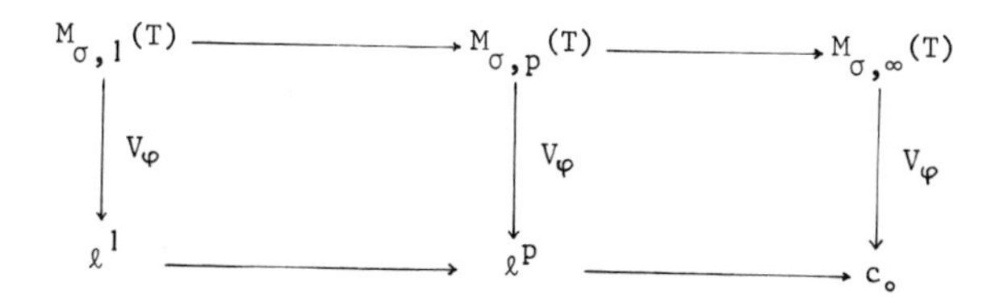

il suffit de vérifier qu'une partie $K \subset M_\sigma(T)$ qui est, soit étroitement relativement compacte, soit précompacte dans $M_{\sigma,\infty}(T)$, est en fait relativement compacte dans $M_{\sigma,1}(T)$. Dans le premier cas chaque image $V_\varphi(K)$ est faiblement relativement compacte dans l'espace ℓ^1, ce qui ramène encore au lemme de Schur et montre que K est relativement compacte dans $M_{\beta,1}(T)$. Dans le second cas la partie bornée K vérifie la condition (C_∞) puisque chaque image $V_\varphi(K)$ est précompacte dans c_0, et K est encore relativement compacte dans $M_{\beta,1}(T)$. On termine en remarquant que $M_{\sigma,1}(T)$ est fermé dans $M_{\beta,1}(T)$. □

REMARQUE

Lorsque $1 < p \leqslant +\infty$ l'espace $M_{\sigma,p}(T)$ n'est pas complet en général. On montre d'ailleurs facilement (voir [4]) que $M_{\sigma,p}(T)$ est quasi-complet, pour une valeur fixée $p > 1$, si et seulement si T est pseudocompact, ce qui revient encore à dire que $M_\sigma(T) = M_\beta(T)$. On en déduit donc que, chaque fois que T n'est pas pseudocompact, l'espace $M_{\sigma,p}(T)$, $1 < p \leqslant +\infty$, fournit un exemple d'elc non quasi-complet dans lequel tout précompact est relativement compact. Cette famille de tels exemples est d'ailleurs notablement plus simple que la plupart des exemples analogues connus. En particulier il est possible, avec $T = \mathbb{N}$, de construire sur l'espace $M_\sigma(T) = \ell^1$, de multiples topologies d'elc non quasi-complets dont les précompacts sont relativement compacts. Pour des développements plus importants sur cette question nous renvoyons à [4].

Le théorème (2.3) fournit donc dans l'espace $M_\sigma(T)$ un critère de compacité étroite particulièrement intéressant. En utilisant l'interprétation de toute $\mu \in M_\sigma(T)$ comme une mesure de Baire sur T, nous allons en déduire un nouveau critère de compacité étroite, mis sous forme ensembliste.

(2.4) THEOREME

Pour qu'une partie bornée $K \subset M_\sigma^+(T)$ *soit étroitement relativement compacte il faut et il suffit que, pour toute suite* (U_n) *de conoyaux de* T, *disjointe et localement finie, on ait*

$$\sup_{\mu \in K} \mu(U_n) \to 0$$

PREUVE

a) La condition est nécessaire : supposons qu'elle ne soit pas vérifiée. Il existe alors $\varepsilon > 0$, une suite $\mu_n \in K$ et une suite (U_n) de conoyaux, disjointe et localement finie, telles que $\mu_n(U_n) > \varepsilon$. Pour chaque n, il existe une fonction continue φ_n telle que $0 \leqslant \varphi_n \leqslant 1_{U_n}$ et $\mu_n(\varphi_n) > \varepsilon$. La fonction $\psi = \sum_{n \geqslant 1} \varphi_n$ est continue et majorée par 1 d'après les hypothèses sur la suite (U_n), ce qui permet, en posant $\varphi_0 = 1 - \psi$, de construire une pcu $\varphi = (\varphi_n)$ telle que $\mu_n(\varphi_n) > \varepsilon$ pour $n \geqslant 1$, ce qui met en défaut la condition (C_1).

b) La condition est suffisante : supposons qu'il existe $\varepsilon > 0$, une suite $\mu_n \in K$

et une pcu $\varphi = (\varphi_n)$ telles que $\mu_n(\varphi_n) > 2\varepsilon$. On supposera de plus que K est contenue dans la boule unité de $M_\sigma(T)$. Pour chaque n soit $\psi_n = \mathrm{Sup}(\varphi_n,\varepsilon) - \varepsilon$; on a alors $0 \leqslant \psi_n \leqslant \varphi_n \leqslant \psi_n + \varepsilon$, d'où l'on tire $\mu_n(\varphi_n - \psi_n) \leqslant \varepsilon$ et par suite $\mu_n(\psi_n) > \varepsilon$, avec aussi la condition $\psi_n \leqslant 1$. Soit $U_n = \mathrm{Coz}(\psi_n) = \{t \; ; \; \psi_n(t) > 0\}$; on obtient là une suite de conoyaux telle que $\mu_n(U_n) > \varepsilon$, puisque $\psi_n \leqslant 1_{U_n}$.

On va maintenant montrer que cette suite (U_n) est localement finie. Pour cela introduisons les conoyaux

$$S_n = \{t \; ; \; \sum_{k \geqslant n} \varphi_k(t) < \varepsilon\}$$

qui constituent une suite croissante telle que $T = \cup S_n$. De plus pour $t \in S_n$ et $k \geqslant n$ on a $\varphi_k(t) < \varepsilon$, donc $\psi_k(t) = 0$, et par suite $S_n \cap U_k = \emptyset$. De ce fait, la suite (S_n) formant un recouvrement ouvert de T, la suite (U_n) est localement finie. Malheureusement elle n'est pas disjointe, ce qui va compliquer maintenant les choses.

Pour tout entier p, la suite $S_n \cap U_p$ est croissante et de réunion U_p. On peut donc définir une fonction N(.), sur les entiers p, en choisissant pour N(p) le plus petit entier n tel que $\mu_p(S_n \cap U_p) > \varepsilon$. On a évidemment $N(p) > p$ puisque $S_p \cap U_p = \emptyset$. Définissons alors la suite strictement croissante d'entiers (n_k) par $n_1=1$ et $n_{k+1} = N(n_k)$. On a :

$$\alpha) \quad \mu_{n_k}(S_{n_{k+1}} \cap U_{n_k}) > \varepsilon$$

$$\beta) \quad j > k \Longrightarrow n_j \geqslant n_{k+1} \Longrightarrow S_{n_{k+1}} \cap U_{n_j} = \emptyset.$$

Posons enfin $V_k = S_{n_{k+1}} \cap U_{n_k}$ et $\nu_k = \mu_{n_k}$. On obtient une suite (V_k) de conoyaux, qui est disjointe et localement finie, et une suite $\nu_k \in K$ telles que $\nu_k(V_k) > \varepsilon$, ce qui contredit la condition du théorème. □

Pour terminer ce paragraphe 2, montrons que les topologies $\mathscr{T}_p$, $1 \leqslant p \leqslant +\infty$, peuvent aisément se substituer à la topologie étroite dans l'étude des espaces $M_\sigma(T)$. On a vu déjà en effet qu'elles fournissent les mêmes parties compactes. De plus :

(2.5) PROPOSITION

Sur le cône positif $M_\sigma^+(T)$ *les topologies* $\mathscr{T}_p$, $1 \leqslant p \leqslant +\infty$, *coïncident avec la topologie étroite.*

PREUVE

Donnons de ce fait une preuve analogue à celle de la proposition (2.2.1) de [2]. Il suffit bien entendu de prouver que si $\mu_\alpha \to \mu$ étroitement dans $M_\sigma^+(T)$, alors la convergence a lieu en réalité dans $M_{\sigma,1}(T)$. Fixons donc $\varphi = (\varphi_n) \in \Phi$ et soit $g_n = \sum_{k \geqslant n} \varphi_k$. La suite (g_n) est telle que $g_n \downarrow 0$, de sorte que, pour tout $\varepsilon > 0$ fixé, il existe un entier p tel que $\mu(g_p) < \varepsilon$. On peut alors trouver un indice α_0, tel que l'on ait, pour $\alpha \geqslant \alpha_0$

$$\sum_{n < p} |\mu_\alpha(\varphi_n) - \mu(\varphi_n)| < \varepsilon.$$

On a, par ailleurs, les inégalités

$$\sum_{n \geqslant p} |\mu_\alpha(\varphi_n) - \mu(\varphi_n)| \leqslant \mu_\alpha(g_p) + \mu(g_p) \leqslant \varepsilon + \mu_\alpha(g_p)$$

de sorte que, pour $\alpha \geqslant \alpha_o$, on obtient

$$\|\mu_\alpha - \mu\|_{\varphi,1} = \sum_n |\mu_\alpha(\varphi_n) - \mu(\varphi_n)| \leqslant 2\varepsilon + \mu_\alpha(g_p)$$

ce qui permet de terminer en remarquant que $\mu_\alpha(g_p) \to \mu(g_p)$ et que $\mu(g_p) < \varepsilon$. □

BIBLIOGRAPHIE

[1] A. Badrikian, *Séminaire sur les fonctions aléatoires linéaires et les mesures cylindriques*, Lecture Notes n° 139, (1970).

[2] J. Berruyer et B. Ivol, *Espaces de mesures et compactologies*, Publ. Dép. Math. Lyon, 9-1, (1972), p. 1-35.

[3] H. Buchwalter, *Topologies et compactologies*, Publ. Dép. Math. Lyon, 6-2, (1969), p. 1-74.

[4] H. Buchwalter, *Quelques curieuses topologies sur* $M_\sigma(T)$ *et* $M_\beta(T)$, Ann. Inst. Fourier, 27-3, (1977), à paraître.

[5] G. De Marco et R.G. Wilson, *Realcompactness and partitions of unity*, Proc. Amer. Math. Soc., 30-1, (1971), p. 189-194.

[6] L. Gillman et M. Jerison, *Rings of continuous functions*, Van Nostrand, (1960), New-York.

[7] M. Rome, *L'espace* $M^\infty(T)$, Publ. Dép. Math. Lyon, 9-1, (1972), p. 36-60.

[8] D. Sentilles et R.F. Wheeler, *Linear functionals and partitions of unity in* $C_b(X)$, Duke Math. J., 41, (1974), p. 483-496.

[9] V.S. Varadarajan, *Measures on topological spaces*, Amer. Math. Soc. Transl., (2), 48, (1965), p. 161-228.

K.-D. Bierstedt, B. Fuchssteiner (eds.)
Functional Analysis: Surveys and Recent Results

AN EXTENSION METHOD OF THE DUALITY THEORY OF LOCALLY CONVEX SPACES WITH APPLICATIONS TO EXTENSION KERNELS AND THE OPERATIONAL CALCULUS

Bernhard Gramsch
Mathematisches Institut der Universität
Kaiserslautern

A starting point of this article is the observation that the classical weak-strong theorem of Grothendieck [15], II, § 3.3 contains, with an appropriate interpretation and refinement, a fundamental extension theorem for vector valued analytic functionals, (ultra-) distributions, and measures. Among other things we give here some applications of this observation. The method of proof of Grothendieck (a combination of duality theory with closed graph theorems and topological tensor products) is generalized in paragraph 2 by an elementary start and sharpened for several applications ([12], [13], [14]). The article is based on the papers [12], [13], and [14].

An inspection of [15], II, § 3.3, leads easily to the following statement:

0.1 . Let $\mathcal{H}$ be a semireflexive locally convex space of scalar valued functions on a non empty set Ω, the topology of $\mathcal{H}$ being finer than the topology of pointwise convergence on Ω. Furthermore every closed linear mapping from any Banach space into $\mathcal{H}$ is assumed to be continuous. Let $\Delta \subset \Omega$ be a subset with the identification property with respect to $\mathcal{H}$ (i.e.: $h \in \mathcal{H}$ and $h(\Delta) \equiv 0$ implies $h \equiv 0$) and $\varphi : \Delta \to F$ an application into the complete locally convex vector space F such that for any y' of the dual F' there exists an $h_{y'} \in \mathcal{H}$ with $h_{y'}(\delta) = y' \circ \varphi(\delta)$ for all $\delta \in \Delta$ (weak $\mathcal{H}$-extension property of φ). Then there exists (exactly one) extension $\tilde{\varphi} \in \mathcal{H}(\Omega,F) := \mathcal{L}(\mathcal{H}'_\tau, F)$ of φ, where $\mathcal{H}'_\tau$ denotes $\mathcal{H}'$ with the Mackey topology τ.

If we specialize in 0.1 the space $\mathcal{H}$ for example to the space of holomorphic functions on a region Ω of the complex plane and F to a Banach space then 0.1

contains already an additional remark to well-known continuation theorems of Gelfand and Schilow, Nachbin [22], Bogdanowicz and Horváth (cf. [18], Th. 1.3) since the "interpolation" set Δ in 0.1 does not need to contain an open subset of Ω, but only a sequence with limit point in Ω. If we add to the proof of 0.1 a lifting which is motivated by a remark in [15], II, p. 88/89 to the Whitney extension theorem, then we obtain:

0.2. Let $\mathcal{H}$ be as in 0.1 and additionally nuclear; furthermore assume F and $\mathcal{H}$ to be (F)-spaces. If Δ is an arbitrary (non empty) subset of Ω and if $\varphi : \Delta \to F$ has the weak extension property as in 0.1, then there exists (at least) one extension $\tilde{\varphi} : \Omega \to F$ of φ with $\tilde{\varphi} \in \mathcal{H} \hat{\otimes}_\pi F$.

For the nuclear (DF)-space of locally holomorphic functions on a compact set ([20], § 27.4.(4)) and a Banach space F an analogous theorem is due to Waelbroeck [26] (cf. [13], [14]). Since the "interpolation" set Δ in 0.1. is an abstract set, it is convenient to identify $E := \mathcal{H}(\Omega)'_\tau$ with a space of functions and Δ with an appropriate subset of E. This remark leads to:

0.3. Let Δ be an arbitrary (non empty) subset of the locally convex space E and $\varphi : \Delta \to F$ a mapping into the complete locally convex space F such that for every $y' \in F'$ there exists an $x' \in E'$ with

$$\langle \delta, x' \rangle_{E,E'} = \langle \varphi(\delta), y' \rangle_{F,F'}$$

for all $\delta \in \Delta$. Furthermore assume that the closed linear hull E_o of Δ in E is infrabarreled and that every closed linear mapping from any Banach space into $(E_o)'_\beta$ is continuous.

1) Then there exists exactly one continuous linear mapping $\varphi_o : E_o \to F$ which extends φ.

2) If the linear mapping φ_o extends to a continuous linear mapping $\tilde{\varphi} : E \to F$, then we have an analogue to the statement 0.2.

The preceding statement 0.3. is already of interest if Δ is a total subset of a function space E. Examples: a) $E = C^k[0,1]$, $\Delta = \{t^n : n = 0,1,\dots\}$.
b) $E = C^\infty(\mathbb{R}^n)$, $\Delta = \{e_\zeta = e^{i\langle\zeta,t\rangle} : t \in \mathbb{R}^n, \zeta \in \mathbb{C}^n\}$, $F = \mathcal{L}(X)$, X a Banach space, $T = (T_1,\dots,T_n)$ a commuting n-tupel of operators $T_j \in \mathcal{L}(X)$, $j = 1,\dots,n$, $\varphi(e_\zeta) = e^{i\langle\zeta,T\rangle}$. In this connection we are looking for integral representations: $x \in E$,

$$\tilde{\varphi}\, x = \int_\Delta \varphi(\delta)\, d\mu_x(\delta),$$

which is an analogue of the representation of the operational calculus by the vector valued Fourier transform

$$\tilde{\varphi}\, x = \frac{1}{(2\pi)^{n/2}} \int_{\mathbb{R}^n} e^{i<\xi,T>} \hat{x}(\xi)\, d\xi .$$

In connection with [4] we can derive a constructive extension procedure for hypoelliptic sheaves using Hilbert space methods in the form of an orthogonalization for the "interpolation" set Δ.

§ 1 NOTATIONS

The vector spaces considered here are spaces over the field $\mathbb{R}$ or $\mathbb{C}$. By $\mathcal{L}(X,Y)$ we denote the vector space of all continuous linear mappings $v : X \to Y$ of the locally convex vector spaces X and Y. E'_c resp. E'_τ resp. E'_β is the dual space of the separated locally convex space E (lcs for short) equipped with the topology of uniform convergence on all absolutely convex compact resp. weakly compact resp. bounded sets of E. For a dual system $<E_1,E_2>$ we denote by $\sigma(E_1,E_2)$ resp. $\sigma(E_2,E_1)$ the weak topology on E_1 resp. E_2. $\mathcal{L}_b(X,Y)$ resp. $\mathcal{L}_e(E'_\tau,Y)$ denotes $\mathcal{L}(X,Y)$ resp. $\mathcal{L}(E'_\tau,Y)$ endowed with the topology of uniform convergence on all bounded subsets of X resp. on all equicontinuous subsets of E'. E^* is the algebraic dual of E. For a subset m of a vector space, $[m]$ denotes the linear hull of m. For a set Δ and a vector space F the symbol F^Δ is the vector space of all mappings from Δ into F; F^Δ carries the topology of pointwise convergence on Δ if F is a topological vector space, i.e. a lcs.

By $\hat{\Gamma}$ we denote the family of all ordered pairs (X,Y) of lcs for which every closed linear map from X into Y is continuous. For a class $\mathfrak{X}$ of lcs $(\Gamma\mathfrak{X})$ resp. $(\mathfrak{X}\,\Gamma)$ denotes the class of all lcs Y such that $(Y,X) \in \hat{\Gamma}$ resp. $(X,Y) \in \hat{\Gamma}$ for all $X \in \mathfrak{X}$. By $\mathfrak{B}$, $\mathfrak{F}$, $\mathcal{L}\mathfrak{F}$, $\mathfrak{U}$ resp. $\mathfrak{T}$ is denoted the class of Banach-, (F)-, (LF)-, ultrabornological resp. barreled lcs. In the same way $\mathfrak{P}$ resp. $\mathfrak{D}\mathfrak{F}$ resp. $\mathfrak{W}$ is the class of Ptak- (cf. [17]) resp. complete (DF)-resp. of spaces of de Wilde [9], ch. I, II with webs of type $\mathfrak{C}$. $\mathfrak{S}\mathfrak{W}$ is the subclass of $\mathfrak{W}$ with a strict web [9], ch. III. 2. $\mathfrak{S}\mathfrak{W}$ is of interest for <u>lifting problems</u> [9], ch. III. 5 (cf. 2.6, 2.7, 2.8). We have $\mathcal{L}\mathfrak{F} \subset \mathfrak{S}\mathfrak{W}$ and $\mathfrak{D}\mathfrak{F} \subset \mathfrak{S}\mathfrak{W}$. There exist numerous closed graph theorems, some due to Ptak, Raikov, Robertson-Robertson and de Wilde.

The following relations hold:

$$(1.1) \qquad (\mathfrak{T}\,\Gamma) \supset \mathfrak{P} \quad \text{and} \quad (\mathfrak{U}\,\Gamma) \supset \mathfrak{W}\,.$$

$\mathfrak{T} = (\Gamma\,\mathfrak{B}) = (\Gamma\,\mathfrak{P})$ and $(\mathfrak{B}\,\Gamma) = (\mathfrak{U}\,\Gamma)$ are well-known. Recently there have been proved more general closed graph theorems [1]). Concerning the permanence properties of $\mathfrak{P}$ and $\mathfrak{W}$ resp. $\mathfrak{T}\mathfrak{W}$ we refer to [17] and [9].

§ 2 EXTENSION AND DUAL SYSTEMS

For certain mappings $\varphi : \Delta \to F_1$ from a set Δ into a vector space F_1 we would like to find, by the embedding of Δ into an appropriate vector space E_1, linear mappings $\tilde{\varphi} : E_1 \to F_1$ which extend φ and which are continuous with respect to suitable topologies on E_1 resp. F_1.

2.1. Definition. Let $\langle E_1, E_2\rangle$ resp. $\langle F_1, F_2\rangle$ be dual systems of vector spaces and Δ an arbitrary (non empty) subset of E_1. A mapping $\varphi : \Delta \to F_1$ has the weak extension property with respect to the preceding dual systems if for each $y_2 \in F_2$ there exists at least one $x_2 \in E_2$ such that for all $\delta \in \Delta$

$$(2.1) \qquad \langle \delta, x_2\rangle_{E_1,E_2} = \langle \varphi(\delta), y_2\rangle_{F_1,F_2}$$

is fulfilled. The family of these mappings φ is denoted by $\phi(\Delta, E_1, E_2; F_1, F_2)$; ϕ for short.

2.2. ϕ is a linear subspace of F_1^{Δ}. If we equip F_1 with a separated locally convex topology, i.e. with $\sigma(F_1,F_2)$ or with a finer one, then ϕ, with the topology of pointwise convergence on Δ induced by F_1^{Δ}, is a separated locally convex vector space. With

$$\Delta^{\perp} = \{x_2 \in E_2 : \langle \delta, x_2\rangle = 0 \ \forall\, \delta \in \Delta\}$$

we have

---

[1]) W. Robertson, Proc. Lond. Math. Soc. 24, 692-738 (1972)
M.H. Powell, Trans. Am. Math. Soc. 211, 391-426 (1975)

(2.2) $$\phi\,(\Delta,\ E_1,E_2;\ F_1,F_2) \cong \phi\,(\Delta,\ [\Delta],\ E_2/\Delta^{\perp};\ F_1,F_2)$$

in a canonical manner.

If E_1 resp. F_1 carries an appropriate locally convex topology which is finer than $\sigma(E_1,E_2)$ resp. $\sigma(F_1,F_2)$, then it is the subject of the following remarks to establish the existence of an (algebraic) extension isomorphism or monomorphism

(2.3) $$\mathcal{E}x : \phi\,(\Delta,\ E_1,E_2;\ F_1,F_2) \to \mathcal{L}(E_1,F_1)/\mathcal{L}_o,$$

$\mathcal{L}_o = \{u \in \mathcal{L}(E_1,F_1) : u(\Delta) = 0\}$; this means, for every $\varphi \in \phi$ there exists a $\tilde{\varphi} \in \mathcal{L}(E_1,F_1)$ with $\tilde{\varphi}|\Delta = \varphi$, $\mathcal{E}x(\varphi) = \tilde{\varphi} + \mathcal{L}_o$. Then the restriction mapping $\mathcal{R}$ to Δ

(2.4) $$\mathcal{R} : \mathcal{L}^{\Phi}(E_1,F_1) \to \phi\,(\Delta,\ E_1,E_2;\ F_1,F_2) \subset F_1^{\Delta}$$

is surjective for an appropriate subspace $\mathcal{L}^{\Phi}(E_1,F_1)$ of $\mathcal{L}(E_1,F_1)$.

<u>2.3. Remark</u>. It is well-known that weakly continous functions on the interval $[0,1]$ with values in a Hilbert space are in general not continuous; furthermore weakly real analytic functions on $[0,1]$ with values in Fréchet spaces are in general not real analytic, this means they do not have locally convergent power series. Consider for instance $f(t,x) = \frac{1}{t - ix}$, $x \in [0,1]$, $t \in (0,1)$ (open interval), $[0,1] \ni x \to \frac{1}{t - ix} \in C(0,1)$ ((F)-space). Therefore the existence of (2.3) in special cases ([20], § 27.4) must depend on special assumptions for the underlying spaces E_j,F_j, $j=1,2$ ([15], II, Prop. 12, and § 3.3).

<u>2.4. Definition</u>. A subset $G \subset \phi$ has the property (b), if for each $y_2 \in F_2$ and for each x_1 of the $\sigma(E_1,E_2)$-closure of $[\Delta]$ in E_1 the set of numbers $M(x_1,y_2,G) :=$

(2.5) $$\{ \langle x_1,x_2\rangle : x_2 \in E_2 \text{ and } \exists\, \varphi \in G \text{ with } \langle\delta,x_2\rangle = \langle\varphi(\delta),y_2\rangle \ \ \forall\delta\in\Delta\}$$

is bounded.

<u>2.5. Theorem</u>. Let F and F'_β be complete lcs and $F_1=F$, $F_2=F'_\beta$ or $F_1=F'_\beta$, $F_2=F$. Assume $E \in \mathcal{T}$ and $E'_\beta \in (\mathcal{B}\Gamma)$ or $(F_2,E'_\beta) \in \hat{\Gamma}$; furthermore let Δ be a total subset of E.

1) Then there exists

$$\mathcal{E}x : \phi\,(\Delta,\ E,E';\ F_1,F_2) \to \mathcal{L}(E,F_1)$$

as an algebraic isomorphism and maps the system of sets $G \subset \phi$ with property (b) onto the system of equicontinuous sets.

2) If $\{\varphi_\alpha\} \subset G$, $G \subset \phi$ with the property (b), is a directed Cauchy system in $\phi \subset F_1^\Delta$, then $\{\mathfrak{E}x(\varphi_\alpha)\}$ converges to some $\tilde{\varphi} \in \mathcal{L}(E,F_1)$ uniformly on every precompact subset of E. The complete hull of subsets $G \subset \phi$ with the property (b) is contained in the separated lcs ϕ.

Proof. 1) Since Δ is total in E, $\varphi \in \phi$ induces a linear mapping

$$(2.7) \qquad \varphi' : F_2 \to E'$$

which is a closed mapping by (2.1) with respect to the weak topologies and therefore also with respect to all finer topologies. Let B be an absolutely convex, $\sigma(F_2,F_1)$-closed equicontinuous (for $F_2=F'_\beta$) resp. bounded (for $F_2=F$) subset of F_2. Then φ', restricted to the Banach space $[B]$, is a continuous linear mapping into E'_β since $E'_\beta \in (\mathfrak{B}\Gamma)$; because of $E \in \mathcal{T}$ the set $\varphi'(B)$ is equicontinuous, thus the transposed mapping

$$(2.7') \qquad {}^t\varphi' : E \to F_2^*$$

of φ' into the algebraic dual F_2^* of F_2 is continuous, if we endow F_2^* with the topology of uniform convergence on all equicontinuous (for $F_2=F'_\beta$) resp. on all bounded (for $F_2=F$) subsets of F_2. In the case $(F_2,E'_\beta) \in \hat{\Gamma}$ it is enough to consider F_2' instead of F_2^*, and we conclude in the same way. It follows

$$(2.8) \qquad {}^t\varphi'(\delta) = \varphi(\delta), \quad \forall\, \delta \in \Delta$$

from

$$(2.8') \qquad \langle {}^t\varphi'(\delta), y_2\rangle_{F_2^*,F_2} = \langle \delta, \varphi'(y_2)\rangle_{E_1,E_2} = \langle \varphi(\delta), y_2\rangle_{F_1,F_2}$$

$\forall\, y_2 \in F_2$. F_1 carries the topology induced by the topology of F_2^* (resp. F_2'), furthermore since Δ is total in E and since F_1 is complete we obtain from the continuity of ${}^t\varphi' : E \to F_2^*$

$$(2.8'') \qquad {}^t\varphi' : E \to F_1$$

as a mapping into F_1 such that we have obtained a uniquely determined extension

$$(2.9) \qquad \tilde{\varphi} := {}^t\varphi' \in \mathcal{L}(E,F_1) \quad \text{for} \quad \varphi \in \phi.$$

The statement (2.6) follows since image $(\mathfrak{R})$ is obviously contained in ϕ.

If $G \subset \phi$ has the property (b), then, in account of $\overline{[\Delta]} = E$,

$\mathcal{E}x(G)$ is pointwise $\sigma(F_1,F_2)$-bounded in F_1 by the relation

$$\langle\tilde{\varphi}(x_1),y_2\rangle = \langle x_1,\varphi'(y_2)\rangle = \langle x_1,x_2^{\varphi}\rangle \tag{2.10}$$

for $\varphi \in \phi$; this follows from $\varphi \in G$, 2.4, (2.5). Because F is complete, it also follows in the case $F_1=F'_\beta$ that the set $\mathcal{E}x(G)$ is pointwise bounded and therefore equicontinuous because of $E \in \mathcal{T}$.

2) Follows immediately from Schaefer [23], ch. III, § 4, applied to $\mathcal{E}x(G)$. Restricted to the equicontinuous sets, the continuous mapping $\mathcal{R}: \mathcal{L}_c(E,F_1) \to \phi$ is a homeomorphism where the index c means the topology of precompact convergence.

Now we pass to the case where Δ is an arbitrary non empty subset of E_1, i.e. not necessarily $\sigma(E_1,E_2)$-total.

2.6. Lemma. Let X,X_o,Y,Y_o be lcs; let $\psi: X \to X_o$ be a surjective, continuous linear mapping into the sequentially complete space X_o; let X be a space with strict web, $X \in \mathcal{SW}$, and Y_o a subspace of Y with the induced topology, $j: Y_o \to Y$. Assume $T_\alpha : Y_o \to X_o$ to be a family of mappings with a nuclear representation

$$T_\alpha y_o = \sum_{k=1}^{\infty} \lambda_k \langle y_o,a_k^{(\alpha)}\rangle x_k, \; \alpha \in A, \; A \text{ index set,}$$

$\lambda_k > 0$, $\sum_{k=1}^{\infty} \lambda_k < \infty$, $\{x_k\}_{k=1}^{\infty}$ bounded in X_o and $\{a_k^{(\alpha)}: \alpha \in A, k=1,2,\ldots\}$ an equicontinuous subset of Y'_o. Then there exists a family of mappings $\tilde{T}_\alpha: Y \to X$ with $T_\alpha = \psi \circ \tilde{T}_{\alpha|Y_o}$ and with a representation

$$\tilde{T}_\alpha y = \sum_{k=1}^{\infty} \tilde{\lambda}_k \langle y, \tilde{a}_k^{(\alpha)}\rangle \tilde{x}_k,$$

where $\tilde{\lambda}_k > 0$, $\sum_{k=1}^{\infty} \tilde{\lambda}_k < \infty$, and $\tilde{x}_k$ is nullsequence in X and $\{\tilde{a}_k^{(\alpha)}: \alpha \in A, k=1,2,\ldots\}$ is an equicontinuous subset of Y'.

Proof. With an appropriate zero sequence $\beta_k > 0$ we obtain $\sum_{k=1}^{\infty} \tilde{\lambda}_k < \infty$, $\tilde{\lambda}_k = \lambda_k \beta_k^{-1}$. The closed absolutely convex hull B of the sequence $\sqrt{\beta_k}\, x_k$ generates a Banach space $X_B \subset X_o$ since X_o is sequentially complete and $\beta_k x_k$ is a nullsequence in X_B. $\{\beta_k x_k\}$ is (following L. Schwartz, cf. [9], ch. III. 4) a "very convergent" null sequence; because of $X \in \mathcal{SW}$ by de Wilde [9], ch. III. 5 this sequence can be lifted to a null sequence $\tilde{x}_k$, $\psi(\tilde{x}_k) = \beta_k x_k$. The equicontinuous set $a_k^{(\alpha)}$ can be extended to an equicontinuous set $\tilde{a}_k^{(\alpha)}$ on Y by the Hahn-Banach theorem; the assertion follows.

We observe especially $\mathfrak{DF} \subset \mathfrak{YW}$.

2.7. Remark. Let $\mathfrak{N}$ be a saturated system ([20], 21.1) of $\sigma(F_2,F_1)$-bounded subsets of F_2 covering F_2. Let F_1 be complete equipped with the polar topology ν associated to $\mathfrak{N}$. Besides the topology $\sigma(E_2,E_1)$ let E_2 carry the complete locally convex topology μ finer than $\sigma(E_2,E_1)$. Furthermore we assume that the ordered pair (F_1,E_2) has the c-lifting-property, this means: for every continuous linear mapping $v : (F_1)'_c \to Q$ into a complete quotient Q of E_2, $q : E_2 \to Q$, there exists $\hat{v} \in \mathcal{L}((F_1)'_c, E_2)$ with $v = q\circ\hat{v}$. Assume that $\varphi' : F_2 \to E_2/\Delta^\perp$, for $\varphi \in \phi$ and $E_2/\Delta^\perp$ complete, transforms the elements of $\mathfrak{N}$ into compact subsets of $E_2/\Delta^\perp$ (this is proved in some cases for Schwartz spaces E_2 and $Q \in (\mathfrak{B}\Gamma)$ with the closed graph theorem). Then for each $\varphi \in \phi$ there exists $\tilde{\varphi} \in \mathcal{L}((E_2)'_c, F_1)$ which extends φ.

Proof. The transposed mapping ${}^t\varphi' : (E_2/\Delta^\perp)'_c \to F_2{}^*$ of $\varphi' : F_2 \to E_2/\Delta^\perp$ is continuous if we endow the algebraic dual F_2^* of F_2 with the polar topology of the system $\mathfrak{N}$. Because of $<\delta,x_2> = <\delta,x_2 + \Delta^\perp>$ the set Δ is embedded into $(E_2/\Delta^\perp)'_c$ as a total set $\overline{\Delta}$($\overline{\Delta}$ is total by reason of $(E'_c)' = E$). As in (2.8') it follows ${}^t\varphi'(\delta) = \varphi(\delta)$, $\delta \in \Delta$. Therefore we obtain from the continuity of ${}^t\varphi'$ and from the completeness of F_1 the statement ${}^t\varphi' \in \mathcal{L}((E_2/\Delta^\perp)'_c, F_1)$. By assumption the transposed mapping $v: (F_1)'_c \to E_2/\Delta^\perp$ has a lifting $\hat{v} \in \mathcal{L}((F_1)'_c, E_2)$. It remains to show that $\tilde{\varphi} := {}^t\hat{v} : (E_2)'_c \to F_1$, restricted to Δ, coincides with φ. Assume $y_2 \in F_2$ and $Q = E_2/\Delta^\perp$; $F_2 \subset (F_1)'$ is clear. Then we have

$$\begin{aligned}
(2.11) \qquad & <\tilde{\varphi}(\delta),y_2>_{F_1,F_1'} &=& <\delta, \hat{v}(y_2)>_{E_2',E_2} &=\\
& <\overline{\delta},q\circ\hat{v}(y_2)>_{Q',Q} &=& <\overline{\delta}, v(y_2)>_{Q',Q} &=\\
& <{}^t\varphi'(\overline{\delta}),y_2>_{F_1,F_1'} &=& <\overline{\delta}, \varphi'(y_2)>_{Q',Q} &=\\
(2.11') \qquad & <\delta,x_2 + \Delta^\perp>_{E_1,E_2} &=& <\varphi(\delta),y_2>_{F_1,F_2} &
\end{aligned}$$

because of 2.1 for all $y_2 \in F_2$. By (2.11) and (2.11') the statement $\tilde{\varphi}(\delta) = \varphi(\delta) \quad \forall\, \delta \in \Delta$ is proved.

2.7'. Remark. The ordered pair of lcs (F,E) has the c-lifting property in the following cases:

1) E and $F \in \mathfrak{F}$ or E and $F \in \mathfrak{DF}$; assume furthermore E or F nuclear (cf. Grothendieck [15], I, 3.2, Prop. 16; II, 3.1, Prop. 12.2.6. Lemma),

2) $E \in \mathfrak{F}$, $F = \mathcal{C}(\Omega)$, Ω locally compact and countable at infinity (cf. [11], § 1),

3) E a (LS)-space and $F = \mathcal{C}(\Omega)$, Ω compact,

4) $E \in \mathfrak{F}$ and $F = \mathcal{CH}(\Lambda)$ defined as in [11], or $F = \mathcal{GH}(\Lambda)$, mixed spaces as defined in [4].

In this connection (2.7') the extension resp. lifting theorems of Whitney for differentiable functions, of Dugundji for continuous functions and of Dunford-Pettis for L^∞-functions are of interest. For further lifting theorems see the work of Kaballo and Vogt in these lecture notes. (Cf. also W. Kaballo, Habilitationsschrift, Kaiserslautern 1976.)

2.8. Theorem. Let the lcs E_j, F_j, $j=1,2$ be of the class $\mathfrak{F} \cup \mathfrak{DF}$ and Δ an arbitrary nonempty subset of E_1. The following conditions a, a', b and c are assumed to be fulfilled:

a) : a_o) $E_1 = (E_2)'_\beta$ or a_1) $E_2 = (E_1)'_\beta$

a') : a'_o) $F_1 = (F_2)'_\beta$ or a'_1) $F_2 = (F_1)'_\beta$

b) : The ordered pair (E_1,F_1) is b_o) of type $(\mathfrak{F},\mathfrak{DF})$ or b_1) of type $(\mathfrak{DF},\mathfrak{F})$ (we assume the same for (E_1,E_2) and (F_2,F_1)).

1) Then for each $\varphi \in \phi(\Delta, E_1,E_2; F_1,F_2)$ there exists a continuous linear mapping $\tilde{\varphi} : E_1 \to F_1$ which extends $\varphi : \Delta \to F_1$.

2) In the cases a_1 (and c_o) $\mathcal{E}x : \phi \to \mathcal{L}(E_1,F_1)/\mathcal{L}_o$ ((2.3)) is bijective $(\mathcal{L}_b(E_1,F_1) = (E_1)' \hat{\otimes}_\pi F_1 = \mathcal{L}_e((F_1)'_c, (E_1)'_\beta))$.

3) If $E_2/\Delta^\perp$ is complete also in the cases $b_o c_1$ (for example E_2 a Ptak space), then in the canonical algebraic sense we have

$$\phi \cong (E_2/\Delta^\perp) \hat{\otimes}_\pi F_1 = (E_2 \hat{\otimes}_\pi F_1)/(\Delta^\perp \hat{\otimes}_\pi F_1).$$

4) Assume c_o). Then there exists for each subset $G \subset \phi$ with the property (b) (cf. 2.4) in the case $E_2 \in \mathfrak{F}$ a rapidly decreasing sequence $\lambda_k > 0$, resp. in the case $E_2 \in \mathfrak{DF}$ for every $r > 1$ a sequence $\lambda_k \le k^{-r}$, $k = 1,2,\ldots$, and a sequence $x_k \in E_2$ and a sequence of linear mappings $v_k : [G] \to F_1$, such that the set $\{v_k(\varphi) : \varphi \in G, k=1,2,\ldots\}$ is bounded in F_1 and that

$$\tilde{\varphi} := \sum_{k=1}^{\infty} \lambda_k\, x_k \otimes v_k(\varphi) \in \mathcal{L}(E_1,F_1) \tag{2.12}$$

extends φ for every $\varphi \in G$ (without loss of generality we may assume G absolutely convex and complete in $\phi \subset F_1^\Delta$). Therefore there exists a linear lifting $(\mathcal{R} \circ \mathcal{E}x = Id_{[G]})$

$$\tilde{\mathfrak{E}}x : [G] \to \mathcal{L}_b(E_1,F_1) \cong E_2 \hat{\otimes}_\pi F_1.$$

<u>Proof</u>. We are going to prove only the statements 1) and 4) in the cases c_o (where E_1 and E_2 are nuclear). - For the proof of the remaining statements we refer to [13].- 1): Let H be the closed subspace of E_1 generated by $[\Delta]$; this space H is again a nuclear (F)- or complete (DF)-space and therefore barreled, too. Because of

$$(2.13) \qquad \phi(\Delta,\ H,\ E_2/H^\perp;\ F_1,F_2) \cong \phi(\Delta,\ E_1,E_2;\ F_1,F_2)$$

we apply 2.5 to the left hand side : For nuclear spaces from $\mathfrak{F} \cup \mathfrak{DF}$ the quotients are again of this type and therefore especially reflexive (F)-spaces or strong duals of reflexive (F)-spaces; it follows $E_2/\Delta^\perp \in \mathfrak{P}$. In account of $\mathfrak{P} \subset (\mathfrak{B}\Gamma)$ the assumptions of 2.5 are satisfied; $H'_\beta = E_2/\Delta^\perp$ is evident. By reason of $\mathcal{L}_b(H,F_1) \cong H'_\beta \hat{\otimes}_\pi F_1$ it remains only to combine [15], II, Prop. 12 with 2.6 .

4): Because of (2.13) and since H is barreled, in account of 2.4 and 2.5.1) the mapping $\mathfrak{E}x : \phi(\Delta,\ H,\ E_2/H^\perp;\ F_1,F_2) \to \mathcal{L}(H,F_1)$ transforms the set G into an equicontinuous subset of $\mathcal{L}(H,F_1)$ whose closed absolutely convex hull $\hat{G}$ is complete in $\mathcal{L}_e(H,F_1)$ ([23], ch. III, § 4). By means of the isomorphism $\mathcal{L}_e(H,F_1) \cong (E_2/\Delta^\perp) \hat{\otimes}_\pi F_1$ the set $\hat{G}$ corresponds to a set $\hat{\hat{G}}$ ([23], ch. IV, 9.1, 9.4). Now we apply Grothendieck [15], II, § 3.1, Prop. 12.2) to $\hat{\hat{G}}$ and we obtain an equicontinuous sequence $u_k : [B] \to F_1$ from the Banach subspace $[B] \subset (E_2/\Delta^\perp) \hat{\otimes}_\pi F_1$, whose unit ball B contains $\hat{\hat{G}}$, such that we have the representation

$$(2.14) \qquad \varphi = \sum_{k=1}^{\infty} \lambda_k\ x'_k \otimes u_k(\hat{\hat{\varphi}}) \text{ for all } \hat{\hat{\varphi}} \in \hat{\hat{G}}$$

$\lambda_k > 0$, $\sum_{k=1}^{\infty} \lambda_k < \infty$, $x'_k \in E_2/\Delta^\perp$, $x'_k \to 0$. An inspection of the proof of Prop. 12 in [15], II, shows that we can choose λ_k, $k=1,2,\dots$, as in the assertion of 2.8.4) (cf. [15], II, p. 75) by reason of [15], II, § 2.4 and by an appropriate iteration of nuclear mappings ([15], II, § 1). Since in the correspondence $\varphi \to \hat{\varphi} \to \hat{\hat{\varphi}}$ we are dealing with algebraic monomorphisms, we have obtained a sequence $v_k : [G] \to F_1$ as asserted in 2.8.4). By the lifting of the null sequence x'_k to a null sequence $x_k \in E_2$ this yields the assertion of 2.8.4).

The following theorem 2.8' is a special case of 2.8 in a formulation for function spaces; this is important in applications.

<u>2.8'. Theorem</u>. Let $\mathfrak{H} \in \mathfrak{F} \cup \mathfrak{DF}$ be a nuclear space of scalar valued functions on a (non empty) set Λ whose topology is finer than the topology of pointwise

convergence on Λ. Furthermore let $F_j \in \mathcal{F} \cup \mathcal{DF}$, j=1,2, be with $F_2 = (F_1)'_\beta$ or $F_1 = (F_2)'_\beta$; assume that the spaces $\mathcal{H}$ and F_1 are both from the class $\mathcal{F}$ or both from the class $\mathcal{DF}$. Let there be defined a mapping $\varphi : \Delta \to F_1$ on the arbitrary (non empty) subset Δ of Λ with the weak $\mathcal{H}$-extension property, this means for each $y_2 \in F_2$ there exists (at least one) $h_{y_2} \in \mathcal{H}$ such that

$h_{y_2}(\lambda) := \langle \delta_\lambda, h_{y_2} \rangle = \langle \varphi(\lambda), y_2 \rangle$, $\forall\ \lambda \in \Delta$, i.e. the function $\langle \varphi(\lambda), y_2 \rangle$ defined on Δ has at least one extension $h_{y_2}(\lambda)$ defined on Λ.

Then there exist null sequences $h_k \in \mathcal{H}$ and $y_k \in F_1$, k=1,2,... such that $\tilde{\varphi} : \Lambda \to F_1$, defined by

$$\tilde{\varphi}(\lambda) = \sum_{k=1}^{\infty} \alpha_k\, h_k(\lambda) y_k, \tag{2.15}$$

is an extension of $\varphi : \Delta \to F_1$, where the sequence $\alpha_k > 0$ is rapidly decreasing in the case $\mathcal{H} \in \mathcal{F}$ resp. α_k can be chosen $0 < \alpha_k < k^{-r}$ for any given $r > 0$ in the case $\mathcal{H} \in \mathcal{DF}$.

Proof. If $\mathcal{H}$ separates the points of Λ then the set Λ can be embedded into $\mathcal{H}'$ and therefore the subset Δ, too. Otherwise we decompose Λ into a family $\hat{\Lambda}$ of equivalence classes : $\lambda_1 \sim \lambda_2$ if $h(\lambda_1) = h(\lambda_2)$ for all $h \in \mathcal{H}$. Λ can be considered as a subset of $\mathcal{H}'$. Now we can apply 2.8.

§ 3 EXTENSION KERNELS

Let $\mathcal{H}(\Lambda)$ be for example a nuclear (F)- or (DF)-space of scalar valued functions on a set Λ and Δ a "thin" subset of Λ with a function space $\mathcal{A}(\Delta)$ such that every element $a \in \mathcal{A}(\Delta)$ can be extended to an element of $\mathcal{H}(\Lambda)$ in a not necessarily unique manner. We are looking for a kernel $K: \mathcal{A}(\Delta) \to \mathcal{H}(\Lambda)$ which provides a simultaneous extension procedure for all elements of $\mathcal{A}(\Delta)$. This intention is motivated by remarks of Grothendieck to the extension theorem of Whitney [15], II, § 3.3 and by an application of theorem B to coherent sheaves (Bungart [7], 13., 18.). For sake of simplicity let $\mathcal{H}(\Lambda)$ separate the points of Λ (otherwise we consider equivalence classes) and let $\mathcal{H}(\Lambda)$ and $\mathcal{A}(\Delta)$ be function spaces with a locally convex topology finer than the pointwise convergence on Λ respectively Δ. Then we have a mapping

(3.1) $$K' : \mathcal{A}(\Delta) \to \mathcal{H}(\Lambda)/\Delta^{\perp},$$

if every element of $\mathcal{A}(\Delta)$ extends to an element of $\mathcal{H}(\Lambda)$, where $\Delta^{\perp} = \{h \in \mathcal{H}(\Lambda) : h(\Delta) = 0\}$. Obviously the mapping K' in (3.1) is closed, therefore $(\mathcal{A}(\Delta), \mathcal{H}(\Lambda)/\Delta^{\perp}) \in \hat{\Gamma}$ (1.1) implies the continuity of K'. We have obtained a lifting problem which connects the search for a kernel $K : \mathcal{A}(\Delta) \to \mathcal{H}(\Lambda)$ to 2.6. and the theorem 2.8. Since under the previous assumptions Δ can be considered as a subset of $\mathcal{H}(\Lambda)'$, our considerations lead to an element $\varphi_{\mathcal{A}}$ or

(3.1') $$\varphi \in \phi(\Delta, \mathcal{H}', \mathcal{H} ; \mathcal{A}', \mathcal{A})$$

which under suitable assumptions (2.6., 2.8. theorem) can be extended to an element of $\mathcal{A}'_{\beta} \hat{\otimes}_{\pi} \mathcal{H}$; this provides the first step to an extension kernel.

As an example let us assume furthermore ([12], 2.1, 2.9) that Δ is an open subset of a compact metrizable Hausdorff space S such that $\bar{\Delta} = S$ and that every element $a \in \mathcal{A}(\Delta)$ has a (unique) extension to a continuous function on S. Then $\mathcal{A}$ can be considered as a subspace of $C(S)$ and we assume $\mathcal{A}$ closed in the uniform topology on S. Now we have $C(S)'/\mathcal{A}^{\perp} \cong \mathcal{A}'$ and the Choquet boundary M of $\mathcal{A}$ in S is a Borel set. Let $\mathcal{M}$ be the subspace of measures in $C(S)'$ concentrated on M; then by a result of Hinrichsen (cf. [12]) the restriction $q : \mathcal{M} \to \mathcal{A}'$ of the quotient map $q_o : C(S)' \to C(S)'/\mathcal{A}^{\perp} = \mathcal{A}'$ is surjective if $\mathcal{A}$ separates the points of S. If $\mathcal{H}(\Lambda)$ is a nuclear (F)- or (DF)-space then the extension $\tilde{\varphi} \in \mathcal{A}'_{\beta} \hat{\otimes}_{\pi} \mathcal{H}$ of $\varphi \in \Phi(\Delta, \mathcal{H}', \mathcal{H}; \mathcal{A}', \mathcal{A})$ has a representation $\tilde{\varphi} = \sum_{j=1}^{\infty} \lambda_j \, a'_j \otimes h_j$ with $\Sigma|\lambda_j| < \infty$, $a'_j \to 0$, $a'_j \in \mathcal{A}'_{\beta}$, $h_j \to 0$, $h_j \in \mathcal{H}(\Lambda)$, $j=1,2,\ldots$ There exists a sequence $\mu_j \in \mathcal{M}$, $q(\mu_j) = a'_j$, $\mu_j \to 0$, such that we have obtained a "kernel"

(3.2) $$K = \sum_{j=1}^{\infty} \lambda_j \, \mu_j \otimes h_j .$$

Another argument (see [12], 2.8, 2.5) shows that there exists a positive measure $\Theta \in \mathcal{M}$, $\|\Theta\| = 1$, and a pointwise defined kernel $k(\lambda,\zeta)$, $\lambda \in \Lambda$, $\zeta \in M$, $k(\lambda,\zeta) \in \mathcal{H}(\Lambda) \hat{\otimes}_{\pi} L^{\infty}(M,\Theta)$, such that

(3.3) $$\tilde{a}(\lambda) = \int_M k(\lambda,\zeta) \, a(\zeta) \, d\Theta(\zeta)$$

provides an extension of $a \in \mathcal{A}$ to an element $\tilde{a} \in \mathcal{H}(\Lambda)$ for all $a \in \mathcal{A}$. From the above example we come to the following theorem:

<u>3.1. Theorem</u>. 1) Let $\mathcal{A}, \mathcal{B}, \mathcal{H}, \mathcal{M}$ be in the class $\mathfrak{F} \cup \mathfrak{D}\mathfrak{F}$.
2) The spaces $\mathcal{A}'_{\beta}$, $\mathcal{B}'_{\beta}$, $\mathcal{H}$ and $\mathcal{M}$ are assumed to be all in the class $\mathfrak{F}$ and

the spaces $\mathcal{A}$, $\mathcal{B}$, $\mathcal{H}'_\beta$ (and $\mathcal{M}'_\beta$) are assumed to be all in the class $\mathcal{DF}$ or vice versa.

3) Let $\mathcal{A}$ be topologically isomorphic to a closed subspace of $\mathcal{B}$ such that $\mathcal{A}'_\beta$ is topologically isomorphic to $\mathcal{B}'_\beta/\mathcal{A}^\perp$. Let furthermore $\mathcal{M}$ be a closed subspace of $\mathcal{B}'_\beta$ such that the restriction $q : \mathcal{M} \to \mathcal{A}'_\beta$ of the quotient map $q_o : \mathcal{B}'_\beta \to \mathcal{A}'_\beta$ is surjective.

4) Let $\mathcal{H}$ be a nuclear space of scalar valued functions on a set Λ whose topology is finer than the pointwise convergence on Λ.

5) Let Δ be an arbitrary non-empty subset of Λ and $\mathcal{H}_\Delta$ the space of restrictions of $\mathcal{H}$ to Δ equipped with the topology of pointwise convergence on Δ.

6) Assume $T : \mathcal{A} \to \mathcal{H}_\Delta$ to be a continuous linear mapping (for example a restriction), i.e. this means: the scalar valued function $\lambda \to (Ta)(\lambda)$, $a \in \mathcal{A}$, has at least one $\mathcal{H}$-extension to all of Λ.

Then there exists a nuclear mapping (an extension kernel) $K : \mathcal{B} \to \mathcal{H}$, $K \in \mathcal{H} \hat{\otimes}_\pi \mathcal{M}$ $(\subset \mathcal{L}(\mathcal{B}, \mathcal{H}))$, with the property

$$(Ka)(\lambda) = (Ta)(\lambda)$$

for all $\lambda \in \Delta$ and $a \in \mathcal{A}$.

Proof. Let $\Delta' = \{\delta_\lambda ; \lambda \in \Delta\} \subset \mathcal{H}'$; obviously the mapping $T : \mathcal{A} \to \mathcal{H}_\Delta$ induces an element $\varphi \in \Phi(\Delta', \mathcal{H}', \mathcal{H} ; \mathcal{A}', \mathcal{A})$. By means of 2.8 and 2.8' we obtain a nuclear mapping $K_o \in \mathcal{A}'_\beta \hat{\otimes}_\pi \mathcal{H}$ which extends φ. Since $\mathcal{M}$ is a closed subspace of $\mathcal{B}'_\beta \in \mathcal{DF} \subset \mathcal{SW}$ (2.6, [16], ch. III.5), $\mathcal{M}$ is also contained in the class $\mathcal{SW}$. Because of 2.6 we obtain the desired kernel $K \in \mathcal{H} \hat{\otimes}_\pi \mathcal{M}$.

3.1'. Remark. The extension kernel $K : \mathcal{B} \to \mathcal{H}$ defines (in some sense) a continuous right inverse $K_{|\mathcal{A}} : \mathcal{A} \to \mathcal{H}$ of the restriction map $r : \mathcal{H} \to \mathcal{H}_\Delta$, if $\mathcal{A}$ is a space of scalar valued functions and if $(Ta)(\lambda) = a(\lambda)$, $\lambda \in \Delta$; $r \circ K_{|\mathcal{A}} = Id_{\mathcal{A}}$.

§ 4 AN APPLICATION TO SPECTRAL THEORY

4.1. The locally convex spaces $\mathcal{O}$, E and F are assumed to be topological algebras with separately continuous multiplication. We start with a multiplicative map $\varphi \in \mathcal{L}(\mathcal{O}, F)$, which may be considered as a usual operational calculus in several variables. We would like to extend φ to a "bigger" algebra E with a multiplicative mapping $\tilde{\varphi} \in \mathcal{L}(E,F)$. Let $\rho : \mathcal{O} \to E$ be a multiplicative map (in general a "restriction" map, sometimes injective). The search for an extension $\tilde{\varphi} : E \to F$ of $\varphi : \mathcal{O} \to F$ amounts to a factorization $\varphi = \tilde{\varphi} \circ \rho$. E may be considered as an algebra with a partition of unity or partly with a partition of unity. The above factorization depends on φ (kernel ρ) = 0 and on 1) and 2):

1) Density problem of $\rho(\mathcal{O})$ in E;

2) the weak continuity problem in the form:

$\forall y' \in F' \ \exists x' \in E' \ \forall \delta \in \mathcal{O} (=\Delta)$

$$(6.1) \qquad < \rho\delta, x' >_{E,E'} = < \varphi(\delta), y' >_{F,F'}.$$

4.2. Let $\varphi \in \phi(\Delta, E, E'; F, F')$ (cf. 2.1., 2.5.) and Δ total in E. Furthermore assume E quasibarreled, $E'_\beta \in (\mathcal{B}\Gamma)$ and F a complete locally convex space (or E metrizable and F sequentially complete). Then there exists exactly one extension $\tilde{\varphi} \in \mathcal{L}(E,F)$ of φ. If E and F are chosen as in 4.1. and if the linear extension $\overline{\varphi}$ on $[\Delta]$ is multiplicative on a set dense in the linear hull $[\Delta]$ of Δ or on a dense subalgebra of E then $\tilde{\varphi} : E \to F$ is multiplicative.

The following theorem uses the theory of ultradistributions of Roumieu and Björk [6], Th. 18. 14 and Komatsu [19], Th. 9.1, a generalization of the Paley-Wiener-Schwartz theorem. These results can be applied easily to locally m-convex algebras F (F has a fundamental system of submultiplicative seminorms). The motivation for 4.4 goes back to Albrecht [1], 3.3. Other related work has been done by Ljubic and Mazaev [21], Tillmann [24], Waelbroeck [25], Wermer [27], Cioranescu [8] and Bierstedt and Meise [5] p. 337, 8. Satz.

4.3. Let ω be a real valued function of $\mathbb{R}^n$ (cf. Björk [6], 1.3. 22) with $\omega(\xi) = \omega(|\xi|)$ and

(α) $0 = \omega(0) = \lim_{\gamma\to 0} \omega(\gamma) \leq \omega(\xi+\eta) \leq \omega(\xi) + \omega(\eta)$, $\xi, \eta \in \mathbb{R}^n$;

(β) $\int_{\mathbb{R}^n} \omega(\xi) \ (1+|\xi|)^{-n-1} \, d\xi < \infty$;

(γ) $\omega(\xi) \geq c \log (1+|\xi|)$ for some $c > 0$, $\xi \in \mathbb{R}^n$.

(For example: $\omega(\xi) = |\xi|^\alpha = (\sum_{j=1}^n |\xi_j|^2)^{\alpha/2}$, $0 < \alpha < 1$)

In analogy to $\mathcal{E}(\mathbb{R}^n) = C^\infty(\mathbb{R}^n)$, $\omega(\xi) = \log(1+|\xi|)$, Björk [6], 1.5, attaches to each such function ω a locally m-convex (F)-algebra $\mathcal{E}_\omega(R^n)$, which e.g. for $\omega(\xi) = |\xi|^\alpha$, $0 < \alpha < 1$, is smaller than $C^\infty(\mathbb{R}^n)$ but which has a partition of unity because of (β) and which is nuclear by means of (γ) (Berenstein, Dostal [3]).

4.4. Theorem. Let a_j, $j=1,\dots,n$, be pairwise commuting elements of a sequentially complete locally m-convex (topological) algebra F with unit element I. Furthermore we assume that for each $y' \in F'$ there exists $\lambda > 0$ and a constant $C(y',\lambda) > 0$ such that for all $\xi \in \mathbb{R}^n$ the following estimate holds:

$$|\langle e^{i\langle\xi,a\rangle}, y' \rangle_{F,F'}| \leq C(y',\lambda)\, e^{\lambda\omega(\xi)}, \tag{4.2}$$

where $\langle\xi,a\rangle := \sum_{j=1}^{n} \xi_j a_j \in F$.

Then there exists exactly one multiplicative mapping $\tilde{\varphi} \in \mathcal{L}(\mathcal{E}_\omega(\mathbb{R}^n),F)$ with the properties $\tilde{\varphi}(t_j) = a_j$ and $\tilde{\varphi}(e^{\langle\zeta,t\rangle}) = e^{\langle\zeta,a\rangle}$ $\forall\, \zeta \in \mathbb{C}^n$. Furthermore for each $\zeta = (\zeta_1,\dots\zeta_n) \notin \mathbb{R}^n$ there exist pairwise commuting elements $b_j \in F$, $j=1,\dots,n$, which commute with the elements a_j, such that $\sum_{j=1}^{n} (\zeta_j - a_j)\, b_j = I \in F$ is fulfilled.

Proof. In order to apply 4.2 we have to define the mapping $\varphi : \Delta \to F$. Let e_ζ be the function $t \to e^{\langle\zeta,t\rangle}$ defined for every $\zeta \in \mathbb{C}^n$ on all of $\mathbb{R}^n \ni t$. $\Delta = \{e_\zeta,\ \zeta \in \mathbb{C}^n\}$ is a total subset of the nuclear Fréchet algebra $\mathcal{E}_\omega(\mathbb{R}^n)$ (cf. Björk, [6], 1.8.11, 1.8.14). We define

$$\varphi(e_\zeta) := e^{\langle\zeta,a\rangle}, \quad \zeta \in \mathbb{C}^n; \tag{4.3}$$

therefore $\varphi(e_{\pm i\xi}) = e^{\pm i\langle\xi,a\rangle}$, $\xi \in \mathbb{R}^n$. The right hand side of (4.3) is defined, since F is locally m-convex and sequentially complete. Obviously $E'_\beta \in (\mathfrak{B}\Gamma)$ is satisfied. By reason of [6], 1.8.14., a form of the Paley-Wiener theorem for ultra-distributions, we obtain for each $y' \in F'$ the existence of a unique element $u_{y'} \in (\mathcal{E}_\omega(\mathbb{R}^n))'$ (see (4.2)) such that

$$\langle e^{i\langle\zeta,\cdot\rangle}, u_{y'} \rangle_{E,E'} = \langle e^{i\langle\zeta,a\rangle}, y' \rangle_{F,F'} \tag{4.4}$$

is fulfilled. By (4.2) there follows the existence of $\tilde{\varphi} \in \mathcal{L}(\mathcal{E}_\omega(\mathbb{R}^n), F)$ extending $\varphi : \Delta \to F$; obviously the mapping $\tilde{\varphi}$ is multiplicative on the linear hull $[\Delta]$ of Δ; $[\Delta]$ being dense in $\mathcal{E}_\omega(\mathbb{R}^n)$, $\tilde{\varphi} \in \mathcal{L}(\mathcal{E}_\omega(\mathbb{R}^n),F)$ is multiplicative. Since the differential quotient $\frac{\partial}{\partial\zeta_j} e^{\langle\zeta,t\rangle}$ exists pointwise on $\mathcal{E}_\omega(\mathbb{R}^n)'$ ($\mathcal{E}_\omega(\mathbb{R}^n)'$ is barreled), this differential quotient exists on $\mathcal{E}_\omega(\mathbb{R}^n)'_\beta$ (by nuclearity). Now from the continuity of $\tilde{\varphi} : \mathcal{E}_\omega(\mathbb{R}^n) \to F$ follows $\tilde{\varphi}(t_j\, e^{\langle\zeta,t\rangle}) = a_j\, e^{\langle\zeta,a\rangle}$ (F is locally m-convex). For $\zeta = 0$ we get $\tilde{\varphi}(t_j) = a_j$. Since $\mathcal{E}_\omega(\mathbb{R}^n)$ contains all real analytic functions (Björk [6], 1.5.15) we put

$$b_j = \tilde{\varphi}\,((\overline{\zeta_j} - t_j)\,(\sum_{j=1}^{n} |\zeta_j - t_j|^2)^{-1}) \qquad \text{(cf. [1])},$$

$t = (t_1,\ldots,t_n) \in \mathbb{R}^n$, $\zeta = (\zeta_1,\ldots,\zeta_n) \notin \mathbb{R}^n$; this implies $\sum_{j=1}^{n} (\zeta_j - a_j)\, b_j = I$.

4.5. Remark. The integration of the relation 4.4 with $f \in \mathcal{D}_\omega(\mathbb{R}^n)$ (Björk [6], 1.4.1) and $\hat{f}$ the Fourier transform of f

$$(4.4') \qquad \frac{1}{(2\pi)^{n/2}} \int_{\mathbb{R}^n} \hat{f}(\xi)\, \langle e^{i\langle\xi,\cdot\rangle}, u_{y'}\rangle_{E,E'}\, d\xi = \frac{1}{(2\pi)^{n/2}} \int_{\mathbb{R}^n} \hat{f}(\xi)\, \langle e^{i\langle\xi,a\rangle}, y'\rangle_{F,F'}\, d\xi$$

suggests (leads to) an integral representation of $\tilde{\varphi}$ on $\mathcal{D}_\omega(\mathbb{R}^n)$

$$(4.5) \qquad \tilde{\varphi}\,(f) = \frac{1}{(2\pi)^{n/2}} \int_{\mathbb{R}^n} \hat{f}(\xi)\, e^{i\langle\xi,a\rangle}\, d\xi.$$

For further applications see [14].

LITERATURE

[1] Albrecht, E.: Funktionalkalküle in mehreren Veränderlichen für stetige lineare Operatoren auf Banachräumen. Manuscripta Math. 14, 1-40 (1974)

[2] Barros-Neto, J.: Spaces of vector valued real analytic functions. Trans. Am. Math. Soc. 112, 381-391 (1964)

[3] Berenstein, C.A., M.A. Dostal: Topological properties of analytically uniform spaces. Trans. Am. Math. Soc. 154, 493-513 (1971)

[4] Bierstedt, K.D., B. Gramsch, R. Meise: Approximationseigenschaft, Lifting und Kohomologie bei lokalkonvexen Produktgarben. Manuscripta Math. 19, 319-364 (1976)

[5] Bierstedt, K.D., R. Meise: Distributionen mit Werten in topologischen Vektorräumen II. Manuscripta Math. 10, 313-357 (1973)

[6] Björk, G.: Linear partial differential operators and generalized distributions. Arkiv f. Mat. 6, 351-407 (1966)

[7] Bungart, L.: Holomorphic functions with values in locally convex spaces and applications to integral formulas. Trans. Am. Math. Soc. 111, 317-344 (1964)

[8] Cioranescu, I.: Operator-valued ultradistributions in the spectral theory. Math. Ann. 223, 1-12 (1976)

[9] De Wilde, M.: Réseaux dans les espaces à semi-normes. Mem. Soc. Roy. Sci. Liége 18, fasc. 2 (1969)

[10] Gramsch, B.: Funktionalkalkül mehrerer Veränderlichen in lokalbeschränkten Algebren. Math. Ann. 174, 311-344 (1967).

[11] Gramsch, B.: Inversion von Fredholmfunktionen bei stetiger und holomorpher Abhängigkeit von Parametern. Math. Ann. 214, 95-147 (1975)

[12] Gramsch, B.: Über das Cauchy-Weil-Integral für Gebiete mit beliebigem Rand. Archiv der Math. 28, 409-421 (1977)

[13] Gramsch, B.: Ein Schwach-Stark-Prinzip der Dualitätstheorie lokalkonvexer Räume als Fortsetzungsmethode. To appear in Math. Zeitschrift

[14] Gramsch, B.: Über eine Fortsetzungsmethode der Dualitätstheorie lokalkonvexer Räume. Ausarbeitung, Kaiserslautern 1976

[15] Grothendieck, A.: Produits tensoriels topologiques et espaces nucléaires Am. Math. Soc. Mem. 16 (1955)

[16] Grothendieck, A.: Topological vector spaces. Gordon and Breach, New York 1973

[17] Horváth, J.: Topological vector spaces and distributions. Addison Wesley 1966

[18] Horváth, J.: Finite parts of distributions. Proc. Conf. Linear Operators and Approximation. Birkhäuser-Verlag, Basel 1972, S. 142-158

[19] Komatsu, H.: Ultradistributions, I. Structure theorems and characterization. J. Fac. Sci. Tokyo, Sec. IA 20, 25-105 (1973)

[20] Köthe, G.: Topologische lineare Räume. Springer-Verlag 1960

[21] Ljubic, J.I., V.I. Mazaev: On operators with separable spectrum. AMS Transl. Ser. 2, 47, 89-129 (1972)

[22] Nachbin, L.: On vector valued versus scalar valued holomorphic continuation. Indag. Math. 35, 352-354 (1973).

[23] Schaefer, H.H.: Topological vector spaces. McMillan, New York 1966

[24] Tillmann, H.G.: Eine Erweiterung des Funktionalkalküls für lineare Operatoren. Math. Ann. 151, 424-430 (1963)

[25] Waelbroeck, L.: Calcul symbolique lié à la croissance de la résolvante. Rend. Sem. Math. Fis. Milano 34, 3-24 (1964)

[26] Waelbroeck, L.: Weak analytic functions and the closed graph theorem. Proc. on infinite dimensional holomorphy, Kentucky 1973. Springer Lecture Notes in Math. 364, 97-100 (1974)

[27] Wermer, J.: The existence of invariant subspaces. Duke Math. J. 19, 615-622 (1952)

[28] Wrobel, C.: Extension du calcul fonctionnel holomorphe et application à l'approximation. C. R. Acad. Sc. Paris 275, 175-177 (1972)

K.-D. Bierstedt, B. Fuchssteiner (eds.)
Functional Analysis: Surveys and Recent Results

LIFTING THEOREMS FOR VECTOR VALUED FUNCTIONS AND THE ε - TENSOR PRODUCT

Winfried Kaballo

Fachbereich Mathematik
Universität Kaiserslautern
D-675 Kaiserslautern, Pfaffenbergstraße 95

Lifting problems for vector valued functions and questions of parameter dependence of solutions of linear equations are studied. This is done by developing a rather general theory of lifting for ε - tensor products, since many spaces of vector valued functions can be interpreted as the complete ε - tensor product of the corresponding space of scalar functions and the locally convex space in which the values are taken. Applications are given to partial differential equations and to a weak-strong extension problem.

INTRODUCTION

The following lifting problem is studied: Let E and Q be locally convex spaces and let $\pi: E \to Q$ be a linear, continuous and surjective map. Let Λ be a set and $f: \Lambda \to Q$ be a function which belongs to a given function space $F(\Lambda, Q)$. The problem is to find a function $g: \Lambda \to E$, belonging to another given function space $G(\Lambda, E)$, that is a lifting of f, i.e. satisfies $\pi g(\lambda) = f(\lambda)$ for all $\lambda \in \Lambda$.

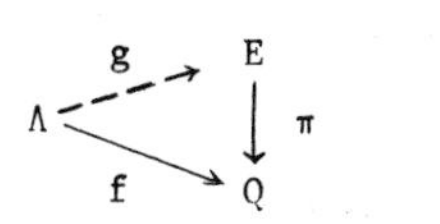

Such problems occur in the investigation of parameter dependences of solutions of linear equations, for instance of differential equations. So π may be a surjective differential operator $P(D): \mathcal{D}'(\Omega) \to \mathcal{D}'(\Omega)$, while $F(\Lambda, \mathcal{D}'(\Omega)) = G(\Lambda, \mathcal{D}'(\Omega)) = C(\Lambda, \mathcal{D}'(\Omega))$

are the spaces of continuous $\mathcal{D}'(\Omega)$ - valued functions. Lifting problems also arise in the theory of Fredholm functions (cf. [8], [11]), integral representation formulas (cf. [7],[9]) and of "weak-strong extension" problems (cf. [10]).

First results concerning the lifting of continuous functions in the case of Banach and (F)-spaces are due to R.G. Bartle - R.M. Graves [1] and E. Michael [23]. A. Grothendieck proved general lifting theorems in connection with nuclearity [13] and the metric theory of tensor products [12]. More recent results are due to B. Gramsch [8] and K. Bierstedt - B. Gramsch - R. Meise [2] for "mixed" functions, P.A. Kučment [19] for holomorphic functions with continuous boundary values, B. Gramsch [9] for bounded holomorphic functions and D. Vogt [30], [31] for tensor products of type (F)-(DF). J. Leiterer [20] developed a method which can be used to prove lifting theorems for variable surjections $\pi(\lambda): E \to Q$ (cf. 4.).

This article is based on parts of the author's Habilitationsschrift [17].

1. (εL) - SPACES

If F and E are locally convex spaces, we denote by $F \varepsilon E = L_e(E'_c, F)$ their ε-product and by $F \hat{\otimes}_\varepsilon E$ their complete ε-tensorproduct, here by $'_c$ the topology of uniform convergence on all compact absolutely convex subsets of E is meant (cf. [27],§1). By a theorem of L. Schwartz [27],§1, a complete space F has the (Schwartz) approximation property (a.p.), iff $F \varepsilon E = F \hat{\otimes}_\varepsilon E$ for every complete space E.

Since for many function spaces we have $F(\Lambda, Q) = F(\Lambda) \hat{\otimes}_\varepsilon Q$ or $= F(\Lambda) \varepsilon Q$ (this is true for instance for all spaces occuring in 1.6.), the lifting problem stated in the introduction can often be restated as follows:

<u>1.1. Problem.</u> Let F, E, Q be locally convex spaces and let $\pi: E \to Q$ be a surjection. When are the induced maps $id \hat{\otimes}_\varepsilon \pi: F \hat{\otimes}_\varepsilon E \to F \hat{\otimes}_\varepsilon Q$ and $id\, \varepsilon\, \pi : F \varepsilon E \to F \varepsilon Q$ surjective, too?

1.2. Definition. A locally convex space F is called (εL)-space, if for every surjection $\pi: E \to Q$ of Banach spaces E and Q $id \hat{\otimes}_\varepsilon \pi: F \hat{\otimes}_\varepsilon E \to F \hat{\otimes}_\varepsilon Q$ is surjective, too.

To characterize the Banach-(εL)-spaces we start with a well known

1.3. Lemma. Let F, E and Q be Banach spaces and $\pi: E \to Q$ a surjection. Suppose that there exists $C > o$ such that for every $x \in F \otimes Q$ there is $y \in F \otimes E$ satisfying $(id \otimes \pi)y = x$ and $||y|| \leq C||x||$. Then $id \hat{\otimes}_\varepsilon \pi: F \hat{\otimes}_\varepsilon E \to F \hat{\otimes}_\varepsilon Q$ is surjective.

The following characterization of Banach-(εL)-spaces is closely related to the well known extension- and lifting theorems for compact operators (cf. [21], [22],II.5.f.). We first recall the definition of $\mathcal{L}_p$ - spaces (cf. [22],II.5.):

1.4. Definition. Let $1 \leq p \leq \infty$ and let $\lambda \geq 1$. A Banach space X is said to be a $\mathcal{L}_{p,\lambda}$ - space if for every finite dimensional subspace B of X there is a finite dimensional space C such that $B \subseteq C \subseteq X$ and $d(C, \ell_p^r) \leq \lambda$, $r = \dim C$; here $d(X,Y) = \inf \{ ||T|| \, ||T^{-1}|| \mid T: X \to Y$ is an isomorphism $\}$ denotes the Banach-Mazur distance of two Banach spaces.
X is said to be a $\mathcal{L}_p$ - space if it is a $\mathcal{L}_{p,\lambda}$ - space for some $\lambda < \infty$.

1.5. Theorem. A Banach space F is an (εL)-space iff it is a $\mathcal{L}_\infty$-space.

Proof. a) Let F be a $\mathcal{L}_{\infty,\lambda}$ - space, $\lambda \geq 1$; then F has the a.p.. Take $x \in F \otimes Q \subseteq L_e(Q_c', F)$; then $\dim R(x) < \infty$. So there is a subspace C of F such that $R(x) \subseteq C$, $\dim C < \infty$, and an isomorphism $T: C \to \ell_\infty^k$ satisfying $||T|| \, ||T^{-1}|| \leq \lambda+\alpha$, $\alpha > o$ arbitrary. Define $\delta_n: \ell_\infty^k \to \mathbb{K}$ ($\mathbb{K} = \mathbb{R}$ or $\mathbb{C}$) by $\delta_n(\xi_j)_{j=1}^k = \xi_n$. Since $(Q_c')' = Q$, the functional $\delta_n Tx: Q_c' \to \mathbb{K}$ may be represented by a vector $q_n \in Q$ as follows: $\delta_n Tx(q') = \langle q_n, q' \rangle$ for all $q' \in Q'$. Take $e_n \in E$ such that $\pi e_n = q_n$ and $||e_n|| \leq (\omega+\alpha)||q_n||$, where ω is the norm of the isomorphism $\tilde{\pi}^{-1}: Q \to E/N(\pi)$. By $z(e') := (\langle e_n, e' \rangle)_{n=1}^k$ a map $z: E' \to \ell_\infty^k$ satisfying $z|_{Q'} = Tx$ is defined; $y := T^{-1}z$ then satisfies $y \in F \otimes E$, $||y|| \leq (C+\alpha)(\omega+\alpha)||x||$ and $(id \otimes \pi)y = x$, hence the assertion follows by Lemma 1.3..

b) Now suppose that F is an (εL)-space. Claim: There is $\lambda > o$ such that for every quotient map $\pi: Y \to X$ of finite dimensional spaces X and Y every $x \in F \otimes_\varepsilon X$ can be lifted to $y \in F \otimes_\varepsilon Y$ such that $||y|| \leq \lambda||x||$. If not, there would exist $\pi_n: Y_n \to X_n$ and $x_n \in F \otimes X_n$ such that $||x_n|| = 1$ and every lifting $y_n \in F \otimes Y_n$

of x_n satisfies $||y_n|| \geq n^3$. Put $Y = (\bigoplus_n Y_n)_{\ell_1}$ and $X = (\bigoplus_n X_n)_{\ell_1}$; then $\pi: Y \to X$, $\pi(\eta_n) := (\pi_n \eta_n)$, is a quotient map. Define $x: F'_c \to X$ by $x(f') := (\frac{1}{n^2} x_n(f'))$; then $x \in F \hat{\otimes}_\varepsilon X$, so by assumption there is a lifting $y \in L_e(F'_c, Y)$ of x. If now $P_n: Y \to Y_n$ is the canonical projection, the map $n^2 P_n y: F'_c \to Y_n$ is a lifting of x_n with norm $\leq n^2||y||$; but this is a contradiction for large n.

Now by [12],p.11, for every finite dimensional Banach space X there is an isometry $(F \otimes_\varepsilon X)' = F' \otimes_\pi X'$; therefore dualization of the above claim implies the following: For every isometric inclusion $\iota: X \to Y$ of finite dimensional spaces and every $x \in F' \otimes X$ the inequality $||x||_{F' \otimes_\pi X} \leq \lambda ||\iota x||_{F' \otimes_\pi Y}$ holds. By definition of the π-norm, this inequality then holds for every isometric inclusion $\iota: X \to Y$ of arbitrary Banach spaces, which means that $(id \otimes_\pi \iota): F' \otimes_\pi X \to F' \otimes_\pi Y$ is always a topological inclusion. However, this implies that F' is a $\mathcal{L}_1$-space, which was noted in [28] and [6]: Another dualization yields the surjectivity of the restriction maps $L(Y,F'') \to L(X,F'')$, which means that F'' is an injective Banach space. Therefore F' is a $\mathcal{L}_1$-space, and thus F is a $\mathcal{L}_\infty$-space (cf. [22],II.5.).

1.6. Remarks and Examples.

(i) An inspection of part b) of the above proof shows that it is sufficient to assume that for every surjection $\pi: E \to Q$ of Banach spaces every $x \in F \hat{\otimes}_\varepsilon Q$ can be lifted to $F \varepsilon E$.

(ii) Examples of $\mathcal{L}_\infty$-spaces are the spaces C(K) of all continuous functions on a compact Hausdorff space K and A(S) of all affine continuous functions on a compact Choquet-Simplex S ([22],p. 199 and 164).

(iii) If Ω is a compact C^∞-manifold of dimension ≥ 2, then for $1 \leq k < \infty$ $C^k(\Omega)$, the space of all k times continuously differentiable functions on Ω, is not a $\mathcal{L}_\infty$-space, as may be deduced from results of G.M. Henkin [14] (cf. [17],§4). However, the spaces $\lambda^{k,\alpha}(\Omega), 0<\alpha<1$, of all C^k-functions whose derivatives of order k satisfy a o-Hölder condition of order α on Ω are $\mathcal{L}_\infty$-spaces, as was shown by R. Bonic, J. Frampton and A. Tromba [3], [5].

(iv) The disc algebra $\mathcal{A}(D)$ is not a $\mathcal{L}_\infty$-space (A. Pelczynski [24]), which implies a negative answer to a question of B. Gramsch [8], p.103, concerning the lifting of holomorphic functions with continuous boundary values. This negative answer has also been obtained by P.A. Kučment and V.I. Ovčinnikov (private communication).

Concerning the lifting of holomorphic functions with growth conditions it can be shown (cf. [17],§5) that the spaces $H_o^\alpha(B) := \{ f \in H(B) \mid \lim_{|z| \to 1} |f(z)| \operatorname{dist}(z, \partial B)^\alpha = o \}$, $o < \alpha < \infty$, are $\mathcal{L}_\infty$-spaces; here B denotes the unit ball in $\mathbb{C}^n$.

Next it will be shown that for rather general locally convex spaces E and Q and surjections $\pi: E \to Q$ the ε-(tensor)product with an identity on a $\mathcal{L}_\infty$-space is surjective, too. A space E is said to be K-complete if the absolutely convex hull of every compact subset of E is compact.

<u>1.7. Theorem.</u> Let E be a K-complete space having a strict web in the sense of M. De Wilde [4], and suppose that in Q every compact subset is very compact ([4],III.4.). Then, if $\pi: E \to Q$ is a surjection, $(id \,\varepsilon\, \pi): F \,\varepsilon\, E \to F \,\varepsilon\, Q$ is surjective, too, for every $\mathcal{L}_\infty$-space F.

<u>Proof.</u> Take $x \in F \,\varepsilon\, Q = L_e(F'_c, Q)$. Since $U^o := \{ f' \in F' \mid \|f'\| \leq 1\}$ is compact in F'_c, $x(U^o)$ is compact in Q, so by assumption it is even very compact. This means ([4],III.4.8.) that there is an absolutely convex compact set $K \subseteq Q$ such that $x(U^o)$ is compact in Q_K. By De Wilde's lifting theorem ([4],III.5.) there exists an absolutely convex compact set $C \subseteq E$ such that $\pi C = K$; hence $\pi: E_C \to Q_K$ is a quotient map. Since Q and Q_K induce the same topology on $x(U^o)$, the restrictions of $x: F'_c \to Q_K$ to the equicontinuous subsets of F' are continuous; but then $x: F'_c \to Q_K$ is continuous, since the $'_c$-topology is the strongest locally convex topology on F' that coincides with the $'_c$- (or weak*-) topology on all equicontinuous subsets of F'. Thus we even have $x \in L_e(F'_c, Q_K) = F \,\varepsilon\, Q_K = F \hat{\otimes}_\varepsilon Q_K$, and by Theorem 1.5. we get a lifting $y \in F \,\varepsilon\, E_C \subseteq F \,\varepsilon\, E$ of x.

<u>1.8. Remarks and Examples.</u>

(i) Theorem 1.7. is only a special case of more general results of the same type (cf. [17],§1).

(ii) The class of spaces with strict web contains all (F)- and sequentially complete (DF)-spaces and is closed under the formation of countable projective and inductive limits (cf. [4]).

(iii) The condition on Q means precisely that Q'_c be a Schwartz space. It is satisfied by (F)- and (DFS)-spaces, or more generally by strong duals of quasinormable infrabarreled spaces. Moreover, it is inherited by compact-regular inductive limits and by countable projective limits with partition of unity (cf. [17],§1).

(iv) In particular, Theorem 1.7. applies to the surjection $P(D)\colon \mathcal{D}'(\Omega) \to \mathcal{D}'(\Omega)$ mentioned in the introduction.

Now (F)- and (DF)-(εL)-spaces will be considered:

1.9. Theorem. A countable dense projective limit $F = \varprojlim_n F_n$ of $\mathcal{L}_\infty$-spaces F_n is an (εL)-space. Even for surjections $\pi\colon E \to Q$ of (F)-spaces $\mathrm{id}\,\hat{\otimes}_\varepsilon\,\pi\colon F\,\hat{\otimes}_\varepsilon\,E \to F\,\hat{\otimes}_\varepsilon\,Q$ is surjective.

Proof. The (F)-space F has the a.p. since all F_n do (cf. [26],p.109). Let $\rho_n\colon F \to F_n$ and $\rho_{n,n+k}\colon F_{n+k} \to F_n$ be the canonical maps and let $x \in F\,\hat{\otimes}_\varepsilon\,Q = F\,\varepsilon\,Q$ be given. An application of Theorem 1.7. to $\rho_n x \in L_e(Q'_c,F_n)$ produces extensions $z_n \in L_e(E'_c,F_n)$. First put $y_1 := z_1$. Since $y_1 - \rho_{12}z_2$ vanishes on Q', we have $y_1-\rho_{12}z_2 \in L(E'_c/Q',F_1)$, hence $y_1-\rho_{12}z_2 \in L_e(U'_c,F_1) = F_1\,\hat{\otimes}_\varepsilon\,U$, where U is the kernel of π. Now let $\{p_j\}$ be a non decreasing system of semi-norms generating the topology of E and put $r_j := \|\ \|_j\,\hat{\otimes}_\varepsilon\,p_j$. Since by assumption $\rho_{12}F_2$ is dense in F_1, $\rho_{12}F_2 \otimes U$ is dense in $F_1\,\hat{\otimes}_\varepsilon\,U$; hence there exists $u_2 \in F_2 \otimes U$ such that $r_1(y_1-\rho_{12}z_2-\rho_{12}u_2) \le \delta_1$, $\delta_1 > 0$ arbitrary. Then $y_2 := z_2+u_2$ satisfies $(\mathrm{id}\,\hat{\otimes}_\varepsilon\,\pi)y_2 = \rho_2 x$ and $r_1(y_1-\rho_{12}y_2) \le \delta_1$. Now we continue in the same way: If $\|\rho_{n,n+1}\| \le C_n$, we put $\delta_n := 2^{-n}(\prod_{i=1}^{n} C_i)^{-1}$ and find $y_n \in F_n\,\hat{\otimes}_\varepsilon\,E$ such that $(\mathrm{id}\,\hat{\otimes}_\varepsilon\,\pi)y_n = \rho_n x$ and $r_n(y_n-\rho_{n,n+1}y_{n+1}) \le \delta_n$ for all $n \in \mathbb{N}$. But then for every fixed $m \in \mathbb{N}$ the series

$$y_m + (\rho_{m,m+1}y_{m+1} - y_m) + (\rho_{m,m+2}y_{m+2} - \rho_{m,m+1}y_{m+1}) + \ldots$$

converges in $F_m\,\hat{\otimes}_\varepsilon\,E = L_e(E'_c,F_m)$; call the limit w_m. Obviously, for every $m \in \mathbb{N}$ we have $(\rho_{m,m+1}\,\hat{\otimes}_\varepsilon\,\mathrm{id})w_{m+1} = w_m$, hence an element $w \in L_e(E'_c,F) = F\,\hat{\otimes}_\varepsilon\,E$ is defined which satisfies $w|_{Q'} = x$, i.e. $(\mathrm{id}\,\hat{\otimes}_\varepsilon\,\pi)w = x$.

1.10. Examples and Remarks.

(i) Let Ω be a region in $\mathbb{R}^n$. Then the space $C(\Omega)$ of all continuous functions on Ω and the spaces $\lambda^{k,\alpha}(\Omega)$ of all C^k-functions on Ω whose derivatives of order k satisfy o-Hölder conditions on compact subsets of Ω (cf. 1.6.(iii)) satisfy the condition of Theorem 1.9. Other examples are certain spaces of continuous or holomorphic functions with growth conditions (cf. [17]).

(ii) Theorem 1.9. also covers the case of nuclear spaces F, since for instance every Hilbert-Schmidt map $T\colon \ell_2 \to \ell_2$ with dense range may be densely factored through c_o. Of course, the theorem is well known in this case, however, perhaps the proof is interesting since it avoids the π-tensor product and also covers the different looking cases of (i).

(iii) There are (FS)-spaces with a.p. which are not (εL)-spaces (cf. [17],§4), though by a result of D. Randtke [25] every (FS)-space is the countable projective limit of spaces isomorphic to c_o; of course, this limit is not dense.

<u>1.11. Theorem.</u> Let $F = \underset{n\longrightarrow}{\lim} F_n$ be a separated countable compact inductive limit of the $\mathcal{L}_\infty$-spaces F_n. Let E be a K-complete space with strict web and let Q be a (DF)-space. Then, if $\pi: E \to Q$ is a surjection, $(id\ \varepsilon\ \pi): F\ \varepsilon\ E \to F\ \varepsilon\ Q$ is surjective, too.

<u>Proof.</u> Since $F = \underset{n\longrightarrow}{\lim} F_n$ is a compact inductive limit, we have $F'_c = F'_b = \underset{\longleftarrow}{\lim}_n F'_n$. If $x \in F\ \varepsilon\ Q = L_e(F'_c, Q)$ is given, then x is a bounded mapping, since F'_c is an (F)-space and Q has a countable fundamental family of bounded sets. So, if the canonical maps of the inductive limit are denoted by $\iota_n: F_n \to F$, we get a factorization $x = \xi\iota'_m$, where $\xi: F'_m \to Q$ is a bounded linear map, for some $m \in \mathbb{N}$. Let $\iota_{m,m+1}: F_m \to F_{m+1}$ be the canonical maps, and put $V = \{ f' \in F'_{m+1} \mid \|f'\| \le 1 \}$. Then $\iota'_{m,m+1}(V)$ is compact in F'_m, hence

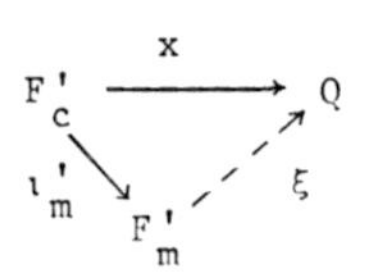

$\xi\iota'_{m,m+1}(V)$ is in Q very compact. Since E is a space with strict web, by De Wilde's lifting theorem [4],III.5., already used in the proof of 1.7., there is an absolutely convex compact set $C \subseteq E$ such that $\pi C = \xi\iota'_{m,m+1}(V) =: K$, and thus $\pi: E_C \to Q_K$ is a quotient map. Now $\xi\iota'_{m,m+2}: F'_{m+2} \to Q_K$ is compact, and since F'_{m+2} is a $\mathcal{L}_1$-space it may be lifted to a compact mapping $\eta: F'_{m+2} \to E_C$ (cf. [22],p. 221). We then put $y := \eta\iota'_{m+2}: F'_c \to E_C$ and find $\pi y = \pi\eta\iota'_{m+2} = \xi\iota'_{m,m+2}\iota'_{m+2} = \xi\iota'_m = x$, so $y \in F\ \varepsilon\ E_C \subseteq F\ \varepsilon\ E$ is a lifting of x.

<u>1.12. Remarks.</u> Of course, Theorem 1.11. covers the case of complete nuclear (DF)-spaces F. This case essentially has been treated by B. Gramsch [10],2.6., by use of a nuclear series development. Finally we note that in Theorem 1.11. the condition on Q is milder than it is in Theorem 1.7., a fact which is due to the compactness of the inductive limit, of course, and may be useful in concrete cases (cf. [17], 3.7., 4.9., 5.12.).

2. (εL) - TRIPLES

Until now only conditions on the space F ensuring the surjectivity of $id \hat{\otimes}_\varepsilon \pi$ for rather large classes of surjections π have been considered. Now, conversely, conditions on the surjection π will be investigated that imply the surjectivity of $id \hat{\otimes}_\varepsilon \pi$ for hopefully large classes of spaces F. Since these conditions may depend on properties of E, Q or the kernel U of π, the following definition turns out to be sensible:

2.1. Definition. A short exact sequence $o \to U \to E \overset{\pi}{\to} Q \to o$ is said to be an (εL)-triple, if for every Banach space F $id \hat{\otimes}_\varepsilon \pi : F \hat{\otimes}_\varepsilon E \to F \hat{\otimes}_\varepsilon Q$ is surjective, too.

The Banach-(εL)-triples were characterized by A. Grothendieck [12], p.27, 76 5), as the author has been told by P. Michor; cf. also the lecture notes of V. Lohsert and P. Michor on Grothendieck's work [12].

2.2. Theorem. A short exact sequence $o \to U \to E \overset{\pi}{\to} Q \to o$ **of Banach spaces is an (εL)-triple iff $U^o = Q' \subseteq E'$** is complemented in E'.

Proof. a) Suppose that $U^o = Q'$ is complemented in E' and let P be a projection of E' onto Q'. In order to show that for an arbitrary Banach space F $(id \hat{\otimes}_\varepsilon \pi): F \hat{\otimes}_\varepsilon E \to F \hat{\otimes}_\varepsilon Q$ is surjective, it suffices to prove that the adjoint map $(id \hat{\otimes}_\varepsilon \pi)': (F \hat{\otimes}_\varepsilon Q)' \to (F \hat{\otimes}_\varepsilon E)'$ is a topological injection (cf. [26], IV.7.9.). However, $(F \hat{\otimes}_\varepsilon Q)' = I(F,Q')$ is even complemented in $I(F,E') = (F \hat{\otimes}_\varepsilon E)'$ (here I denotes the ideal of integral operators, cf. [12], [13]), since $\hat{P}: I(F,E') \to I(F,Q')$, $\hat{P}(A) := PA$, is a projection.

b) Now suppose that $o \to U \to E \overset{\pi}{\to} Q \to o$ is an (εL)-triple. Claim: There is $\lambda > o$ such that for every finite dimensional space X and every tensor $x \in X \otimes_\varepsilon Q$ there is a lifting $y \in X \otimes_\varepsilon E$ such that $||y|| \leq \lambda ||x||$. If not, there would exist spaces X_n and tensors $x_n \in X_n \otimes_\varepsilon Q$, $||x_n|| = 1$, such that every lifting of x_n has norm $\geq n^3$. Put $X := (\bigoplus_n X_n)_{\ell_2}$, then $X' = (\bigoplus_n X_n')_{\ell_2}$; let $P_n : X' \to X_n'$ be the canonical projection. Define $x: X' \to Q$ by $x(x') := \sum_{n=1}^{\infty} \frac{1}{n^2} x_n(P_n x')$; then $x \in X \hat{\otimes}_\varepsilon Q$ can be lifted to $y \in X \hat{\otimes}_\varepsilon E \subseteq K(X',E)$. But then $n^2 y P_n : X_n' \to E$ is a lifting of x_n which has norm $\leq n^2 ||y||$, which gives a contradiction for large n.

So our claim is proved. Now by [12], p.11, for every finite dimensional space X $(X \hat{\otimes}_\varepsilon Y)'$ is isometric to $X' \hat{\otimes}_\pi Y'$, which means precisely the accessibility of the π-norm. Therefore, dualization of the claim gives the following: For every finite dimensional space X $(id \otimes_\pi \pi'): X \otimes_\pi Q' \to X \otimes_\pi E'$ is an isomorphism such that $\| \ \|_{X \otimes_\pi Q'} \le \lambda \| \ \|_{X \otimes_\pi E'}$. Since $\| \ \|_{F \otimes_\pi E'} = \inf\{\| \ \|_{X \otimes_\pi E'} \mid X \subseteq F, \dim X < \infty\}$, this implies that for every Banach space F $(id \otimes_\pi \pi'): F \otimes_\pi Q' \to F \otimes_\pi E'$ is a topological isomorphism. Another dualization yields the surjectivity of the restriction $\rho: L(E',F') \to L(Q',F')$, and the assertion follows by taking $F = Q$.

In order to find interesting examples of Banach-(εL)-triples the following concept, suggested by Kučment's work [19], is useful:

2.3. Definition. Let U be a subspace of the Banach space E. The inclusion map $\iota: U \to E$ is said to be approximatively left invertible (a.l.i.), if there is a constant $C > 0$ and a net $\{\ell_\alpha\}_{\alpha\in A}$ of operators $\ell_\alpha: E \to U$ such that $\|\ell_\alpha\| \le C$ and $\ell_\alpha u \to u$ for all $u \in U$.

2.4. Examples.

(i) If U is a $\mathcal{L}_\infty$-space, then every inclusion $\iota: U \to E$ is a.l.i.. For, if U is a $\mathcal{L}_{\infty,\lambda}$-space and B is a finite dimensional subspace of U such that $d(B,\ell_\infty^k) \le \lambda+1$, there is a projection P_B of E onto B with norm $\le \lambda+1$, since ℓ_∞^k is a $\mathcal{P}_1$-space. Then the net $\{P_B\}$ satisfies the requirements of 2.3. (cf. Definition 1.4.).

(ii) Let X and Y be Banach spaces such that X' or Y has the bounded approximation property (b.a.p.). Then the inclusion $\iota: K(X,Y) \to L(X,Y)$ of the space of compact operators into the space of all operators from X to Y is a.l.i..

First, suppose that Y has the b.a.p.. Then there is a net $\{F_\alpha\}$ of finite dimensional operators on Y such that $\|F_\alpha\| \le C$ and $F_\alpha y \to y$ for all $y \in Y$. Then we just define $\ell_\alpha: L(X,Y) \to K(X,Y)$ by $\ell_\alpha(A) := F_\alpha A$ and get $\|\ell_\alpha\| \le C$ and $\ell_\alpha(K) = F_\alpha K \to K$ for all $K \in K(X,Y)$ because of $F_\alpha \to id_Y$ in $L_c(Y)$.

Now suppose that X' has the b.a.p.. Then there is a net $\{G_\alpha\}$ of finite dimensional operators on X' such that $\|G_\alpha\| \le C$ and $G_\alpha x' \to x'$ for all $x' \in X'$. There exist finite dimensional operators H_α on X such that $H_\alpha' = G_\alpha$. Define $\ell_\alpha: L(X,Y) \to K(X,Y)$ by

$\ell_\alpha(A) := AH_\alpha$; then $||\ell_\alpha|| \le C$ and $||K - KH_\alpha|| = ||K' - H_\alpha' K'|| = ||K' - G_\alpha K'|| \to o$ for all $K \in K(X,Y)$.

(iii) If X is a Banach space with b.a.p., then the inclusion $\iota: X \to X''$ is a.l.i..
For, if $\{F_\alpha\}$ is as in (ii), we just put $\ell_\alpha := F_\alpha'': X'' \to X$, and the conditions of 2.3. are satisfied.

<u>2.5. Proposition.</u> If $\iota: U \to E$ is an a.l.i. inclusion and $Q := E/U$, then the triple $o \to U \to E \to Q \to o$ is an (εL)-triple.

<u>Proof.</u> Consider the net $\{\ell_\alpha'\}$ in $L(U',E')$. Since $||\ell_\alpha'|| \le C$, there is a subnet $\{\ell_\beta'\}$ of $\{\ell_\alpha'\}$ converging pointwise weak* to a limit $\rho \in L(U',E')$. Then for all $u \in U$ and $u' \in U'$ we have
$\langle u, \iota'\rho u'\rangle = \langle \iota u, \rho u'\rangle = \lim \langle \iota u, \ell_\beta' u'\rangle = \lim \langle \ell_\beta u, u'\rangle = \langle u, u'\rangle$,
hence $\iota'\rho = \mathrm{id}$. So $U^o = Q' = N(\iota')$ is complemented in E', and the assertion follows from Theorem 2.2..

<u>2.6. Remarks.</u>

(i) Dually to 2.3. there is a notion of approximatively right invertible surjections $\pi: E \to Q$ (a.r.i.). If $\pi: E \to Q$ is a.r.i., then $o \to U \to E \to Q \to o$ is an (εL)-triple, and π is always a.r.i. if Q is a $\mathcal{L}_1$-space, cf. [17], §2.

(ii) Proposition 2.5. and the corresponding assertion for a.r.i. surjections may also be proved very elementarily, without use of Grothendieck's theory of integral operators (cf. [17],2.5.).

(iii) Example 2.4.(ii) in combination with Proposition 2.5. is important for the regularization of Fredholm functions (cf. [17], §8).

As it is suggested by 2.4.(iii), the difference of the notions "a.l.i." and "(εL)-triple" is essentially due to the possible absence of the b.a.p.. This and a "concrete" lifting test are established in the next

<u>2.7. Proposition.</u> Let $o \to U \xrightarrow{\iota} E \xrightarrow{\pi} Q \to o$ be a short exact sequence of Banach spaces such that U has the b.a.p.. Then the following assertions are equivalent:

(1) $\iota: U \to E$ is a.l.i.

(2) $U^o = Q'$ is complemented in E'.

(3) $(\mathrm{id}\ \hat{\otimes}_\varepsilon\ \iota'): U\ \hat{\otimes}_\varepsilon\ E' \to U\ \hat{\otimes}_\varepsilon\ U'$ is surjective.

<u>Proof.</u> We only have to show the implication (3) $\to$ (1). Since U has the b.a.p., there is a net $\{h_\alpha\} \subseteq U' \otimes U$ such that

$||h_\alpha|| \leq C$ and $h_\alpha u \to u$ for all $u \in U$. Since the canonical map $E' \hat{\otimes}_\varepsilon U \to U' \hat{\otimes}_\varepsilon U$ is surjective, there is a constant $M \geq 1$ and there exist $\ell_\alpha \in E' \hat{\otimes}_\varepsilon U$ such that $||\ell_\alpha|| \leq MC$ and $(\iota' \hat{\otimes}_\varepsilon id)\ell_\alpha = h_\alpha$. But then $\ell_\alpha \in K(E,U)$ and $\ell_\alpha u = h_\alpha u \to u$ for all $u \in U$.

For some more equivalences we refer to [17], 2.9.. Proposition 2.7. can for instance be applied to the inclusion $\mathcal{A}(D) \to C(\partial D)$ of the disc algebra into the space of continuous functions on the unit circle: By a theorem of A. Pelczynski [24] $\mathcal{A}(D)$ is not a $\mathcal{L}_\infty$-space, and so (2) cannot be true, since $C(\partial D)$ is. Then by (3) we get the result that $\mathcal{A}(D) \hat{\otimes}_\varepsilon C(\partial D)' \to \mathcal{A}(D) \hat{\otimes}_\varepsilon \mathcal{A}(D)'$ is not surjective.

Next it will be shown that in case of (εL)-triples the tensor product of π with an identity on a more general space than a Banach space remains surjective:

<u>2.8. Proposition.</u> Let $o \to U \to E \overset{\pi}{\to} Q \to o$ be an (εL)-triple of Banach spaces and let F be a complete locally convex space. Then $(id \hat{\otimes}_\varepsilon \pi): F \hat{\otimes}_\varepsilon E \to F \hat{\otimes}_\varepsilon Q$ is surjective, too, if one of the following conditions holds:

(1) F is an (F)-space.

(2) $F = \varinjlim_\alpha F_\alpha$ is a compact-regular inductive limit of (F)-spaces F_α with a.p.

(3) Q has the a.p., and every compact subset of F is very compact.

<u>Proof.</u> Take $x \in F \hat{\otimes}_\varepsilon Q \subseteq L_e(Q'_c,F)$ and put $V^o = \{q' \in Q' \mid ||q'|| \leq 1\}$; then $x(V^o)$ is compact in F. In case (3) there is a Banach space B, continuously embedded into F, such that $x(V^o)$ is compact in B. Then as in the proof of 1.7. it follows $x \in L_e(Q'_c,B) = B \hat{\otimes}_\varepsilon Q$, since Q has the a.p., but then x can be lifted to $B \hat{\otimes}_\varepsilon E \subseteq F \hat{\otimes}_\varepsilon E$. In case (2) it is shown analogously that x is in fact an element of a fixed $F_\alpha \hat{\otimes}_\varepsilon Q$, and so the assertion will follow from a proof of (1). So let F be an (F)-space. Then there is a sequence (x_n) in $F \otimes Q$ converging to x in $F \hat{\otimes}_\varepsilon Q$. Obviously $K := x(V^o) \cup \bigcup_{n=1}^{\infty} x_n(V^o)$ is relatively compact in F, hence K - K is contained in a very compact set. If B is a Banach space, continuously embedded into F, such that K - K is relatively compact in B, then x_n converges to x in $L_e(Q'_c,B)$ which implies $x \in B \hat{\otimes}_\varepsilon Q$. But then x can be lifted to $B \hat{\otimes}_\varepsilon E \subseteq F \hat{\otimes}_\varepsilon E$.

An interesting example of condition (3) is the following: Let H be a Hilbert space and $A \subseteq L(H)$ be a C^*-algebra containing K(H) such that $A/K(H)$ is commutative. Then $A/K(H) \cong C(M)$, where M is the space of maximal ideals of $A/K(H)$, has the a.p., and $o \to K(H) \to A \to A/K(H) \to o$ is an (εL)-triple (cf. [17], 2.13.).

Now various types of (F)- and (DF)-(εL)-triples can be found by using the previous results and forming countable projective and inductive limits, cf. [17], §2. However, here we restrict ourselves to the following

2.9. Theorem. Let $o \to U \xrightarrow{\iota} E \xrightarrow{\pi} Q \to o$ be a short exact sequence of (F)-spaces such that $U = \varprojlim_n U_n$ is a dense countable projective limit of $\mathcal{L}_\infty$-spaces U_n. Then $o \to U \xrightarrow{\iota} E \xrightarrow{\pi} Q \to o$ is an (εL)-triple; even for every (F)-space F $(id \hat{\otimes}_\varepsilon \pi): F \hat{\otimes}_\varepsilon E \to F \hat{\otimes}_\varepsilon Q$ is surjective.

Proof. If $\rho_n: U \to U_n$ are the canonical maps, the semi-norms $p_n := \| \ \|_n \circ \rho_n$ form a non decreasing sequence of semi-norms that generates the topology of U. Let $r_1 \leq r_2 \leq \ldots$ be a fundamental system of semi-norms on E such that $r_n|_U \leq p_n \leq r_{n+1}|_U$ for all $n \in \mathbb{N}$. Put $V_n = \{ e \in E \mid r_n(e) \leq 1 \}$ and $W_n = \{u \in U \mid p_n(u) \leq 1\}$; then $V_{n+1} \cap U \subseteq W_n \subseteq V_n \cap U$, hence $(V_n \cap U)^o \subseteq W_n^o \subseteq (V_{n+1} \cap U)^o$, where the polars are taken relative to $\langle U,U' \rangle$. Then by $q_n(e) := \sup \{ |\langle e,e' \rangle| \mid e' \in V_{n+1}^o \text{ and } \iota' e' \in W_n^o \}$ we define a semi-norm on E satisfying $r_n \leq q_n \leq r_{n+1}$ and $q_n|_U = p_n$. Now we have $E = \varprojlim_n \hat{E}_{q_n}$ and $Q = \varprojlim_n Q_n$, where $Q_n := \hat{E}_{q_n}/\hat{U}_{p_n}$; recall that $\hat{U}_{p_n} \cong \hat{U}_n$ is a $\mathcal{L}_\infty$-space.
Now take any $x \in F \hat{\otimes}_\varepsilon Q \subseteq L_e(F'_c, Q)$. If $\sigma_n: Q \to Q_n$ is the canonical map, consider $\sigma_n x \in F \hat{\otimes}_\varepsilon Q_n$ which can be lifted to $y_n \in F \hat{\otimes}_\varepsilon \hat{E}_{q_n}$, since $o \to \hat{U}_{p_n} \to \hat{E}_{q_n} \to Q_n \to o$ is an (εL)-triple by 2.4.(i) and 2.5.. But then the "Mittag-Leffler-technique" already used in the proof of Theorem 1.9. completes the proof of the theorem.

2.10. Remarks and Examples.

(i) Theorem 2.9. is new even for the case of a nuclear space U. Recall that the same assertion can easily be proved by use of the π-tensor product, if Q is nuclear; however, that method seems to be not applicable in the case of a nuclear subspace U.

(ii) We mention the following application of Theorem 2.9.: Let P(D) be a hypoelliptic linear partial differential operator with constant coefficients and $\Omega \subseteq \mathbb{R}^n$ be an open and P(D)-convex set (cf. [15], 3.5.,4.1.). By a theorem of L. Hörmander [15],p.82,

$P(D): B^{loc}_{p,k\tilde{P}}(\Omega) \to B^{loc}_{p,k}(\Omega)$ is surjective, where the spaces $B^{loc}_{p,k}(\Omega)$ are local Sobolev-spaces (cf. [15], 2.3.). Obviously, the spaces $B^{loc}_{p,k}(\Omega)$ are not nuclear, however, since $P(D)$ is hypoelliptic, its kernel $N(P(D))$ is contained in $C^\infty(\Omega)$ and thus nuclear. Note that by a result of A. Grothendieck (cf. [29],p.544) $N(P(D))$ is, at least for elliptic operators $P(D)$, not complemented in $B^{loc}_{p,k\tilde{P}}(\Omega)$; nevertheless $o \to N(P(D)) \to B^{loc}_{p,k\tilde{P}}(\Omega) \to B^{loc}_{k,p}(\Omega) \to o$ is an (εL)-triple by Theorem 2.9.. So, if for instance a $B^{loc}_{p,k}(\Omega)$ - valued C^ℓ-function $f(x,t)$ is given, the parameter dependent equation $P(D)g(x,t) = f(x,t)$ has a solution g which is a C^ℓ-function in the parameter t, though C^ℓ-functions cannot be lifted in general, cf. 1.6.(iii).

3. A WEAK - STRONG EXTENSION THEOREM

In [10] B. Gramsch essentially sharpened and generalized the classical weak-strong theorem of A. Grothendieck [13], II §3.3., cf. [10], 2.8., and Gramsch's lecture at this conference.
The method of proof is based on duality theory, closed graph - and lifting theorems; for the existence of certain liftings nuclearity of the corresponding spaces is assumed in most cases. So it is clear that more general theorems can be proved by use of more general (εL)-spaces and (εL)-triples.
This will be illustrated in one particular case where Theorem 2.9. can be applied:

3.1. Definition. Let Ω be an open subset of $\mathbb{R}^n$ and let $P(x,D)$ be a hypoelliptic operator of order m defined on Ω. If ω is an open subset of Ω, we define $\mathcal{H}_\omega(\Omega) := \{ f \in C^m(\Omega) \mid P(x,D)f(x) = o$ for all $x \in \omega \}$.

Obviously, $\mathcal{H}_\omega(\Omega)$ is a closed subspace of $C^m(\Omega)$ which is not nuclear if $\Omega \setminus \overline{\omega} \neq \emptyset$. We put $\Delta := \Omega \setminus \omega$ and prove:

3.2. Theorem. Let F be an (F)-Montel space with a.p. and consider a function $\phi: \Delta \to F$ with the "weak $\mathcal{H}_\omega(\Omega)$-extension property", which means that for every $f' \in F'$ there exists $h_{f'} \in \mathcal{H}_\omega(\Omega)$ satisfying $\langle\phi(\delta), f'\rangle = h_{f'}(\delta)$ for all $\delta \in \Delta$.
Then ϕ has an extension $\tilde{\phi} \in \mathcal{H}_\omega(\Omega,F) := \mathcal{H}_\omega(\Omega) \hat{\otimes}_\varepsilon F$.

<u>Proof.</u> Put $\Delta^\perp = \{ g \in \mathcal{H}_\omega(\Omega) \mid g|_\Delta = o \}$; then by $\phi'(f') = h_{f'} + \Delta^\perp$ a map $\phi' : F' \to \mathcal{H}_\omega(\Omega)/_{\Delta^\perp}$ is defined. Obviously, $\phi' : F'_\sigma \to \mathcal{H}^f_\omega(\Omega)/_{\Delta^\perp}$ is closed; since $F'_c = F'_b$ is ultrabornological, $\phi' : F'_c \to \mathcal{H}_\omega(\Omega)/_{\Delta^\perp}$ is continuous. Since F has the a.p. we even get $\phi' \in F \hat{\otimes}_\varepsilon \mathcal{H}_\omega(\Omega)/_{\Delta^\perp}$. Now for $g \in \Delta^\perp$ we have $g(x) = o$ for $x \in \Delta$ and $P(x,D)g(x) = o$ for $x \in \omega$, hence $P(x,D)g(x) = o$ for all $x \in \Omega$. So $\Delta^\perp$ is contained in $N(P(x,D)) \subseteq C^\infty(\Omega)$, and therefore $\Delta^\perp$ is nuclear. But then by Theorem 2.9. there is a lifting $\tilde{\phi}$ to $F \hat{\otimes}_\varepsilon \mathcal{H}_\omega(\Omega)$ of ϕ'. Since for all $\delta \in \Delta$ and $f' \in F'$ we have $\langle\tilde{\phi}(\delta), f'\rangle = \langle\delta, \tilde{\phi}'f'\rangle = \langle\delta, \phi'f'\rangle = \langle\delta, h_{f'}\rangle = \langle\phi(\delta), f'\rangle$, $\tilde{\phi}$ is an extension of ϕ.

The theorem is false for Banach spaces F as can be seen by taking $\omega = \emptyset$. In general the space $\Delta^\perp$ is not trivial, since by a result of Pliš (cf.[10], 1.3.) there exist elliptic operators with C^∞-coefficients the kernels of which contain functions with compact support.

4. A LIFTING THEOREM FOR VARIABLE SURJECTIONS

Until now always fixed surjections $\pi : E \to Q$ of locally convex spaces have been considered. Of course, the case of variable surjections $\pi(\lambda) : E \to Q$ is also very interesting. The lifting problem is then the following: Given a function $f : \Lambda \to Q$, find a function $g : \Lambda \to E$ such that $\pi(\lambda)g(\lambda) = f(\lambda)$ for all $\lambda \in \Lambda$.
For continuous parameter dependence the problem was treated in [1] and [23], for holomorphic dependence and Banach spaces E, Q the problem was solved by J. Leiterer [20]; his method may be applied to many other cases, cf. [17], §9. Essentially the result is this: Given a fixed parameter dependence the variable lifting problem is always solvable if (and only if) the lifting problem is solvable for all fixed surjections and for all pointwise right invertible functions $\pi(\lambda)$. To make this principle more explicit we prove the following

<u>4.1. Theorem.</u> Let X and Y be Banach spaces and let Ω be a region in $\mathbb{R}^n$ respectively in $\mathbb{C}^n$. Let $\pi : \Omega \to L(X,Y)$ be a function that is
(a) continuous,
(b) of class $\lambda^{k,\alpha}$ (cf. 1.6., 1.10.),
(c) holomorphic,
such that $\pi(\lambda)$ is surjective for all $\lambda \in \Omega$.

Then if $f: \Omega \to Y$ is a
(a) continuous,
(b) $\lambda^{k,\alpha}$- or
(c) holomorphic
function, and if in case (c) Ω is a domain of holomorphy, there is a
(a) continuous,
(b) $\lambda^{k,\alpha}$- or
(c) holomorphic
function $g: \Omega \to X$ such that $\pi(\lambda)g(\lambda) = f(\lambda)$ for all $\lambda \in \Omega$.

Proof. There exist ℓ_1-spaces $\ell_1(A)$ and $\ell_1(B)$ such that X and Y are quotient spaces of them; let $\rho: \ell_1(A) \to X$ and $\sigma: \ell_1(B) \to Y$ be the quotient maps. Consider the function $\pi(\lambda)\rho: \ell_1(A) \to Y$ which can pointwise be lifted to $\ell_1(B)$, since $\ell_1(A)$ is a projective Banach space. This means that the map $\sigma\circ: L(\ell_1(A),\ell_1(B)) \to L(\ell_1(A),Y)$ is surjective. But then $\pi\circ\rho$ has a continuous, $\lambda^{k,\alpha}$- or holomorphic lifting $\pi_o: \Omega \to L(\ell_1(A),\ell_1(B))$,

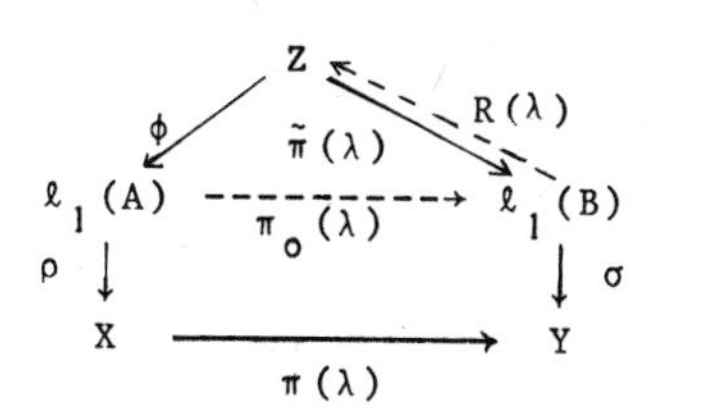

i.e. we have $\sigma\pi_o(\lambda) = \pi(\lambda)\rho$ for all $\lambda \in \Omega$. Now we put $Z := \ell_1(A) \oplus N(\sigma)$ and define $\tilde{\pi}(\lambda): Z \to \ell_1(B)$ by $\tilde{\pi}(\lambda)\ (a,n) = \pi_o(\lambda)a + n$. Then $\tilde{\pi}$ is continuous, $\lambda^{k,\alpha}$- or holomorphic and is pointwise surjective. However, since $\ell_1(B)$ is a projective Banach space, $\tilde{\pi}$ is even pointwise right invertible. But then there exists a continuous, $\lambda^{k,\alpha}$- or holomorphic global right inverse R of $\tilde{\pi}$, i.e. a function $R: \Omega \to L(\ell_1(B),Z)$ satisfying $\tilde{\pi}(\lambda)R(\lambda) = id$ for all $\lambda \in \Omega$. The existence of R is clear locally, and in cases (a) and (b) its global existence follows from the existence of a C^∞- partition of unity. In case (c) the global existence of R follows from a "paraalgebra version" [16] of a theorem of G. R. Allan; here it is essential that Ω is a domain of holomorphy.

Now let a function $f: \Omega \to Y$ of type (a), (b) or (c) be given. Then there is a lifting $\tilde{f}: \Omega \to \ell_1(B)$ of f which is also continuous, $\lambda^{k,\alpha}$- or holomorphic. If $\phi: Z \to \ell_1(A)$ is the projection onto the first component, we put $g(\lambda) := \rho\phi R(\lambda)\tilde{f}(\lambda)$ for $\lambda \in \Omega$. Then $g: \Omega \to X$ is continuous, $\lambda^{k,\alpha}$- or holomorphic, too, and satisfies $\pi(\lambda)g(\lambda) = \pi(\lambda)\rho\phi R(\lambda)\tilde{f}(\lambda) = \sigma\pi_o(\lambda)\phi R(\lambda)\tilde{f}(\lambda) = \sigma\tilde{\pi}(\lambda)R(\lambda)\tilde{f}(\lambda) = \sigma\tilde{f}(\lambda) = f(\lambda)$ for all $\lambda \in \Omega$. So g is a lifting of f, and the proof of the theorem is complete.

4.2. Remarks.

(i) In the holomorphic case, the theorem is not true for arbitrary regions Ω in $\mathbb{C}^n$, even for pointwise right invertible functions.

(ii) A similar lifting theorem holds for functions satisfying rather general Lipschitz conditions (cf. [17], §4, §9), for functions of mixed type on regular domains (cf. [8], [2], [17], §7, §9) and for holomorphic functions on polynomially convex regions in nuclear (DF)-spaces with Schauder bases. This last result is proved by using a nuclearity theorem of P.J. Boland and L. Waelbroeck and a result of [11] on maximal ideal spaces; details may be found in [17], §9.

REFERENCES

1. Bartle, R.G., and Graves, R.M. (1952). Mappings between function spaces. Transactions Am. Math. Soc. 72, 4oo-413.

2. Bierstedt, K., Gramsch, B., and Meise, R. (1976). Approximationseigenschaft, Lifting und Kohomologie bei lokalkonvexen Produktgarben. Manuscripta Math. 19, 319-364.

3. Bonic, R., Frampton, J., and Tromba, A. (1969). Λ - manifolds. Journal of Funct. Anal. 3, 31o-32o.

4. De Wilde, M. (1969). Réseaux dans les espaces linéaires à semi-normes. Mém. Soc. Roy. Sci. Liège 18.

5. Floret, K. (1973). $\mathcal{L}_1$ - Räume und Liftings von Operatoren nach Quotienten lokalkonvexer Räume. Math. Z. 134, 1o7-117.

6. Frampton, J., and Tromba, A. (1972). On the classification of spaces of Hölder continuous functions. Journal of Funct. Anal. 1o, 336-345.

7. Gleason, A.M. (1962). The abstract theorem of Cauchy-Weil. Pac. Journal of Math. 12, 511-525.

8. Gramsch, B. (1975). Inversion von Fredholmfunktionen bei stetiger und holomorpher Abhängigkeit von Parametern. Math. Ann. 214, 95-147.

9. Gramsch, B. (1975). Über das Cauchy-Weil-Integral für Gebiete mit beliebigem Rand. To appear in Arch. Math.

1o. Gramsch, B. (1975). Über eine Fortsetzungsmethode der Dualitäts= theorie lokalkonvexer Räume. To appear.

11. Gramsch, B., and Kaballo, W. (1976). Regularisierung von Fredholmfunktionen. To appear.

12. Grothendieck, A. (1952). Résumé de la théorie métrique des produits tensoriels topologiques. Bol. Soc. Mat. Sao Paulo 7.
13. Grothendieck, A. (1955). Produits tensoriels topologiques et espaces nucléaires. Am. Math. Soc. Mem. 16.
14. Henkin, G.M. (1967). Impossibility of a uniform homeomorphism between spaces of smooth functions of one and of n variables ($n \geq 2$). Math. USSR - Sbornik 3, 551-561.
15. Hörmander, L. (1963). Linear partial differential operators. Springer, Berlin - Heidelberg - New York.
16. Kaballo, W. (1974). Über holomorphe und meromorphe einseitige Inverse. Manuscripta Math. 13, 1-13.
17. Kaballo, W. (1976). Lifting-Sätze für Vektorfunktionen und das ε - Tensorprodukt. Habilitationsschrift Kaiserslautern.
18. Köthe, G. (1966). Topologische lineare Räume I. Springer, Berlin - Heidelberg - New York.
19. Kučment, P.A. (1975). Lifting of functions with values in Banach spaces. Mat. Issled. (Kishinev) I 35, 185-193 (Russian).
2o. Leiterer, J. (1976). Banach coherent analytic Fréchet sheaves. To appear in Math. Nachr.
21. Lindenstrauß, J. (1964). Extension of compact operators. Am. Math. Soc. Mem. 48.
22. Lindenstrauß, J., and Tzafriri, L. (1973). Classical Banach spaces. Springer, Berlin - Heidelberg - New York.
23. Michael, E. (1956). Continuous selections I. Ann. Math. 63, 361-382.
24. Pelczynski, A. (1974). Sur certaines propriétés isomorphiques nouvelles des espaces de Banach de fonctions holomorphes A et H^∞. C. R. Acad. Sci. Paris 279, 9-12.
25. Randtke, D.J. (1973). A structure theorem for Schwartz spaces. Math. Ann. 2o1, 171-176.
26. Schaefer, H.H. (1971). Topological vector spaces. Springer, Berlin - Heidelberg - New York.
27. Schwartz, L. (1957). Théorie des distributions à valeurs vectorielles I. Ann. Inst. Fourier 7, 1-142.
28. Stegall, C.P., and Retherford, J.R. (1972). Fully nuclear and completely nuclear operators with applications to $\mathcal{L}_1$- and $\mathcal{L}_\infty$-spaces. Transactions Am. Math. Soc. 163, 457-492.
29. Trèves, F. (1967). Topological vector spaces, distributions and kernels. Academic Press, New York.

3o. Vogt, D. (1975). Vektorwertige Distributionen als Randver= teilungen holomorpher Funktionen. Manuscripta Math. 17, 267-29o.

31. Vogt, D. (1975). Tensorprodukte von (F)- mit (DF)-Räumen und ein Fortsetzungssatz. To appear.

K.-D. Bierstedt, B. Fuchssteiner (eds.)
Functional Analysis: Surveys and Recent Results

SUBSPACES AND QUOTIENT SPACES OF (s)

Dietmar Vogt

Fachbereich Mathematik
der Gesamthochschule
Wuppertal, Germany

The main intention of this article is to give a survey of results which have been obtained by the author on subspaces and by the author together with M.J. Wagner on quotient spaces of s, the space of rapidly decreasing sequences of real or complex numbers (§§ 1,...,6). The results in these paragraphs are contained in Vogt [20] and Vogt-Wagner [21]. They are presented with only minor changes in arrangement and details of proofs. There is given a complete characterization of subspaces and quotient spaces of s, which in the case of nuclear (F)-spaces with basis (=sequence spaces) has been obtained independently and using other methods by Dubinsky [3] and Dubinsky-Robinson [4]. For the development of the concept of classes (DN) and (Ω) see [18], [19], [22]. In the proof of our results splitting theorems for exact sequences of (F)-spaces (s. §2) play an important role. These are also important in other fields of analysis and functional analysis. The last two paragraphs are concerned with related results and applications of these theorems.

§ 1. DEFINITIONS AND MAIN RESULTS

Let E be a (F)-space, $\| \ \|_1 \leq \| \ \|_2 \leq \dots$ a fundamental system of seminorms and $U_1 \supset U_2 \supset \dots$ a basis of absolutely convex neighbourhoods of zero in E.

1.1. Definition: E has property (DN), if there exists a continuous norm $\| \ \|$ on E, such that for every $k \in \mathbb{N}$ we have a $p \in \mathbb{N}$ and $C > 0$ with

$$\| \ \|_k \leq r \| \ \| + \frac{C}{r} \| \ \|_{k+p}$$

for all $r > 0$

(or equivalently: $\| \ \|_k^2 \leq C \| \ \| \ \| \ \|_{k+p}$).

1.2. Definition: E has property (Ω), if for every $p \in \mathbb{N}$ there exists a $q \in \mathbb{N}$, such that for every $k \in \mathbb{N}$ we have a $n \in \mathbb{N}$ and $C > 0$ with

$$U_q \subset C\, r^n\, U_k + \frac{1}{r} U_p$$

for all $r > 0$.

Both definitions are easily seen to be independent of the choice of the seminorm-system, resp. the basis of neighbourhoods of zero.

These properties will be shown to be characteristic for the subspaces, resp. quotient spaces of s, where s denotes the space of all rapidly decreasing sequences of real or complex numbers. That is

$$s = \{x = (x_1, x_2, \ldots) : \|x\|_k = \sum_j |x_j| \, j^k < +\infty \text{ for all } k\}.$$

s furnished with the norms $\| \, \|_k$ is a nuclear (F)-space. It plays an important role in analysis, since many of the usual spaces of C^∞-functions are isomorphic to s or closely related to s.

We formulate now the main results we are concerned with in this article:

1.3. Theorem: A nuclear (F)-space E is isomorphic to a subspace of s, iff E has property (DN).

1.4. Theorem: A nuclear (F)-space E is isomorphic to a quotient space of s, iff E has property (Ω).

1.5. Theorem: A nuclear (F)-space E is isomorphic to a continuously projected subspace of E, iff E has properties (DN) and (Ω).

One direction of the proof of these theorems follows immediately from the following lemma and the fact that s has properties (DN) and (Ω) (cf §5).

1.6. Lemma: (a) With E also every subspace has property (DN).
(b) With E also every quotient space has property (Ω).
The proof is easy.

It remains now to show, that every nuclear (F)-space
1. with property (DN) is isomorphic to a subspace of s,
2. with property (Ω) is isomorphic to a quotient space of s,
3. with properties (DN) and (Ω) is isomorphic to a continuously projected subspace of s.

For that we need some general results which are contained in the following §§ 2,3. The proofs of 1.3. - 1.5. will be finished in § 4.

§ 2. SPLITTING - THEOREMS

The results contained in this § will show us some of the most important features of the classes (DN) and (Ω). They have various applications in other fields of analysis (s. §8). We first give an equivalent dual formulation of property (DN) which we shall need in the proof of 2.2.:

2.1. Lemma: E has property (DN) iff there exists a neighbourhood U of zero in E, such that to every k we have a p and $C > 0$ such that

$$U_k^o \subset r\, U^o + \frac{C}{r} U_{k+p}^o$$

for all $r > 0$.

The proof is a straightforward dualisation argument.

In the proof of 2.2. we shall also use the following notational conventions: If H is a $\widehat{(F)}$-space, $\| \ \|_1 \leq \| \ \|_2 \leq \ldots$ a fundamental system of seminorms, then $H_k := H/\ker \| \ \|_k$ is the Banach space belonging to the k-th seminorm, $\rho_k : H \to H_k$, $\rho_{n,k} : H_n \to H_k$, $n > k$ are the natural maps. It is then known from the general theory of locally convex spaces that $H = \lim\text{proj}\, H_k$, that means: if we have another locally convex space E and for every k a $\pi_k \in L(E,H_k)$ such that $\rho_{n,k} \circ \pi_n = \pi_k$ for all $n > k$, then there exists an unique $\pi \in L(E,H)$ with $\rho_k \circ \pi = \pi_k$ for all k. L(E,F) always denotes the space of continuous linear maps between the locally convex spaces E and F.

In the proof of 2.2. (and only there) we shall use for s not the norms defined in § 1, but the equivalent norm system $\| x \|_k = \sup_j |x_j| j^k$, hence $s_k = \{x = (x_1, x_2, \ldots) : \| x \|_k := \sup_j |x_j| j^k < +\infty\}$ with the norm $\| \ \|_k$.

2.2. Theorem: If $0 \to H \to \tilde{E} \xrightarrow{q} E \to 0$ is an exact sequence of nuclear (F)-spaces, H is a quotient space of s, E has (DN), then the sequence is split (i.e. there exists a continuous right inverse for q).

Proof: We assume H to be a subspace of $\tilde{E}$ and have to show: H is continuously projected in $\tilde{E}$. H is by assumption a quotient of s, let $Q: s \to H$ be the quotient map. We can further assume that for each k we have an induced quotient map $Q_k : s_k \to H_k$.

The canonical map $H \to H_k$ is nuclear. We can extend it to a map $F_k \in L(\tilde{E},H_k)$. With $\tilde{G}_k := \rho_{k+1,k} \circ F_{k+1} - F_k \in L(\tilde{E},H_k)$ we have $\tilde{G}_k = 0$ on H. So $\tilde{G}_k$ induces a map in $L(E,H_k)$, which is necessarily nuclear and can be lifted to $G_k \in L(E,s_k)$. G_k has the form $G_k(x) = (G_1^k(x), G_2^k(x), \ldots)$, where $G_j^k \in E'$ and $\{j^k G_j^k : j = 1,2,\ldots\}$ is equicontinuous and therefore contained in a U_k^o. We can assume that $U_k \supset U_{k+1}$, $\{U_k\}$ basis of neighbourhoods of zero in E.

Since E has property (DN), we can find a $U \subset E$ fulfilling the conditions of 2.1. By shrinking the U_k if necessary we can arrange that

$$U_k^o \subset r\, U^o + \frac{2^{-k-2}}{r} U_{k+1}^o$$

for all $r > 0$, $k \in \mathbb{N}$. Choosing $r = j\, 2^{-k-1}$ we get after multiplication with $2\, j^{-k}$

$$(*) \qquad 2\, j^{-k} U_k^o \subset j^{-k+1}\, 2^{-k}\, U^o + j^{-(k+1)}\, U_{k+1}^o$$

for all $j,k \in \mathbb{N}$.

We determine now (having j fixed) inductively a sequence $A_j^k \in E'$ with $A_j^k \in j^{-k} U_k^o$. We start with $A_j^o = 0$. If $A_j^k \in j^{-k} U_k^o$ we have $G_j^k + A_j^k \in 2\, j^{-k} U_k^o$. On account of (*) we can find $A_j^{k+1} \in j^{-k-1} U_{k+1}^o$, such that $G_j^k + A_j^k - A_j^{k+1} \in j^{-k+1}\, 2^{-k}\, U^o$. Defining $A_k\, x := (A_1^k(x), A_1^k(x), \ldots)$ we get an $A_k \in L(E, s_k)$, we define $\tilde{A}_k := Q_k \circ A_k \circ q \in L(\tilde{E}, H_k)$, $\pi_k := F_k - \tilde{A}_k \in L(\tilde{E}, H_k)$.

For $x \in \tilde{U} := q^{-1} U$ we have (with $\| \ \|_k$ norm in H_k resp. in s_k, $\rho_{k,k-1}$ omitted):

$$\| \rho_{k+1,k}\, \pi_{k+1} x - \pi_k x \|_{k-1} = \| \tilde{G}_k x + \tilde{A}_k x - \tilde{A}_{k+1} x \|_{k-1}$$

$$= \| (G_k + A_k - A_{k+1})\, qx \|_{k-1} \leq 2^{-k}$$

It follows easily that for each $n \in \mathbb{N}$ and $x \in E$ $\lim_{\substack{k\to\infty \\ k>n}} (\rho_{k,n} \circ \pi_k)\, x$ exists and defines a $\tilde{\pi}_n \in L(\tilde{E}, H_n)$. Since apparently $\rho_{n+1,n} \circ \tilde{\pi}_n = \tilde{\pi}_{n+1}$, there exists $\pi \in L(\tilde{E}, H)$ with $\tilde{\pi}_n = \rho_n \circ \pi$. For $x \in H$ we have $\tilde{\pi}_n\, x = \lim_{\substack{k\to\infty \\ k>n}} \rho_{k,n}\, (F_k x) = \rho_n x$, so $\pi x = x$ and π is a continuous projection from $\tilde{E}$ onto H.

Remark: We did not need really the nuclearity of $\tilde{E}$ and E. We used it only for simplification of the notations in the proof. The proof for a more general version of 2.2. without the assumption of nuclearity for $\tilde{E}$, E is essentially the same.

Now we come to the second splitting theorem involving property (Ω).

2.3. Theorem: If $0 \to E \to \tilde{E} \overset{q}{\to} s \to 0$ is an exact sequence of (F)-spaces and E has property (Ω), then the sequence is split.

Proof: We assume that E is a subspace of $\tilde{E}$. Let $W_k \supset W_{k+1}$ be a basis of absolutely convex neighbourhoods of zero in $\tilde{E}$, $U_k = W_k \cap E$. Then U_k is a neighbourhood basis of zero in E and using condition (Ω) we can arrange that with appropriate $\nu(k) \in \mathbb{N}$ we have

$$U_k \subset r^{\nu(k)}\, U_{k+1} + \frac{1}{r}\, U_{k-1}$$

for all $k \in \mathbb{N}$, $r > 2$.

If e_j is the j-th unit vector in s, then using the canonical norms $\|x\|_k = \sum_j j^k |x_j|$ we have $\|e_j\|_k = j^k$. Therefore for every k there exists a $n(k) \in \mathbb{N}$, $C_k \geq 1$ and a sequence $d_j^k \in E$, such that $d_j^k \in C_k\, j^{n(k)}\, W_k$ for all j,k, and $q(d_j^k) = e_j$. We can assume that $n(k) \leq n(k+1)$, $C_k \leq C_{k+1}$ and that with $m(k) = n(k+1)$ we have

$$(\nu(k) + 1)\, m(k) \leq m(k+1),\ C_k^{\nu(k)+1}\, 2^{(k+1)\nu(k)+1} \leq C_{k+1}$$

for all $k \in \mathbb{N}$.

Multiplying with $2\, C_k\, j^{m(k)}$ and choosing $r = 2^{k+1}\, C_k\, j^{m(k)}$ we obtain

$$(*) \qquad 2\, C_k\, j^{m(k)}\, U_k \subset 2\, C_k^{\nu(k)+1}\, j^{m(k)+\nu(k)m(k)}\, 2^{(k+1)\nu(k)}\, U_{k+1} + 2^{-k}\, U_{k-1}$$

$$\subset C_{k+1}\, j^{m(k+1)}\, U_{k+1} + 2^{-k}\, U_{k-1}.$$

Since $b_j^k := d_j^{k+1} - d_j^k \in 2\ j^{n(k+1)}\ W_k \cap H = 2\ j^{m(k)}\ U_k$, we can choose inductively (j fixed) a sequence $a_j^k \in C_k\ j^{m(k)}\ U_k$ in the following way: $a_j^o = 0$, if $a_j^k \in C_k\ j^{m(k)} U_k$ is chosen, then $b_j^k + a_j^k \in 2C_k\ j^{m(k)}\ U_k$ and according to (*) we can find $a_j^{k+1} \in C_{k+1}\ j^{m(k+1)}\ U_{k+1}$ such that $b_j^k + a_j^k - a_j^{k+1} \in 2^{-k}\ U_{k-1}$.

We define $R_j^k := d_j^k - a_j^k$, then we have

$$R_j^k \in 2\ C_k\ j^{m(k)}\ W_k,$$

$$R_j^{k+1} - R_j^k = b_j^k + a_j^k - a_j^{k+1} \in 2^{-k}\ U_{k-1}$$

for all $j,k \in \mathbb{N}$. It follows that $\lim_{k\to+\infty} R_j^k := R_j$ exists and $R_j \in 2\ C_k\ j^{m(k)}\ W_k + W_{k-1} \subset 3\ C_k\ j^{m(k)}\ W_{k-1}$. So we can define $R\,x = \sum_j x_j\,R_j$ for $x = (x_1, x_2, \ldots) \in s$ and with $\|\ \|_k$ denoting the seminorm in $\tilde{E}$ belonging to W_k resp. the canonical norm in s we obtain

$$\| Rx \|_{k-1} \leq 3\ C_k\ \| x \|_{m(k)}$$

Therefore $R \in L(s,\tilde{E})$ and, because $q(R_j) = \lim_{k\to+\infty} q(R_j^k) = \lim_{k\to+\infty} q(d_j^k) = e_j$, we get $q \circ R = id$.

Remark: In analogy to 2.2. we need also here only that we have an exact sequence $0 \to E \to \tilde{E} \to F \to 0$, where F is a subspace of s (s. [21]).

§ 3. CONSEQUENCES OF THE KOMURA IMBEDDING THEOREM

In 1966 it was shown by T. and Y. Komura proving a conjecture of Grothendieck that every nuclear (F)-space can be imbedded in $s^{\mathbb{N}}$. Using this result we obtain the following propositions which are crucial for the remaining part of the proof of thms. 1.3 - 1.5. We start with the following observation:

3.1. Lemma: There is an exact sequence

$$0 \to s \to s \xrightarrow{\phi} s^{\mathbb{N}} \to 0.$$

Proof: Let ω be the space of all (real or complex) sequences, $\Delta: \mathcal{D}[-1,+1] \to \omega$ the application $\phi \mapsto (\phi(o), \phi'(o), \ldots)$. Then according to a theorem of E. Borel Δ is surjective. The kernel of Δ is isomorphic to $\mathcal{D}[-1,o] \times \mathcal{D}[o,+1]$. Since $\mathcal{D}[a,b] \cong s$ for all $a < b$, and $s \times s \cong s$, the exact sequence

$$0 \to \mathcal{D}[-1,o] \times \mathcal{D}[o,+1] \to \mathcal{D}[-1,+1] \xrightarrow{\Delta} \omega \to 0$$

leads to an exact sequence

$$0 \to s \to s \to \omega \to 0.$$

Since $s \hat{\otimes}_\pi s = s$, $s \hat{\otimes}_\pi \omega = s^{\mathbb{N}}$ we get the desired result by tensoring with s. We use thereby the result of Grothendieck that for any nuclear (F)-space E, $\hat{\otimes}_\pi E$ is an exact functor in the category of nuclear (F)-spaces.

3.2. Proposition: For every nuclear (F)-space E there exists an exact sequence

$$0 \to s \to \tilde{E} \to E \to 0,$$

where $\tilde{E}$ is a subspace of s.

Proof: Using the Komura theorem we imbed E as a subspace into $s^{\mathbb{N}}$. With $\tilde{E} = \phi^{-1}(E)$ (ϕ the application in 3.1.) we obtain an exact sequence

$$0 \to s \to \tilde{E} \xrightarrow{q} E \to 0$$

with $q = \phi|_{\tilde{E}}$.

3.3. Proposition: If E is a subspace of s then there exists
(a) a subspace F of s and an exact sequence

$$0 \to E \to s \to F \to 0,$$

(b) a projected subspace G of s, and an exact sequence

$$0 \to E \to G \to s \to 0.$$

Proof: (a) According to 3.2. we have an exact sequence

$$0 \to s \to F \xrightarrow{h} s/E \to 0.$$

Let H be the following subspace of $F \otimes s$:

$$H = \{(x,y) : hx = Qy\},$$

where Q is the quotient map $s \to s/E$. If p_1, p_2 (resp. i_1, i_2) are the canonical maps $F \oplus s \longrightarrow F$, $F \oplus s \longrightarrow s$ (resp. $F \longrightarrow F \oplus s$, $s \longrightarrow F \oplus s$), we have the following commutative diagram with exact rows and columns:

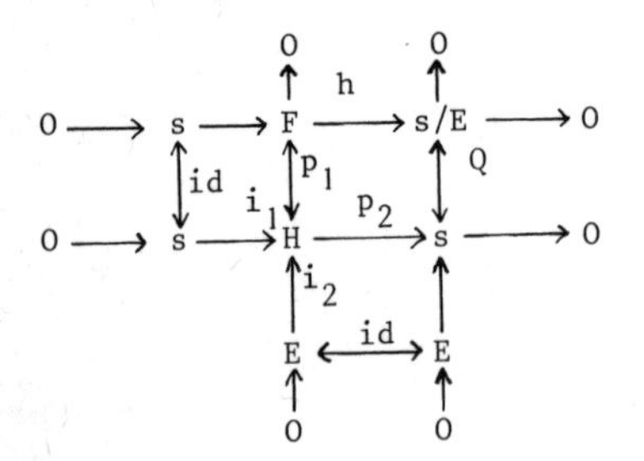

Since the second row is exact, it is (cf.3.2 or 3.3) split, so $H \cong s \oplus s \cong s$. The left column gives the desired result.

(b) We find a subspace $F \subset s$ and an exact sequence

$$0 \to E \to s \xrightarrow{q} F \to 0$$

according to (a).
We apply (a) to the space F, which gives us a $\tilde{F} \subset s$ and an exact sequence

$$0 \to F \to s \to \tilde{F} \to 0.$$

F is a quotient of s, $\tilde{F}$ as a subspace of s has property (DN). Theorem 3.2. tells us that the sequence is split, and therefore that $s \cong F \oplus \tilde{F}$.
With $G = s \oplus \tilde{F}$, $\tilde{q} : G = s \oplus \tilde{F} \xrightarrow{q \oplus id} F \oplus \tilde{F} \cong s$ we get the desired exact sequence

$$0 \to E \to G \xrightarrow{\tilde{q}} s \to 0 .$$

G is a projected subspace of $s \oplus s \cong s$.

§ 4. END OF PROOF OF THE MAIN THEOREMS

We are now prepared to finish the proofs of thms. 1.3 - 1.5:

4.1. Proposition: Every nuclear (F)-space E with property (DN) is isomorphic to a subspace of s.

Proof: According to 3.2. there exists a subspace $\tilde{E} \subset s$ and an exact sequence

$$0 \to s \to \tilde{E} \to E \to 0.$$

Since E has property (DN), this sequence is split (cf. 2.2.). So E is isomorphic to a subspace of $\tilde{E} \subset s$.

The definitions of (DN) and (Ω) for a space E obviously do not use the completeness of E. It is also obvious that if a metrizable locally convex space E has (DN), then also its completion $\hat{E}$. If E is also nuclear, then $\hat{E}$ is isomorphic to a subspace of s, and therefore this is also true for E. We have proved:

4.2. Corollary: A metrizable locally convex space is isomorphic to a subspace of s, iff it has property (DN).

Now we proceed to finish the proof of 1.5., which we shall use to prove 1.4.:

4.3. Proposition: Every nuclear (F)-space E with properties (DN) and (Ω) is isomorphic to a projected subspace of s.

Proof: Using 4.1. we see that E is isomorphic to a subspace of s. Therefore we have by 3.3. a projected subspace G of s and an exact sequence

$$0 \to E \to G \to s \to 0.$$

Since E has property (Ω), this sequence is split, so E is a projected subspace of G and therefore of s.

For the proof of 1.4. we need one more lemma:

4.4. Lemma: If $0 \to F \to \tilde{E} \xrightarrow{\phi} E \to 0$ is an exact sequence of (F)-spaces, F and E have property (Ω), then also $\tilde{E}$ has property (Ω).

Proof: We assume that F is a subspace of $\tilde{E}$, $U_1, U_2, \dots$ a basis of neighbourhoods of zero in E, $p \in \mathbb{N}$. Since F has property (Ω), there exists a $Q \in \mathbb{N}$ such that for every $k \in \mathbb{N}$ we have a $m \in \mathbb{N}$ and $C_1 > 0$ with

$$U_Q \cap F \subset C_1 r^m (U_k \cap F) + \frac{1}{r} (U_p \cap F) \subset C_1 r^m U_k + \frac{1}{r} U_p$$

for all $r > 0$.

Replacing r by r^{k+1} and multiplying with r^n we get

$$r^n (U_Q \cap F) \subset C_1 r^{m(k+1)+n} U_k + \frac{1}{r} U_p$$

for all $n \in \mathbb{N}$, $r > 0$.

Using property (Ω) for E, we get a $q \in \mathbb{N}$ and for every $k \in \mathbb{N}$ a $n \in \mathbb{N}$ and $C_2 > 0$, such that

$$\phi U_q \subset C_2 r^n \phi U_k + \frac{1}{r} \phi U_Q$$

or

$$U_q \subset C_2 r^n U_k + \frac{1}{r} U_Q + F .$$

Assuming without loss of generality $Q \leq q$ it is easily seen that we can replace F in the above inclusion by $(U_q + C_2 r^n U_k + \frac{1}{r} U_Q) \cap F$, which for $k \geq q$, $r \geq 1$ is contained in $(C_2+2) r^n (U_Q \cap F)$. So we get

$$\begin{aligned} U_q &\subset C_2 r^n U_k + \frac{1}{r} U_Q + (C_2+2) r^n (U_Q \cap F) \\ &\subset C_3 r^{m(n+1)+n} U_k + \frac{1}{r} U_Q + (C_2+2) \frac{1}{r} U_p \\ &\subset C_3 r^{\nu} U_k + \frac{C}{r} U_p . \end{aligned}$$

From this the desired result follows easily.

Now the proof of 1.4. can be finished.

4.5. Proposition: Every nuclear (F)-space E with property (Ω) is isomorphic to a quotient space of s.

Proof: 3.2. gives us an exact sequence

$$0 \to s \to \tilde{E} \to E \to 0,$$

where $\tilde{E}$ is a subspace of s. Using 4.4. and the fact that s has property (Ω) we see that with E also $\tilde{E}$ has property (Ω). $\tilde{E}$ has (DN) as a subspace of s. From 4.3. it follows that $\tilde{E}$ is isomorphic to a projected subspace of s. This gives the result.

§ 5. SEQUENCE SPACES

We intend now to use the special case of Köthe sequence spaces to point out the meaning of properties (DN) and (Ω). If $A = (a_{j,k})$ is an infinite matrix with $0 \leq a_{j,k} \leq a_{j,k+1}$, $\sup_k a_{j,k} > 0$ for all j,k, then we define

$$\lambda(A) := \{x = (x_1, x_2, \ldots) : \|x\|_k := \sum_{j=1}^{\infty} a_{j,k} |x_j| < +\infty \text{ for all } k\}.$$

$\lambda(A)$ with the seminorms $\| \ \|_k$ is a (F)-space. The well-known Grothendieck-Pietsch criterion says that $\lambda(A)$ is nuclear if and only if to every k there exists p such that

$$\sum_j \frac{a_{j,k}}{a_{j,k+p}} < +\infty.$$

5.1. Proposition: $\lambda(A)$ has property (DN), iff there exists a k_o such that for every k we have a $p \in \mathbb{N}$, $C > 0$ with

$$a_{j,k}^2 \le C\, a_{j,k_o}\, a_{j,k+p}$$

for all j.

Proof: One direction follows immediately from the definition by inserting the unit vectors and using the fact that $\| \ \|$ can be assumed to be one of the $\| \ \|_k$. The opposite direction is a consequence of the Cauchy-Schwartz inequality:

$$\|x\|_k^2 \le (\sum_j (a_{j,k_o} |x_j|)^{\frac{1}{2}} (a_{j,k+p} |x_j|)^{\frac{1}{2}})^2 \le \|x\|_{k_o} \|x\|_{k+p} .$$

To get the corresponding result for (Ω), we first dualise the definition. Here we denote by $\| \ \|_{-k}$ in E' the Minkowski-norm of U_k^o. ($\|x\|_{-k} = +\infty$ for $x \notin E'_{U_k^o}$.)

5.2. Lemma: E has property (Ω), iff for every p there exists a q such that for every k we have a n and $C > 0$ with

$$\| \ \|_{-q}^{n+1} \le C \| \ \|_{-k} \| \ \|_{-p}^k .$$

Proof: By a straightforward dualisation argument we see that (Ω) is equivalent to: For every p, there exists .... with

$$\| \ \|_{-q} \le C\, r^n \| \ \|_{-k} + \frac{1}{r} \| \ \|_{-p} \quad \text{for all } r > 0.$$

For fixed x we calculate the minimum of the function of r on the right side and get the result with changed C.

5.3. Proposition: $\lambda(A)$ has property (Ω), iff for every p there exists a q such that for every k we have a n and $C > 0$ with

$$C\, a_{j,q}^{n+1} \ge a_{j,k}\, a_{j,p}^n$$

for all j.

Proof: Again one direction of the proof follows easily by inserting the j-th coordinate map in the formula of 5.2. Since $\|x\|_{-k} = \sup \{|x_j|\, a_{j,k}^{-1} : j \in \mathbb{N}\}$ with $\alpha/_0 = +\infty$ for $\alpha>0$, $0/_0 = 0$, the other direction is obvious.

From 5.1. and 5.3. we derive a nice characterisation in the case of sequence spaces. We use thereby the following notation: if $\lambda(A)$ is a sequence space and $B = (b_{j,k})$ another matrix, we say that B is equivalent to A, iff $\lambda(A) = \lambda(B)$.

A necessary and sufficient condition for that are estimates of the form:
$a_{j,k} \leq C_k\, b_{j,\nu(k)},\ b_{j,k} \leq C'_k\, a_{j,\mu(k)}$ for all j,k .

5.4. Theorem: $\lambda(A)$ has property

(1) (DN) iff there exists an equivalent B with $b_{j,k}^2 \leq b_{k,k-1}\, b_{j,k+1}$,

(2) (Ω) iff there exists an equivalent B with $b_{j,k}^2 \geq b_{j,k-1}\, b_{j,k+1}$,

(3) (DN) and (Ω) iff there exists an equivalent B with $b_{j,k}^2 = b_{j,k-1}\, b_{j,k+1}$.

Proof: One direction of the proofs follows from the fact that using the right side in (1) we get

$$\frac{b_{j,k}}{b_{j,1}} = \prod_{i=1}^{k-1} \frac{b_{j,i+1}}{b_{j,i}} \leq \prod_{i=k}^{2k-2} \frac{b_{j,i+1}}{b_{j,i}} \leq \frac{b_{j,2k}}{b_{j,k}}$$

and using the right side in (2):

$$b_{j,p}\left(\frac{b_{j,q}}{b_{j,p}}\right)^{k-p} \geq b_{j,p} \prod_{i=p}^{k-1} \frac{b_{j,i+1}}{b_{j,i}} = b_{j,k} \text{ with } q = p+1$$

and so the conditions in 5.1., resp. 5.3.. The case of zeros is excluded in (1) and handled in (2) quite easily. The right side in (3) includes the conditions in (1), (2), so also here we get the desired result.

Turning to the proof of the opposite direction we see that in the case (1), since there exists a k_o such that for every $k > k_o$ we have a $p \in \mathbb{N}$, $C_k \geq 1$ with

$$b_{j,k} \leq C_k\, a_{j,k_o}\, a_{j,k+p}$$

we can proceed inductively, defining $b_{j,1} := a_{j,k_o}$ and, for $b_{j,\nu} = a_{j,k}$ chosen, defining $b_{j,k+1} := C_k\, a_{j,k+p}$.

Case (2) is less trivial. Again by proceeding inductively ("by forgetting everything between p and q") we can arrange that for every $p \in \mathbb{N}$ the following is true: For $k > p$ there exists $n \in \mathbb{N}$ and $C > 0$ such that

$$C\, a_{j,p+1}^{k+1} \geq a_{j,k}\, a_{j,p}^{k} .$$

By enlarging k and C if necessary we get for each k a $n(k)$ and $C_k \geq 1$ such that

$$C_k\, a_{j,\nu+1}^{n(k)+1} \geq a_{j,k}\, a_{j,\nu}^{n(k)}$$

for all $\nu < k$ and j.

For fixed j we define $b_{j,n}$ inductively beginning with $b_{j,1} = a_{j,1}$, $b_{j,2} = a_{j,2}$. Having defined $b_{j,1}, \ldots, b_{j,k}$ $(k \geq 2)$, we take

$$b_{j,k+1} := \begin{cases} \dfrac{b_{j,k}^2}{b_{j,k-1}} & \text{if } b_{j,k}^2 < a_{j,k+1}\, b_{j,k-1} \\ a_{j,k+1} & \text{if } b_{j,k}^2 \geq a_{j,k+1}\, b_{j,k-1} \end{cases}$$

Note that in the first case $b_{j,k-1} > 0$. By definition we have $b_{j,k}^2 \geq b_{j,k-1}\, b_{j,k+1}$, $b_{j,k} \leq b_{j,k+1}$ and $b_{j,k} \leq a_{j,k}$ for all j and k.

We are ready, if we prove that $C_k\, b_{j,k+n(k)} \geq a_{j,k}$ for all j,k. For that we denote by ν the greatest number $\leq k+n(k)$ such that $b_{j,\nu} = a_{j,\nu}$ and distinguish two cases:

i) $\nu \geq k$, then $b_{j,k+n(k)} \geq b_{j,\nu} = a_{j,\nu} \geq a_{j,k}$;

ii) $\nu < k$, then we have $b_{j,\mu+1} = \dfrac{b_{j,\mu}^2}{b_{j,\mu-1}}$ for all $\mu = \nu, \ldots, n+n(k)-1$ and therefore

$$b_{j,k+n(k)} = b_{j,\nu} \prod_{\mu=\nu}^{k+n(k)-1} \frac{b_{j,\mu+1}}{b_{j,\mu}} = b_{j,\nu} \left(\frac{b_{j,\nu}}{b_{j,\nu-1}}\right)^{k+n(k)-\nu}$$

$$= a_{j,\nu}\left(\frac{a_{j,\nu}}{b_{j,\nu-1}}\right)^{k+n(k)-\nu} \geq a_{j,\nu}\left(\frac{a_{j,\nu}}{a_{j,\nu-1}}\right)^{n(k)} \geq \frac{1}{C_k}\, a_{j,k} ,$$

which gives the result.

To prove the remaining part of (3) we can assume because of (1) that we have $a_{j,k}^2 \leq a_{j,k-1}\, a_{j,k+1}$ for all j,k, furthermore we can attain that to every k there exists $n(k)$, $C_k > 0$ with

$$C_k\, a_{j,2}^{n(k)+1} \geq a_{j,k}\, a_{j,1}^{n(k)} .$$

So by defining $b_{j,k} = a_{j,1} \left(\dfrac{a_{j,2}}{a_{j,1}}\right)^k$ we have

$$b_{j,k} \leq a_{j,1} \prod_{\nu=1}^{k} \frac{a_{j,\nu+1}}{a_{j,\nu}} \leq a_{j,k+1} ,$$

$$a_{j,k} \leq C_k\, b_{j,n(k)+1} ,$$

and the $b_{j,k}$ fulfill the desired equality.

<u>5.5. Corollary:</u> $\lambda(A)$ has properties (DN) and (Ω), iff it is of type $E_{b,+\infty}$.

Here $E_{b,r}$, $b = (b_1, b_2, \ldots)$ denotes a $\lambda(A)$ with $a_{j,k} = b_j^{r_k}$, $r_k \nearrow r$, $b_j \geq 1$. Using the notation $\Lambda_\rho(\alpha) = \lambda(A)$ with $a_{j,k} = \rho_k^{\alpha_j}$ $\rho_k \nearrow \rho$, $\alpha_j \nearrow +\infty$ one gets:

<u>5.6. Corollary</u>: A space $\Lambda_\rho(\alpha)$
(1) has (DN), iff $\rho = +\infty$,
(2) always has (Ω),
(3) has (DN) and (Ω), iff $\rho = +\infty$.

From the above results combined with 1.3.,... 1 5. one derives characterisations of nuclear (F)-spaces with basis which are subspaces, quotient spaces or projected subspaces of s. These have been obtained independently and by other methods by Dubinsky [3], Dubinsky-Robinson [4] and for the case of projected subspaces earlier by Bessaga [1]. In [23] Wagner characterizes the sequence spaces which are subspaces, quotient spaces or projected subspaces of a general power series space Λ in terms of (DN), (Ω) and similar properties, together with types of nuclearity which are related to Λ.

§ 6. QUOTIENT UNIVERSAL SPACES AND MARTINEAU'S CONJECTURE

As mentioned above, in 1966 T. and Y. Komura proved that every nuclear (F)-space is a subspace of $s^{\mathbb{N}}$, but nothing was known about the existence of spaces which have all nuclear (F)-spaces as quotient spaces (quotient universal spaces). It was a natural conjecture (Martineau's conjecture, cf. [12], [16]) that $s^{\mathbb{N}}$ or (since $s^{\mathbb{N}}$ is a quotient of s) equivalently s should have this property. Counter-examples for this conjecture have been given in [22], [21] and independently in [5]. In [21] the nonexistence of a quotient universal nuclear (F)-space has been proved. The following lemma gives a simple method to construct such counter-examples.

<u>6.1. Lemma</u>: If $a_{j,k} \geq 1$, $\sum_{j=1}^{\infty} \frac{a_{j,1}}{a_{j,2}} < +\infty$ and $\lim_{j\to+\infty} \frac{a_{j,k}^n}{a_{j,k+1}} = 0$ for all n and k, then $\lambda(A)$ is nuclear and not isomorphic to a quotient space of s.

<u>Proof</u>: Nuclearity follows inductively from

$$\sum_{j=1}^{\infty} \frac{a_{j,k}}{a_{j,k+1}} \leq C \sum_{j=1}^{\infty} \frac{1}{a_{j,k}} \leq C \sum_{j=1}^{\infty} \frac{a_{j,k-1}}{a_{j,k}}$$

where $C = \sup (a_{j,k}^2 \; a_{j,k+1}^{-1})$, $k \geq 2$.

If $\lambda(A)$ was isomorphic to a quotient space of s, then it would have property (Ω) (s. 1.6.) and therefore (s. 5.3.) we would have a q, n, $C > 0$ with

$$C\, a_{j,q}^{n+1} \geq a_{j,q+1} \; a_{j,1}^n \geq a_{j,q+1}$$

for all j, which contradicts to our assumption.

It is now easy to give concrete counterexamples.

6.2. Theorem: $\lambda(A)$ with $a_{j,k} = e^{k^j}, e^{j^k}, j^{j^k}, \underbrace{\exp(\exp(\ldots j))}_{k\times}$ is nuclear and is not isomorphic to a quotient space of s.

To prove the nonexistence of any nuclear (F)-space E_o (even of a (FS)-space E_o), which has all nuclear (F)-spaces as quotient spaces, we use the following modification of condition (Ω). ϕ denotes here a strictly positive, increasing function on $(0,+\infty)$.

6.3. Definition: E has property (Ω_ϕ), if for every $p \in \mathbb{N}$ there exists a $q \in \mathbb{N}$, such that for every $k \in \mathbb{N}$ we have a $C > 0$ with

$$U_q \subset C\phi(r)\, U_k + \frac{1}{r} U_p$$

for all $r > 0$.

We admit without proof the following lemma, which corresponds to 1.6.

6.4. Lemma: With E every quotient space of E has property (Ω_ϕ).

The line of argumentation is now the following: We shall prove that for every nuclear (F)-space (even for every (FS)-space) E_o there is a ϕ such that E_o has property (Ω_ϕ). If moreover E_o is quotient universal, then every nuclear (F)-space has property (Ω_ϕ). So the proof of nonexistence is complete, if for every ϕ we construct a nuclear (F)-space not having property (Ω_ϕ).

6.5. Lemma: For every (FS)-space E there exists a ϕ such that E has property (Ω_ϕ).

Proof: By assumption to every p there exists a $q = q(p)$ such that U_q is precompact relative to U_p, i.e. to every $n = 1,2,\ldots$ there exists a finite set $e_n \subset E$ with

$$U_q \subset e_n + \frac{1}{n} U_p .$$

We can assume $e_n \subset e_{n+1}$. For $k \in \mathbb{N}$ and $r > 0$ we define

$$\phi_{p,k}(r) = \sup\{\|x\|_k : x \in e_n\},$$

if $n \in \mathbb{N}$ and $n-1 \leq r \leq n$. $\phi_{p,k}$ is then nonnegative increasing, and we have for all $r > 0$, $p,k \in \mathbb{N}$

$$U_{q(p)} \subset \phi_{p,k}(r)\, U_k + \frac{1}{r} U_p .$$

Choosing a positive, increasing ϕ with $\frac{\phi_{p,k}}{\phi}$ bounded we get the result.

Remark: Using the same argumentation we can even find an absolutely convex compact set B and a ϕ such that for every $p \in \mathbb{N}$ there exists $q \in \mathbb{N}$ and $C > 0$ with

$$U_q \subset C\phi(r)\, B + \frac{1}{r} U_p$$

for all $r > 0$.

6.6. Lemma: If $\lambda(A)$ is nuclear and has property (Ω_ϕ), $a_{j,1} \geq 1$ for all j, then there exists a q, such that for every k we have a $C > 0$ with

$$a_{j,k} \leq C\phi(a_{j,q})\, a_{j,q} .$$

The proof is very similar to that of 5.3. We omit it here (cf. [21]).

6.7. Lemma: For every ϕ there exists a nuclear (F)-space which does not have property (Ω_ϕ).

Proof: Beginning with $a_{j,1} = 1$ for all j we choose inductively $a_{j,k+1} \geq a_{j,k}$ in such a way that

$$\lim_j \frac{a_{j,k}}{a_{j,k+1}} \phi(a_{j,k}) = 0 \text{ and } \sum_{j=1}^{\infty} \frac{a_{j,k}}{a_{j,k+1}} < +\infty .$$

The argumentation in the remark after 6.3. gives now the following theorem which proves the nonexistence of a quotient universal nuclear (F)-space.

6.8. Theorem: There exists no nuclear (F)-space (even no (FS)-space) E_o such that every nuclear (F)-space is isomorphic to a quotient space of E_o.

It should be remarked that in the construction of 6.5., we can postulate even stronger nuclearities. So for example there exists no (FS)-space which has all s-nuclear, $\Lambda(\alpha)$-nuclear etc. spaces as quotients. On the other side in [23] it is proved that there exists a separable (F)-space which has all separable (F)-spaces (and therefore all nuclear spaces) as quotient spaces. 6.3. and 6.5. show that this space does not have property (Ω_ϕ) for any ϕ.

§ 7. LIFTING AND EXTENSION OF LINEAR MAPS, TENSORPRODUCTS OF (F)-SPACES WITH (DF)-SPACES

We shall present now some consequences of the results of § 2 without trying to get the highest degree of generalization. The remarks in § 2 show for example that the assumptions of nuclearity can be considerably weakened.
In the following thm. in fact only one of the spaces in the exact sequence has to be assumed to be nuclear.

7.1. Theorem: Let $0 \to E \to F \to G \to 0$ be an exact sequence of nuclear (F)-spaces such that E has (Ω), G has (DN); then the sequence is split.

Proof: This is an immediate consequence of 2.2. together with 1.4.

From this we can derive easily lifting and extension theorems for linear maps.

7.2. Theorem: Let $0 \to E \to F \overset{q}{\to} G \to 0$ be an exact sequence of nuclear (F)-spaces, E having property (Ω), let H be a nuclear (F)-space having (DN), $\phi \in L(H,G)$. Then there exists a $\psi \in L(H,F)$ with $q \circ \psi = \phi$:

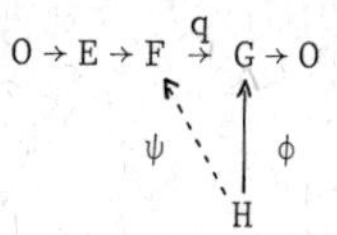

$$\begin{array}{ccccccccc} 0 & \to & E & \to & F & \overset{q}{\to} & G & \to & 0 \\ & & & & & \overset{\psi}{\nwarrow} & \uparrow \phi & & \\ & & & & & & H & & \end{array}$$

Proof: Denote by $X \subset H \oplus F$ the space $X := \{(x,y) : \phi x = qy\}$ and by $p_1 : X \to H$, $p_2 : X \to F$, $i_2 : E \to H \oplus F$ the canonical maps. Then we have an exact sequence

$$0 \to E \overset{i_2}{\to} X \overset{p_1}{\to} H \to 0,$$

which is split according to 7.1. Let r be a right inverse for p_1, then $\psi = p_2 \circ r$ is the desired map, for we have $q \circ \psi = q \circ p_2 \circ r = \phi \circ p_1 \circ r = \phi$.

7.3. Theorem: Let $0 \to E \overset{i}{\to} F \to G \to 0$ be an exact sequence of nuclear (F)-spaces, G having property (DN), let H be a nuclear (F)-space having (Ω), $\phi \in L(E,H)$. Then there exists a $\psi \in L(F,H)$ with $\psi \circ i = \phi$:

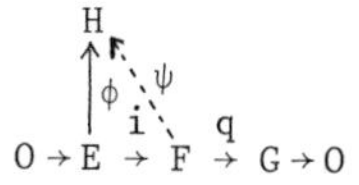

$$\begin{array}{ccccccccc} & & H & & & & & & \\ & & \uparrow \phi & \overset{\psi}{\nwarrow} & & & & & \\ 0 & \to & E & \overset{i}{\to} & F & \overset{q}{\to} & G & \to & 0 \end{array}$$

Proof: Denote by $Y_o \subset H \oplus F$ the space $Y_o := \{(\phi\xi, -i\xi) : \xi \in E\}$, which is closed in $H \oplus F$, and define $Y := H \oplus F / Y_o$. q induces a map $Q \in L(Y,G)$ by $Q(x,y) = q(y)$, whereas the canonical injection $i_1 : H \to H \oplus F$ induces an injection $I : H \to Y$. We get an exact sequence

$$0 \to H \overset{I}{\to} Y \overset{Q}{\to} G \to 0,$$

which is again split according to 7.1. Let P be a left inverse for I, then $\psi = P \circ I_2$ is the desired map, where I_2 is induced by the canonical injection $i_2 : F \to H \oplus F$. For we have $\psi \circ i = P \circ I_2 \circ i = P \circ I \circ \phi = \phi$.

We have seen that in certain situations we can lift linear maps, resp. extend linear maps. We are now going to interprete these results as results giving an affirmative answer under certain general conditions to the following question: Given an exact sequence of nuclear (F)-spaces ((DF)-spaces) $0 \to E \overset{i}{\to} F \overset{q}{\to} G \to 0$ and a nuclear (DF)-space ((F)-space) H, is the sequence

$$0 \to F \hat{\otimes}_\pi H \xrightarrow{i \otimes id} F \hat{\otimes}_\pi H \xrightarrow{q \otimes id} G \hat{\otimes}_\pi H \to 0$$

exact again?

In general this is not true, for take a non split sequence of (nuclear)(F)-spaces and let H = G. The element of $G \hat{\otimes}_\pi H$ which belongs to id_G will not be in the image of $q \otimes id$ (cf. [7]). It should be mentioned that it is well known that the answer is always affirmative if E,F,G,H are nuclear (F)-spaces or complete nuclear (DF)-spaces (cf. [7]).

From a theorem of Grothendieck [7] we know that the only interesting question in this context is the surjectivity of $q\otimes \mathrm{id}$. So only this is to prove in the following theorems.

7.4. Definition: A (DF)-space H has property (A) if its strong dual has property (DN), it has property (DΩ) if its strong dual has property (Ω).

Necessary and sufficient conditions for (A) and (DΩ) in terms of a fundamental system $B_1 \subset B_2 \subset \ldots$ of absolutely convex bounded sets easily can be obtained from 2.1. and 5.2. by replacing the U_k^o by B_k resp. the $\| \ \|_{-k}$ by the Minkowski norms of the B_k.

7.5. Theorem: Let $0 \to E \overset{i}{\to} F \overset{q}{\to} G \to 0$ be an exact sequence of nuclear (F)-spaces, E having property (Ω), H a complete nuclear (DF)-space with property (A), then also

$$0 \to E \hat{\otimes}_\pi H \xrightarrow{i\otimes \mathrm{id}} F \hat{\otimes}_\pi H \xrightarrow{q\otimes \mathrm{id}} G \hat{\otimes}_\pi H \to 0$$

is exact.

Proof: 7.2. tells us that the application $\psi \mapsto q\circ\psi$ is surjective from $L(H',F)$ to $L(H',G)$. This is equivalent to the surjectivity of $q\otimes \mathrm{id}$.

7.6. Theorem: Let $0 \to E \overset{i}{\to} F \overset{q}{\to} G \to 0$ be an exact sequence of nuclear (DF)-spaces, E having property (A), H a nuclear (F)-space with property (Ω), then also

$$0 \to E \hat{\otimes}_\pi H \xrightarrow{i\otimes \mathrm{id}} F \hat{\otimes}_\pi H \xrightarrow{q\otimes \mathrm{id}} G \hat{\otimes}_\pi H \to 0.$$

is exact.

Proof: From 7.3 we see that $\psi \longmapsto \psi\circ q^\tau$ is surjective from $L(F',H)$ to $L(G',H)$. Again this is equivalent to the surjectivity of $q\otimes \mathrm{id}$.

§ 8. EXAMPLES AND APPLICATIONS

For $K \subset \mathbb{R}^n$ compact, $\Omega \subset \mathbb{R}^n$ open we use the following well-known notations:

$\mathcal{E}(\Omega) = C^\infty(\Omega)$ = infinitely often differentiable functions on Ω,

$\mathcal{D}(K) = \{\phi \in \mathcal{E}(\mathbb{R}^n) : \operatorname{supp} \phi \subset K\}$,

$\mathcal{D}(\Omega) = \{\phi \in \mathcal{E}(\mathbb{R}^n) : \operatorname{supp} \phi \subset \Omega,\ \operatorname{supp} \phi \text{ compact}\}$,

$\mathcal{E}(\Omega,K) = \{\phi \in \mathcal{E}(\Omega) : \phi^{(k)}(x) = 0 \text{ all } x \in K,\ k \in \mathbb{N}^n\}$,

$\mathcal{E}(K)$ = Whitney-differentiable functions on K (cf. [11] e.g.).

For a sequence $M_p, M_p > 0$ all p, we denote by $\mathcal{E}^{(M_p)}(\Omega)$ (resp. $\mathcal{E}^{\{M_p\}}(\Omega)$) the set of all $\phi \in \mathcal{E}(\Omega)$ such that to every $K \subset \Omega$, and every $h > 0$ (resp. there exists a $h > 0$ such that)

$$\|\phi\|_{K,h} = \sup_{\substack{x\in K \\ j}} \frac{\phi^{(j)}(x)}{M_{|j|}\, h^{|j|}} < +\infty .$$

$\mathcal{D}^{(M_p)}(K)$, $\mathcal{D}^{(M_p)}(\Omega)$, $\mathcal{D}^{\{M_p\}}(K)$, $\mathcal{D}^{\{M_p\}}(\Omega)$ are then defined as in the case of C^∞- functions.

We assume that M_p is so well behaved that there exist $\mathcal{D}^{(M_p)}$- partitions of unity and all spaces are closed under multiplication and differentation (cf. [9]).

We assume all spaces to be equipped with their natural topologies.

8.1. Proposition: (1) $\mathcal{D}(K)$, $\mathcal{D}^{(M_p)}(K)$ always have property (DN).

(2) $\mathcal{E}(\Omega)$, $\mathcal{E}^{(M_p)}(\Omega)$ have not (DN).

(3) $\mathcal{E}(K)$ sometimes has (DN), sometimes not.

Proof: (1) For $\mathcal{D}(K)$ we can use the following norms: $|||\phi|||_k := \| \phi^{(k_1,\dots,k)} \|_{L_2}$.
By partial integration we get

$$|||\phi|||_k^2 = \int \phi \, \overline{\phi}^{(2k,\dots,2k)} \leq |||\phi|||_o \; |||\phi|||_{2k} .$$

Even simpler we can proceed in the case of $\mathcal{D}^{(M_p)}$:

$$|||\phi|||_k^2 = \sup_{j,x} \frac{|\phi^{(j)}(x)|^2}{M_{|j|}^2 \; h^{2|j|}} \leq \sup_{j,x} \frac{|\phi^{(j)}(x)|}{M_{|j|}} \; \sup_{j,x} \frac{|\phi^{(j)}(x)|}{M_{|j|} \; h^{2|j|}}$$

$$\leq |||\phi|||_1 \; \| \phi \|_{h^2} .$$

(2) Neither $\mathcal{E}(\Omega)$ nor $\mathcal{E}^{(M_p)}(\Omega)$ admit a continuous norm.

(3) $K = \{0\}$ gives $\mathcal{E}(K) = \omega$ which apparently has not (DN). $K = [o,1]$ gives $\mathcal{E}(K) \cong s$ which has (DN).

8.2. Proposition: $\mathcal{E}(\Omega)$, $\mathcal{D}(K)$, $\mathcal{E}(\Omega,K)$, $\mathcal{E}(K)$, $\mathcal{E}^{(M_p)}(\Omega)$, $\mathcal{D}^{\{M_p\}'}(K)$, $\mathcal{D}^{\{M_p\}'}(\Omega)$ have property (Ω).

Proof: Using partition of unity one proves easily that $\mathcal{E}(\Omega)$ is a quotient space of $\prod_j \mathcal{D}(K_j)$ if $\Omega = \bigcup_j \overset{o}{K}_j$. Taking the K_j as cubes and using the fact that $\mathcal{D}(\text{cube}) \cong s$ (cf. [16]), we see that $\mathcal{E}(\Omega)$ is a quotient of $s^{\mathbb{N}}$, which is a quotient of s.
From Whitney's extension theorem we see that $\mathcal{E}(K)$ is a quotient of $\mathcal{E}(\mathbb{R}^n)$, hence quotient of s, so it has (Ω).

In the case of $\mathcal{E}^{(M_p)}(\Omega)$ we proceed as in the case of $\mathcal{E}(\Omega)$ taking instead of $\mathcal{D}(K_j)$, K_j cube, the space $\mathcal{P}^{(M_p)}(\prod_{j=1}^{k} [a_j,b_j])$ of all $\phi \in \mathcal{E}^{(M_p)}(\mathbb{R}^n)$ which are periodic in each variable x_j with period $b_j - a_j$, $j=1 \dots n$. According to a result of Petzsche [15] this space is a power series space (of type $+\infty$).

The same procedure reduces the case of $\mathcal{D}^{\{M_p\}'}(\Omega)$ to $\mathcal{D}^{\{M_p\}'}(K)$. Since $\mathcal{D}^{\{M_p\}'}(K)$ is reflexive, we have apparently that the $\|\ \|_h$ on $\mathcal{D}^{\{M_p\}}(K)$ are the Minkowski norms of a fundamental system of equicontinuous sets in $(\mathcal{D}^{\{M_p\}'}(K))'$. But for this we get for given $h_1 > h_o$, $h > h_1$ and n such that $(\frac{h}{h_1}) \leq (\frac{h_1}{h_o})^n$:

$$\|\phi\|_{h_1}^{n+1} = \sup_{\substack{j\\ x\in K}} \frac{|\phi^{(j)}(x)|}{M_{|j|}\, h^{|j|}} \left(\frac{|\phi^{(j)}(x)|}{M_{|j|}\, h_o^{|j|}}\right)^n \left(\frac{h\, h_o^{\,n}}{h_1^{n+1}}\right)^{|j|} \leq \|\phi\|_h \ \|\phi\|_{h_o}^n .$$

Therefore for every h_o there exists a h_1 $(>h_o)$ such that for every h we have a $n \in \mathbb{N}$ with

$$\|\phi\|_{h_1}^{n+1} \leq \|\phi\|_h \ \|\phi\|_{h_o}^n .$$

This gives the result by 5.2.

It remains to prove property (Ω) for $\mathcal{D}(K)$ and $\mathcal{E}(\Omega,K)$. But $\mathcal{D}(K)$ is a projected subspace of $\mathcal{E}(\mathbb{R}^n,\tilde{K})$ where $\tilde{K} = L \setminus \mathring{K}$ with L compact, $K \subset \mathring{L}$, so the only open case is $\mathcal{E}(\Omega,K)$, which apparently reduces to $\mathcal{E}(\mathbb{R}^n,K)$ by using a $\mathcal{D}$-function which is 1 in a neighbourhood of K.

Let be $\phi\in\mathcal{D}$, supp $\phi \subset \{x: |x| \leq 1\}$, $\int\phi = 1$, $\phi_r(x) := r^{-n}\phi(rx)$, $K_r = \{x: d(x,K) \geq r^{-1}\}$, $\alpha_r(x) := \int_{K_r} \phi_{2r}(x-\xi)d\xi$, $\beta_r(x) = 1 - \alpha_r(x)$.

With

$$\|\phi\|_m = \sup_{\substack{|j|\leq m\\ x\in L_m}} |\phi^{(j)}(x)|$$

where $K \subset \mathring{L}_o \subset L_o \subset \mathring{L}_1 \subset \ldots$, L_m compact, $\bigcup_m L_m = \mathbb{R}^n$, we have for $f\in \mathcal{E}(\mathbb{R}^n)$:

$$\alpha_r f \in \mathcal{E}(\mathbb{R}^n,K),\ \|\alpha_r f\|_m \leq C_1 r^m \|f\|_m ,$$

and for $f\in\mathcal{E}(\mathbb{R}^n,K)$: $\beta_r f \in \mathcal{E}(\mathbb{R}^n,K)$, $\|\beta_r f\| \leq C_2 \frac{1}{r} \|f\|_{m+1}$.

The proof is then completed by the following:

<u>8.3. Lemma</u>: If $F \subset E$ and if for every $r > 0$ there exists $\alpha_r \in L(E,F)$, $\beta_r \in L(F,F)$ such that $\alpha_r + \beta_r = id_F$, $\alpha_{r'} \circ \alpha_r = \alpha_r$ all $r' \geq 3r$ and with an appropriate seminorm system on E : $\|\alpha_r f\|_m \leq C_1 r^m \|f\|_m$, $\|\beta_r f\|_m \leq \frac{C_2}{r} \|f\|_{m+1}$,

then with E also F has property (Ω).

<u>Proof</u>: Let $U_k = \{f: \|f\|_k \leq 1\}$; then, using (Ω) for E, for every $p \in \mathbb{N}$ we have a $q > p$ such that for every k there exist $\nu, C > o$ with

$$U_q \subset C\, r^{\nu}\, U_k + \frac{1}{r} U_p$$

for all $r > 0$. Replacing r by r^{p+q+1} and multiplying by r^q we get

$$r^q\, U_q \subset C\, r^{\mu}\, U_k + \frac{1}{r^{p+1}} U_p\,,$$

where $\mu = \nu\,(2q+1)+q$.

For $f \in F \cap U_q$ we have $f = \alpha_r f + \beta_r f$, $\beta_r f \in \frac{C_2}{r} U_p$, $\alpha_r \in C_1\, r^q\, U_q$. Then $\alpha_r f = \psi_1 + \psi_2$, $\psi_1 \in C_3 r^{\mu}\, U_k$, $\psi_2 \in \frac{C_1}{r^{p+1}} U_p$ and therefore $\alpha_r \phi = \alpha_{3r}\, \psi_1 + \alpha_{3r}\, \psi_2$ with $\alpha_{3r}\, \psi_1 \in C_4\, r^{\mu+k}\, U_k$, $\alpha_{3r}\, \psi_2 \in C_5\, \frac{1}{r} U_p$. From this the desired result follows easily.

Property (Ω) for $\mathcal{E}(\mathbb{R}^n, K)$ was first proved by Tidten [17], who used the following result to give sufficient conditions for the existence of a continuous extension operator $\mathcal{E}(K) \to \mathcal{E}(\mathbb{R}^n)$ ([17]):

8.4. Proposition: $\mathcal{E}(K)$ has (DN), iff there exists a continuous extension operator $\mathcal{E}(K) \to \mathcal{E}(\mathbb{R}^n)$.

Proof: If there exists a continuous extension operator $\mathcal{E}(K) \to \mathcal{E}(\mathbb{R}^n)$, then for L compact, $K \subset L^o$, there exists a continuous extension operator $\mathcal{E}(K) \to \mathcal{D}(L)$. So $\mathcal{E}(K)$ is isomorphic to a (projected) subspace of $\mathcal{D}(L)$, which has (DN).

For the converse use the exact sequence

$$0 \to \mathcal{E}(\mathbb{R}^n, K) \overset{i}{\to} \mathcal{E}(\mathbb{R}^n) \overset{q}{\to} \mathcal{E}(K) \to 0,$$

where q is the restriction map and i is the natural imbedding. The surjectivity of q comes from Whitney's theorem. Since $\mathcal{E}(\mathbb{R}^n, K)$ has always property (Ω) (cf. 8.2.), we get an extension operator (i.e. a right inverse for q) if $\mathcal{E}(K)$ has property (DN).

We finish this § by an application of the tensorproduct-version of the thms. in § 7, which gives a distribution valued version of a theorem of E. Borel.

8.5. Proposition: Given $\Omega \subset \mathbb{R}^n$ open and for every $j \in \mathbb{N}^n$ $T_j \in \mathcal{D}'(\Omega)$, there exists a $f \in C^{\infty}(\mathbb{R}^n, \mathcal{D}'(\Omega))$ such that $f^{(j)}(o) = T_j$ for all j.

Proof: By the theorem of E. Borel we have the following exact sequence

$$0 \to \mathcal{E}(\mathbb{R}^n, \{o\}) \to \mathcal{E}(\mathbb{R}^n) \overset{\Delta}{\to} \omega_n \to 0\,,$$

where ω_n is the space of all sequences $(x_j)_{j \in \mathbb{N}^n}$, and $\Delta f = (f^{(j)}(o))_j$.

Since for every compact set $K \subset \mathbb{R}^n$, $\mathcal{D}(K)$ has property (DN), or equivalently $\mathcal{D}'(K)$ has property (A), we obtain from 7.2. that

$$\Delta : \mathcal{E}(\mathbb{R}^n, \mathcal{D}'(K)) \to (\mathcal{D}'(K))^{\mathbb{N}^n}$$

is surjective.

$\mathcal{D}'(\Omega)$ is a projected subspace of $\prod_\nu \mathcal{D}'(K_\nu)$ with $K_\nu \subset \Omega$ compact, $\bigcup_\nu \overset{\circ}{K}_\nu = \Omega$. Hence also

$$\Delta : \mathcal{E}(\mathbb{R}^n, \mathcal{D}'(\Omega)) \to (\mathcal{D}'(\Omega))^{\mathbb{N}^n}$$

is surjective.

Taking for example $n = 1$, $\Omega = \mathbb{R}^n$ we get, using a regularisation argument:

8.6. Theorem: For every sequence $T_o, T_1, \ldots \in \mathcal{D}'(\mathbb{R}^n)$ there exists a C^∞- function f on $H_+ := \{(t,x) \in \mathbb{R}^{n+1} : t > o\}$ such that f and all derivatives of f are locally slowly increasing and have distributional boundary values, when $t \to o+$,

$$T_k = \lim_{t \to o+} \text{distr} \frac{\partial^k f}{\partial t^k}(x,t).$$

REFERENCES

1 BESSAGA, C.: Some Remarks on Dragilev's Theorem, Studia Math. 31 (1968), 3o7-318

2 DRAGILEV, M.M.: On Regular Bases in Nuclear Spaces, Am. Math. Soc. Transl. (2) 93 (197o), 61-82 (Engl. Translation of Mat. Sb. 68 (11o) (1965), 153-173)

3 DUBINSKY, E.: Basic Sequences in (s), to appear

4 DUBINSKY, E., RAMANUJAN, M.S.: On λ-Nuclearity, Mem. Am. Math. Soc. 128 (1972)

5 DUBINSKY, E., ROBINSON, W.: Quotient Spaces of (s) with Basis, to appear

6 DREWNOWSKI, L., LOHMAN, R.H.: On the Number of Separable Locally Convex Spaces Proc. Am.Math. Soc. 58 (1976), 185-188

7 GROTHENDIECK, A.: Produits tensoriels topologiques et espaces nucléaires, Mem. Am.Math. Soc. 16 (1955)

8 KABALLO, W.: Lifting-Sätze für Vektorfunktionen und das ε-Tensorprodukt, Habilitationsschrift, Kaiserslautern (1976)

9 KOMATSU, H.: Ultradistributions I, Structure Theorems and a Characterization J. Fac. Sc. Univ. Tokyo, Sec. I A, 2o, No. 1, (1973) 25-1o5

1o KOMURA, T. und Y.: Über die Einbettung der nuklearen Räume in $(s)^A$, Math. Ann. 162 (1966), 284-288

11 MALGRANGE, B.: Ideals of Differentiable Functions, Bombay (1966)

12 MARTINEAU, A.: Sur une propriété universelle de l'espace des distributions de M. Schwartz, C.R. Acad. Sci. Paris 259 (1964), 3162-3164

13 MITYAGIN, B.S.: Approximative Dimension and Bases in Nuclear Spaces, Russian Math. Surveys 16 (1961), No. 4, 59-127 (Engl. Translation of Usp. Math. Nauk 16 (1961), No. 4, 63-132)

14 PELCZYNSKI, A.: On the Approximation of S-Spaces by Finite Dimensional Spaces, Bull. Acad. Polon. Sci. 5 (1957), 879-881

15 PETZSCHE, H.J.: Darstellung der Ultradistributionen vom Beurlingschen und Roumieuschen Typ durch Randwerte holomorpher Funktionen, Dissertation, Düsseldorf (1966)

16 PIETSCH, A.: Nukleare lokalkonvexe Räume, Berlin 1969

17 TIDTEN, M.: Fortsetzungen von C^∞-Funktionen, welche auf einer abgeschlossenen Menge in $\mathbb{R}^n$ definiert sind, to appear

18 VOGT, D.: Vektorwertige Distributionen als Randverteilungen holomorpher Funktionen, manuscripta math. 17 (1975), 267-290

19 VOGT, D.: Tensorprodukte von (F)- mit (DF)-Räumen und ein Fortsetzungssatz, to appear

2o VOGT, D.: Charakterisierung der Unterräume von s, to appear

21 VOGT, D., Wagner, M.J.: Charakterisierung der Quotientenräume von (s) und eine Vermutung von Martineau, to appear

22 WAGNER, M.J.: Über zwei spezielle Klassen von Stufenräumen, Diplomarbeit, Mainz (1975)

23 WAGNER, M.J.: Unterräume und Quotienten von Potenzreihenräumen. Dissertation, Wuppertal (1977)

24 ZAHARIUTA, V.P.: On the Isomorphism of Cartesian Products of Locally Convex Spaces, Studia Math. 46 (1973), 2o1-221

K.-D. Bierstedt, B. Fuchssteiner (eds.)
Functional Analysis: Surveys and Recent Results

REPRESENTATION OF DISTRIBUTIONS AND ULTRADISTRIBUTIONS BY HOLOMORPHIC FUNCTIONS

Reinhold Meise
Mathematisches Institut der Universität
Düsseldorf

0. INTRODUCTION

The aim of the present article is to give a survey of recent developments in the theory of representation of distributions and ultradistributions by boundary values of holomorphic functions, as it was initiated by Köthe and Tillmann. We shall begin with some "historical remarks" leading to the formulation of the general problem in the setting of distributions. In the second part we demonstrate in the case of certain $K\{M_p\}$ spaces that it is rather useful to introduce locally convex topologies on the corresponding spaces of holomorphic functions and to apply Grothendieck's theory of topological tensor products. Then we sketch how this method and a refined version of the Mittag-Leffler argument were used by Vogt in the case of arbitrary distributions on $\mathbb{R}^N$. In the last part we shortly introduce ultradistributions (following Komatsu) and give the main results (as well as the ideas of their proofs) concerning the representation of ultradistributions by holomorphic functions. These results were obtained by Komatsu, Körner and Petzsche.

1. FORMULATION OF THE PROBLEM

Only a few years after the introduction of distributions by Schwartz [37], Köthe [20], as a consequence of his results in [21], recognized a relation between distributions on the unit circle and holomorphic functions on its complement. To be more precise, he proved, among other things, the following result: Let $\bar{\mathbb{C}}$ denote the Riemann sphere, S^1 the unit circle, and $H_o(\bar{\mathbb{C}}\setminus S^1)$ the space of functions holomorphic on $\bar{\mathbb{C}}\setminus S^1$, vanishing at infinity and with the property that there exist constants $C > 0$ and $k \in \mathbb{N}$ such that $|f(z)| \leq C|1-|z||^k$

for every $z \in \bar{\mathbb{C}} \setminus S^1$ with $|z| < 1$ or $1 < |z| < 2$. $H_o(\bar{\mathbb{C}} \setminus S^1)$ is endowed with its natural inductive limit topology. Then the mapping $\wedge : C^\infty(S^1)'_b \to H_o(\bar{\mathbb{C}} \setminus S^1)$, $\hat{T} : z \to \frac{1}{2\pi i} \left\langle T_x, \frac{1}{x-z} \right\rangle$, is a topological isomorphism. The inverse of $\wedge$ is the mapping $R : H_o(\bar{\mathbb{C}} \setminus S^1) \to C^\infty(S^1)'_b$, given by

$$R(f) : \varphi \to \lim_{r \to 1-} \int_{S^1} (f(rz) - f(\tfrac{z}{r}))\varphi(z)\,dz.$$

With the appropriate modifications this result remains true for any simply closed analytic curve C instead of S^1.

Then Tillmann [40] generalized Köthe's results in two directions. First he showed that one could also use simply closed curves C passing through the point ∞ and analytic in each finite part; next he investigated the case of an N-fold product of such curves as a subset of $\bar{\mathbb{C}}^N$. In this way he showed that it was possible to represent (as described below) distributions with compact support on $\mathbb{R}^N$ by holomorphic functions on $(\mathbb{C} \setminus \mathbb{R})^N$. Tillmann's results suggested the following general question:

<u>Problem.</u> Let E denote a complete locally convex (l.c.) space (over the field $\mathbb{C}$) and $\mathcal{K}(\mathbb{R}^N)$ a l.c. space of C^∞-functions on $\mathbb{R}^N$ which contains $\mathcal{D}(\mathbb{R}^N)$ as a continuously and densely embedded subspace. Is it then possible to find a vector space $H_{\mathcal{K}}(N,E)$ of holomorphic E-valued functions on $(\mathbb{C} \setminus \mathbb{R})^N$ such that the mapping $R_E^N : H_{\mathcal{K}}(N,E) \to L_b(\mathcal{K}(\mathbb{R}^N), E) := \mathcal{K}'_b(\mathbb{R}^N, E)$ can be defined by

$$R_E^N(f) : \varphi \to \lim_{\varepsilon \downarrow 0} \int_{\mathbb{R}^N} \{ \sum_{\sigma \in \{-1,1\}^N} (\prod_{j=1}^N \sigma_j) f(x + i\sigma\varepsilon) \} \varphi(x)\,dx$$

for every $f \in H_{\mathcal{K}}(N,E)$ and $\varphi \in \mathcal{K}(\mathbb{R}^N)$ (or at least for every $\varphi \in \mathcal{D}(\mathbb{R}^N)$) and such that this mapping is actually surjective? If the answer to this is affirmative, can one then also introduce an appropriate l.c. topology on $H_{\mathcal{K}}(N,E)$ such that R_E^N becomes a topological homomorphism?

Among the authors who dealt with this question we should like to mention the following ones:

$\mathcal{E}(\mathbb{R}^N)$: Tillmann [40], [43], Itano [18], Bierstedt and Meise [5].

$\mathcal{D}_{L^p}(\mathbb{R}^N)$: Tillmann [41], Luszczki and Zielezny [26], Bengel [2], [3].

$\mathcal{S}(\mathbb{R}^N)$: Tillmann [42], Martineau [27], Schmidt [36], Meise [30], [31], Bierstedt and Meise [5], Vogt [48].

$\mathcal{D}_F(\mathbb{R}^N)$: Tillmann [42].

$\mathcal{D}(\mathbb{R}^N)$: Tillmann [42], Martineau [27], Itano [18], Meise [28], Konder [24], Vogt [46], [47], [49].

In the next sections we shall show how Grothendieck's theory of topological tensor products can be used to treat the vector valued and the higher dimensional cases in a unified manner. For this purpose a l.c. topology with nice properties is introduced on the spaces $H_{\mathcal{K}}(N) := H_{\mathcal{K}}(N,\mathbb{C})$, as it was done by Wloka [54], the author [29] and Vogt [46]. This method also works in the case of ultradistributions, as we shall see in section 4.

Since topological tensor products can be applied only on product sets, we shall not deal with distributional boundary values of holomorphic functions in tubular radial domains, as several authors have done (see e.g. Vladimirov [45], Carmichael [8], [9], [10]). Moreover, this article will not contain any information on applications of distributional boundary values in other fields, as given in the books of Bremermann [7], Beltrami and Wohlers [1], Vladimirov [45] and Colojoară and Foiaş [11], where the interested reader can find further references.

2. CERTAIN NUCLEAR $K\{M_p\}$ SPACES

The program sketched above shall now be carried out with almost all details for certain $K\{M_p\}$ spaces, because the general idea will become quite clear in this simple case.

2.1 Throughout this section let $(M_n)_{n\in\mathbb{N}}$ denote a fixed sequence of entire functions in the complex plane which satisfy the following conditions:

(1) For every $n\in\mathbb{N}$ $M_n|[0,\infty)$ is strictly increasing.

(2) For every $n\in\mathbb{N}$ and every $x\in[0,\infty)$ we have $1 \le M_n(x) \le M_{n+1}(x)$.

(3) For every $n\in\mathbb{N}$ there exist $m\in\mathbb{N}$ $(m\geq n)$ and $C_n > 0$ such that for every $x\in[0,\infty)$ we have $(1+x^2)M_n(x) \le C_nM_m(x)$.

(4) For every $n\in\mathbb{N}$ and every $\nu\in\mathbb{N}_0$ there exist $m\in\mathbb{N}$ and $C_n > 0$ such that for every $z\in\mathbb{C}$: $|M_n^{(\nu)}(z)| \le C_n\, M_m(|z|)$.

(5) For every $n\in\mathbb{N}$ there exist $m\in\mathbb{N}$ $(m\geq n)$ and $B_n > 0$ such that for every $z\in\mathbb{C}$ with $|\mathrm{Im}\, z| \le 1$ we have $M_n(|z|) \le B_nM_m(|\mathrm{Re}\, z|)$.

(6) For every $n\in\mathbb{N}$ there exist $m\in\mathbb{N}$ $(m\geq n)$, $r_n > 0$ and $K_n > 0$ such that $\sup_{x\in[0,\infty)} \frac{M_n(x+r_n)}{M_m(x)} \leq K_n$.

<u>2.2 Definition.</u> Let $(M_n)_{n\in\mathbb{N}}$ be as above and $N\in\mathbb{N}$. Then we define:

a) $\mathcal{K}(\mathbb{R}^N) := \{\varphi\in C^\infty(\mathbb{R}^N)$: for every $n\in\mathbb{N}$

$$p_n(\varphi) := \sup_{|\alpha|\leq n} \sup_{x\in\mathbb{R}^N} |D^\alpha\varphi(x)| \prod_{j=1}^{N} M_n(|x_j|) < \infty \},$$

and we give $\mathcal{K}(\mathbb{R}^N)$ the l.c. topology induced by the norms $(p_n)_{n\in\mathbb{N}}$.

b) $v_n : (\mathbb{C}\setminus\mathbb{R})^N \to \mathbb{R}$ by $v_n(z) := \prod_{j=1}^{N} [(M_n(|z_j|))^{-1}\min(1,|\text{Im } z_j|^n)]$,

and $H_n := \{f$: f is a holomorphic function on $(\mathbb{C}\setminus\mathbb{R})^N$ satisfying

$$\|f\|_n := \sup_{z\in(\mathbb{C}\setminus\mathbb{R})^N} |f(z)|v_n(z) < \infty \}.$$

Since $H_n \subset H_{n+1}$ we can define $H_{\mathcal{K}}(N)$ to be the locally convex inductive limit of the spaces H_n.

<u>2.3 Examples.</u> a) The functions $M_n : z \to (1+z^2)^n$ satisfy all the conditions of 2.1 and lead to the well known space $\mathcal{S}(\mathbb{R}^N)$ of rapidly decreasing C^∞-functions on $\mathbb{R}^N$, introduced by Schwartz [37]. The corresponding space $H_{\mathcal{S}}(N)$ was introduced by Tillmann [42] and is called the space of slowly increasing holomorphic functions on $(\mathbb{C}\setminus\mathbb{R})^N$.

b) Let $(a_n)_{n\in\mathbb{N}}$ be a strictly increasing sequence of positive real numbers and let $(k_n)_{n\in\mathbb{N}}$ be an increasing sequence of natural numbers. Then the functions $M_n : z \to \exp(a_n z^{k_n})$ satisfy all the conditions of 2.1.

Taking $a_n = n$ and $k_n = 1$ one gets the space $\mathcal{K}_1(\mathbb{R}^N)$ (introduced by Sebastião e Silva [38]) which, together with its dual (respectively with the convolution operators on it), was investigated by Hasumi [16] and Zielezny [55].

Let $A > 0$ and $k\in\mathbb{N}$ be given. Taking $a_n = \frac{1}{keA^k}(1-\frac{1}{n})$ and $k_n = k$ one obtains the space $S_{1/k,A}(\mathbb{R}^N)$, a special case of the spaces $S_{\alpha,A}$ introduced by Gelfand and Shilov [14].

<u>2.4 Proposition.</u> a) The space $\mathcal{K}(\mathbb{R}^N)$ is a nuclear Fréchet space which contains $\mathcal{D}(\mathbb{R}^N)$ as a continuously and densely embedded subspace.

b) The space $H_{\mathcal{K}}(N)$ is a (DFN)-space, i.e. it is the strong dual of a nuclear Fréchet space.

Proof. a) By 2.1 (1), (4) and (3) the nuclearity follows from Swartz [39], Thm 2. All the other statements are proved in Chapter 2., section 2 of Friedman [13].
b) The conditions given in 2.1 imply that Satz 2. in § 2 of Meise [29] can be applied, which proves b).

2.5. Proposition (Tillmann [42], Wiegner [51]). The mapping $R^1 : H_{\mathcal{K}}(1) \to \mathcal{K}_b'(\mathbb{R})$ is a surjective topological homomorphism with $\ker R^1 = H(\mathbb{C}) \cap H_{\mathcal{K}}(1)$ ($H(\mathbb{C})$ denotes the space of entire functions).

Proof. As in the case $\mathcal{S}(\mathbb{R})$ (see Friedman [13], Thm. 12, Chap. 2.), one proves that any $T \in \mathcal{K}'(\mathbb{R})$ can be represented as $T = (d/dx)^k (M_n(|x|)g(x))$, where k and n are suitable natural numbers and g is a continuous and bounded function in $L^1(\mathbb{R})$. Denoting by $\chi_{[0,\infty)}$ the characteristic function of the intervall $[0,\infty)$, one defines $g_+ := g\chi_{[0,\infty)}$ and $g_- := g(1-\chi_{[0,\infty)})$. Then the functions

$$g_\pm : (\mathbb{C} \setminus \mathbb{R}) \to \mathbb{C},\ g_\pm(z) := \frac{1}{2\pi i} \int_{\mathbb{R}} \frac{g_\pm(x)}{x-z} dx$$

are used to define

$f : (\mathbb{C} \setminus \mathbb{R}) \to \mathbb{C}$ by $f(z) := (d/dz)^k [M_n(z) g_+(z) + M_n(-z) g_-(z)]$.

By use of the special representation of T and of partial integration, one shows $f \in H_{\mathcal{K}}(1)$ and $R^1(f) = T$ as well as the continuity of R^1. Since $H_{\mathcal{K}}(1)$ and $\mathcal{K}_b'(\mathbb{R})$ are (LF)-spaces, continuity and surjectivity of R^1 imply (by the open mapping theorem for (LF)-spaces) that R^1 is an open mapping.
It is trivial that $H(\mathbb{C}) \cap H_{\mathcal{K}}(1)$ is contained in $\ker R^1$. For every $f \in H_{\mathcal{K}}(1)$ there exists an $m \in \mathbb{N}$ such that m-fold primitives $f_+^{[-m]}$ and $f_-^{[-m]}$ in the upper and lower half plane have continuous boundary values on $\mathbb{R}$. (This is seen by estimating the growth near the real axis.) Hence for $f \in \ker R^1$ it follows by partial integration that $(d/dx)^m (f_+^{[-m]}(x+io) - f_-^{[-m]}(x-io)) = 0$. Therefore $f_+^{[-m]}(x+io) - f_-^{[-m]}(x-io)$ is a polynomial p with $\deg p \leq m-1$. This implies that there exists an entire function h such that $h^{(m)} = f$, hence $f \in H(\mathbb{C}) \cap H_{\mathcal{K}}(1)$.

2.6 Proposition. For every complete (DF)-space E the mapping $R^1 \hat{\otimes} id_E : H_{\mathcal{K}}(1) \hat{\otimes}_\pi E \to \mathcal{K}_b'(\mathbb{R}) \hat{\otimes}_\pi E$ is a surjective topological homomorphism with $\ker R^1 \hat{\otimes} id_E = (\ker R^1) \hat{\otimes}_\pi E$.

Proof. This is a consequence of 2.5 and 2.4 b) together with theorems of Grothendieck [15] on the π-tensor product of topological homomorphisms with dense range and on the (completed) π-tensor product of (DF)-spaces. We shall sketch the proof of the surjectivity in order to show how the nuclearity of $\mathcal{K}(\mathbb{R})$ and $H_{\mathcal{K}}(1)$ is used. Because of 2.3 the abstract version of the Schwartz kernel theorem gives $H_{\mathcal{K}}(1)\hat{\otimes}_\pi E = L_b(H_{\mathcal{K}}(1)'_b, E)$ and $\mathcal{K}'_b(\mathbb{R})\hat{\otimes}_\pi E = L_b(\mathcal{K}(\mathbb{R}), E)$. Using this identification, the mapping $R^1\hat{\otimes}\, id_E$ corresponds to the mapping $u \to u\circ {}^tR^1$.

Let $T\in L_b(\mathcal{K}(\mathbb{R}), E)$ be given. Because $\mathcal{K}(\mathbb{R})$ is bornological, T is even a continuous linear mapping from $\mathcal{K}(\mathbb{R}^N)$ into E_{bor}, the bornological space associated with E. Since E is a (DF)-space, it has a fundamental sequence $(B_n)_{n\in\mathbb{N}}$ of bounded, closed, circled and convex sets. For every $n\in\mathbb{N}$ the associated normed space E_{B_n} is then a Banach space and E_{bor} is represented as the inductive limit of the E_{B_n}. Now theorem A of Grothendieck [15], I, p. 16 implies that there exists $n\in\mathbb{N}$ such that $T\in L(\mathcal{K}(\mathbb{R}), E_{B_n})$. The nuclearity of $\mathcal{K}(\mathbb{R})$ implies that T can be represented as $T(x) = \sum_{j=1}^{\infty} \lambda_j \langle x, y_j\rangle e_j$, where $\{y_j : j\in\mathbb{N}\} \subset \mathcal{K}'(\mathbb{R})$ is equicontinuous, $\{e_j : j\in\mathbb{N}\}$ is bounded in E_{B_n} and $(\lambda_j)_{j\in\mathbb{N}}$ is a sequence in l^1. Since by Lemma 5 of Meise [30] it is possible to find a bounded set $\{f_j : j\in\mathbb{N}\} \subset H_{\mathcal{K}}(1)$ such that $R^1(f_j) = y_j$, one can define $S: H_{\mathcal{K}}(1)'_b \to E_{B_n} \hookrightarrow E$ by $S(h') := \sum_{j=1}^{\infty} \lambda_j \langle h', f_j\rangle e_j$. It is easy to see $S\circ {}^tR^1 = T$, hence $R^1\hat{\otimes}\, id_E$ is surjective.

The following proposition will enable us to reduce the N-dimensional case to the one-dimensional vector valued case.

2.7 Proposition. a) For every complete l.c. space E and every $N\in\mathbb{N}$ we have $H_{\mathcal{K}}(N)\hat{\otimes}_\pi E = \{f\in H((\mathbb{C}\setminus\mathbb{R})^N, E) : \text{for every } e'\in E',\ e'\circ f\in H_{\mathcal{K}}(N)\}$ $(=: H_{\mathcal{K}}(N,E))$.

b) $H_{\mathcal{K}}(N)\hat{\otimes}_\pi H_{\mathcal{K}}(M) = H_{\mathcal{K}}(N+M)$ for every $N,M\in\mathbb{N}$.

c) $\mathcal{K}'_b(\mathbb{R}^N)\hat{\otimes}_\pi \mathcal{K}'_b(\mathbb{R}^M) = \mathcal{K}'_b(\mathbb{R}^{N+M})$ for every $N,M\in\mathbb{N}$.

Proof. a) This is a consequence of 2.4 b) and Grothendieck [15], II, Thm. 13., where the (completed) π-tensor product of certain nuclear

function spaces with a complete l.c. space E is described as a space of E-valued functions.
b) This is proved in Meise [29], §3, Satz 2.
c) The proof can be given in the same way as it was done in Treves [44], Cor. to Thm. 51.6.

Now we are able to solve the problem formulated in section 1. for the spaces $\mathcal{K}(\mathbb{R}^N)$ and the class of complete (DF)-spaces, by defining the space $H_{\mathcal{K}}(N,E)$ as in 2.7 a). The following theorem generalizes Satz 2 of Meise [30].

2.8 Theorem. For every complete (DF)-space E and every $N \in \mathbb{N}$ the mapping $R_E^N : H_{\mathcal{K}}(N,E) \to \mathcal{K}_b'(\mathbb{R}^N, E)$ is a surjective topological homomorphism, and $\ker R_E^N = \{f \in H_{\mathcal{K}}(N,E) : f = \sum_{j=1}^{N} f_j$, where $f_j \in H_{\mathcal{K}}(j-1) \hat{\otimes}_\pi \ker R^1 \hat{\otimes}_\pi H_{\mathcal{K}}(N-j) \hat{\otimes}_\pi E \}$.

Proof. First one gives another description of $H_{\mathcal{K}}(N,E)$, e.g. as in §2, Bemerkung 1. in Meise [29], and uses this to show that R_E^N can be defined on $H_{\mathcal{K}}(N,E)$ and that it is continuous. Then it is immediately clear that R_E^N coincides with $R^N \hat{\otimes}\, id_E$. Hence one can use 2.7 b) and c) together with 2.6 to prove the theorem by induction on N. In order to avoid too complicated formulas we will only give the induction step from N = 1 to N = 2: By 2.6, 2.7 b) and c) and by what we said at the beginning,
$R_E^2 = (id_{\mathcal{K}_b'(\mathbb{R})} \hat{\otimes} R^1 \hat{\otimes}\, id_E) \circ (R^1 \hat{\otimes}\, id_{H_{\mathcal{K}}(1) \hat{\otimes}_\pi E})$ is a surjective topological homomorphism. The statement about $\ker R_E^N$ is proved as in the case of $\mathcal{S}(\mathbb{R}^N)$, cf. Meise [31], §3., Satz 2.

Now we can ask the natural question, whether it is possible to have a representation of all distributions in $\mathcal{K}_b'(\mathbb{R}^N, E)$ for a class of l.c. spaces considerably larger than the class of all complete (DF)-spaces. For the space $\mathcal{S}(\mathbb{R}^N)$ of rapidly decreasing C^∞-functions this can be done by enlarging the spaces of holomorphic functions used for the representation.

2.9 Definition. For $n \in \mathbb{N}$ let $v_n : (\mathbb{C} \setminus \mathbb{R})^N \to \mathbb{R}$ be given by
$v_n(z) := \prod_{j=1}^{N} [(1+|z_j|)e^{-|\mathrm{Im}\, z_j|} \min(1,|\mathrm{Im}\, z_j|)]^n$. These $(v_n)_{n\in\mathbb{N}}$ define spaces H_n of holomorphic functions on $(\mathbb{C} \setminus \mathbb{R})^N$ in the same way as in 2.2 b). We denote $\underset{n\to}{\mathrm{ind}}\, H_n$ by $H_{\mathcal{S}}^e(N)$. Then it follows from Meise [29], that 2.4 b) and 2.7 a) and c) remain true.

<u>2.10 Theorem</u> (Meise [31]). For every $N \in \mathbb{N}$ and every complete l.c. space E the mapping $R_E^N : H_{\mathcal{S}}^e(N,E) \to \mathcal{S}_b'(\mathbb{R}^N, E)$ is a surjective topological homomorphism, the kernel of which is of the same form as indicated in 2.8 (with $H_{\mathcal{S}}^e$ instead of $H_{\mathcal{S}}$).

<u>Proof.</u> Since the induction arguments can be applied again, we only have to show that R_E^1 is surjective for every complete l.c. space E. Now we use the Fourier transform $\mathcal{F}$ on $\mathcal{S}_b'(\mathbb{R}, E)$. If $T \in \mathcal{S}'(\mathbb{R}, E)$ is given, set $S := \mathcal{F}^{-1}T$ and decompose $S = S_o + S_+ + S_-$, where $\mathrm{Supp}(S_o)$ is compact, $\mathrm{Supp}(S_+) \subset [0,+\infty)$ and $\mathrm{Supp}(S_-) \subset (-\infty,0]$. Then we define $\tilde{S}_o$ as the Fourier-Laplace transform of S_o (i.e. $\tilde{S}_o(z) :=$ $:= \langle S_{ox}, e^{-ixz} \rangle$), $\tilde{S}_+ : z \to \langle S_{+x}, e^{-ixz} \rangle$ for Im z < 0 and $\tilde{S}_- : z \to \langle S_{-x}, e^{-ixz} \rangle$ for Im z > 0. If we set $f : (\mathbb{C} \setminus \mathbb{R}) \to E$,

$$f(z) := \begin{cases} \tilde{S}_+(z) & \text{for } \operatorname{Im} z < 0 \\ \tilde{S}_-(z) + \tilde{S}_o(z) & \text{for } \operatorname{Im} z > 0, \end{cases}$$

we get a function in $H_{\mathcal{S}}^e(1,E)$ which has the property $R_E^1(f) = \mathcal{F} S = T$.

In [48] Vogt showed that the fact that a representation of all distributions in $\mathcal{K}'(\mathbb{R}, E)$ for every complete l.c. space E is possible has the farreaching consequence that ker R^1 is a continuously projected subspace of $H_{\mathcal{K}}(1)$. This is an immediate corollary of the following proposition.

<u>2.11 Proposition</u> (Vogt [48]). Let E and F be l.c. spaces and let $\alpha \in L(E,F)$ be surjective. If there exists a l.c. space Y for which $Y_c' = F$, then the surjectivity of $\alpha \varepsilon \mathrm{id}_Y : E\varepsilon Y \to F\varepsilon Y$ implies that ker α is a continuously projected subspace of E.

<u>Proof.</u> By definition we have $F\varepsilon Y = L_e(Y_c', F) = L_e(F,F)$. Since $\mathrm{id}_F \in L(F,F)$, there exists $u \in E\varepsilon Y = L_e(Y_c',E) = L_e(F,E)$ such that $\mathrm{id}_F = (\alpha \varepsilon \mathrm{id}_Y)(u) = \alpha \circ u$. Hence $u \circ \alpha \in L(E,E)$ is a projection, because the equality $(u \circ \alpha) \circ (u \circ \alpha) = u \circ \alpha$ holds. The identity $\alpha \circ u = \mathrm{id}_F$ implies that u is injective, hence $\ker \alpha = \ker u \circ \alpha$.

<u>2.12 Remarks.</u> a) The kernel of $R^1 : H_{\mathcal{S}}^e(1) \to \mathcal{S}_b'(\mathbb{R})$ is the space of all Fourier-Laplace transforms of distributions with compact support. By 2.10 and 2.11 this is a continuously projected subspace of $H_{\mathcal{S}}^e(1)$.

b) Using a conformal mapping and the duality result of Köthe [21], Vogt [46] was able to give a concrete representation for $(H_{\mathcal{S}}(1))_b'$.

Then he showed that, in this case, ker R^1 (= all complex polynomials) is not a continuously projected subspace of $H_{\mathcal{S}}(1)$. Hence by 2.11, theorem 2.8 cannot be true for all complete l.c. spaces E. Vogt [46] gave also examples of spaces E for which R^1_E is not surjective, e.g. $\mathcal{S}(\mathbb{R})$, $C^\infty(\mathbb{R})$, $\mathcal{D}(\mathbb{R})$, $\mathcal{D}_{L^p}(\mathbb{R})$, $C^\infty[a,b]$ and $\mathcal{D}[a,b]$. On the other hand it is an easy consequence of 2.8 that R^N_E is surjective for every continuously projected subspace of a product of complete (DF)-spaces. Hence R^N_E is surjective for $\mathcal{D}'_b(\Omega)$, $C^m(\Omega)$ and $(\mathcal{D}^m(\Omega))'_b$ for $0 \leq m < \infty$.

c) For the $K\{M_p\}$ spaces given by 2.1 and 2.2 (except $\mathcal{S}$) it seems to be unknown, whether ker R^1 is continuously projected. And it is also not known, whether an enlargement of the spaces of holomorphic functions leads to a similar result as for $\mathcal{S}$.

3. THE SPACE $\mathcal{D}(\mathbb{R}^N)$

Because of the structure of $\mathcal{D}(\mathbb{R}^N)$ - which is an (LFN)-space but not an (FN)-space - the representation of arbitrary distributions on $\mathbb{R}^N$ is more complicated than for the distributions considered in section 2. The first result in this direction was given by Tillmann [42] and reads as follows.

3.1 Theorem (Tillmann [42]). Let $H_{\mathcal{D}}(1)$ denote the space of all holomorphic functions on $(\mathbb{C}\setminus\mathbb{R})$ with the property that for every $n\in\mathbb{N}$ there exists an $m\in\mathbb{N}$ such that $\sup\{|f(z)||\mathrm{Im}\, z|^m : |\mathrm{Re}\, z| \leq n, 0 < |\mathrm{Im}\, z| \leq 1\} < \infty$. Then $R^1 : H_{\mathcal{D}}(1) \to \mathcal{D}'(\mathbb{R})$ is surjective and $\ker R^1 = H(\mathbb{C})$.

Proof. The proof is similar to the proof of the classical Mittag-Leffler theorem. Let $T \in \mathcal{D}'(\mathbb{R})$ be given. Then T can be represented as a locally finite sum $T = \sum_{n\in\mathbb{Z}} T_n$ of distributions T_n with $\mathrm{Supp}(T_n) \subset (n-1,n+1)$. Because of Tillmann's result on the representation of distributions with compact support (mentioned in section 1.), there exists $f_n\in H_{\mathcal{D}}(1)$ with $R^1(f_n) = T_n$. Since f_n is holomorphic in $\{z\in\mathbb{C} : |z| \leq |n|-1\}$, one can find polynomials p_n such that $\sum_{n\in\mathbb{Z}}(f_n-p_n)$ converges uniformly on every compact subset of $(\mathbb{C}\setminus\mathbb{R})$ and such that the limit defines a function $f\in H_{\mathcal{D}}(1)$. It is easy to show that $R^1(f) = T$. The statement concerning the kernel is proved by applying the arguments in the proof of 2.5 locally.

Remarks. a) The proof given above applies also to distributions with values in an (F)-space. This fact was noticed by Itano [18] and the author [28].
b) In the case of $\mathcal{D}(\mathbb{R}^N)$, $N > 1$, the Mittag-Leffler argument works only for distributions with support in certain cones, as the author [28] demonstrated.

Itano [18] showed that it is not possible to represent all distributions in $\mathcal{D}'(\mathbb{R}, E)$ for every complete l.c. space E. Soon after Itano's counterexample appeared Vogt constructed a very simple and illustrating example which we shall give now.

3.2 Example (Vogt [75]). Let E be the space $\bigoplus_{n\in\mathbb{N}} \mathbb{C}$ of all finite sequences endowed with the finest l.c. topology and define $T \in \mathcal{D}'(\mathbb{R}, E)$ by $T := \sum_{n\in\mathbb{N}} e_n \cdot \delta_n$, where $e_n = (\delta_{jn})_{j\in\mathbb{N}} \in E$ and δ_n denotes the evaluation at the point n. Then T cannot be represented by a holomorphic function. This is proved by the following arguments:
Let K_+ resp. K_- denote compact subsets with interior points which are contained in the upper resp. lower half plane. Then for every $f \in H(\mathbb{C}\setminus\mathbb{R}, E)$, $f(K_+ \cup K_-)$ is a compact subset in E, and hence contained in a finite dimensional subspace E_o of E. By the theorem of Hahn-Banach and analytic continuation it follows that $f(\mathbb{C}\setminus\mathbb{R}) \subset E_o$. If $R^1_E(f)$ is defined, then the range if $R^1_E(f)$ is contained in E_o, because E_o is closed. Since the range of T equals E, T cannot be in the range of R^1_E.

Following the general program we now define a l.c. space $H_{\mathcal{D}}(N)$ of holomorphic functions on $(\mathbb{C}\setminus\mathbb{R})^N$, introduced by Vogt [46], [47].

3.3 Definition. Let N be any natural number. For $n,m \in \mathbb{N}$ denote by G_{nm} the set $G_{nm} := \{z\in\mathbb{C} : |\mathrm{Re}\, z| < n + 1/m,\ 0 < |\mathrm{Im}\, z| < n + 1/m\}$ and define $v_m : \mathbb{C}^N \to \mathbb{R}$ by $v_m(z) := \prod_{j=1}^{N} |\mathrm{Im}\, z_j|^m$. Then we define

$H_{nm} := \{f\in H(G^N_{nm}) : \|f\|_{n,m} := \sup\{|f(z)|v_m(z) : z\in G^N_{nm}\} < \infty\}$, endowed

with the norm $\|\cdot\|_{n,m}$. For fixed n we can form $\operatorname{ind}_{m\to} H_{nm}$. Taking the projective limit over n, we get a l.c. space of holomorphic functions on $(\mathbb{C}\setminus\mathbb{R})^N$, called the space of locally slowly increasing functions $H_{\mathcal{D}}(N) := \operatorname{proj}_{\leftarrow n} \operatorname{ind}_{m \to} H_{nm}$.

The following proposition was proved in Vogt [47], 1.4, 1.5 and 1.6.

<u>3.4 Proposition</u> (Vogt [47]). a) $H_{\mathcal{D}}(N)$ is a complete nuclear space for every $n \in \mathbb{N}$.

b) $H_{\mathcal{D}}(N) \hat{\otimes}_{\pi} H_{\mathcal{D}}(M) = H_{\mathcal{D}}(N+M)$ for every $N, M \in \mathbb{N}$.

c) For every complete l.c. space E and every $N \in \mathbb{N}$ we have $H_{\mathcal{D}}(N) \hat{\otimes}_{\pi} E = \{f \in H((\mathbb{C} \setminus \mathbb{R})^N, E) : \text{for every } e' \in E', e' \circ f \in H_{\mathcal{D}}(N)\}$ $(=: H_{\mathcal{D}}(N,E))$.

<u>3.5 Theorem</u> (Vogt [47]). For every $N \in \mathbb{N}$ the mapping $R^N : H_{\mathcal{D}}(N) \to \mathcal{D}'_b(\mathbb{R}^N)$ is a surjective topological homomorphism, the kernel of which is of the same form as indicated in 2.8.

<u>Proof:</u> We only sketch the main ideas of the proof:
Using a special partition of unity by C^∞-functions one shows that $\mathcal{D}'_b(\mathbb{R}^N)$ is a continuously projected subspace of the space $(s'_b)^{\mathbb{N}}$, where s denotes the (FN)-space of rapidly decreasing sequences. By a more complicated induction argument than the one that we used in the proof of 2.8, the proof of the surjectivity of R^N is then reduced to the surjectivity of the mappings R^1_E for the spaces $E = H_{\mathcal{D}}(N,s')$ (see 3.4 c)). A deep analysis of the structure of the spaces $H_{\mathcal{D}}(N,s')$ led Vogt to the observation that one can find an (F)-space H_o which is continuously embedded in $H_{\mathcal{D}}(N,s')$ and has the following property (see Vogt [47], Lemma 2.3):

For every holomorphic function f on $D_r = \{z \in \mathbb{C} : |z| < r\}$ $(r > 0)$ with values in $H_{\mathcal{D}}(N,s')$ and any given neighbourhood U of zero in H_o, there exists a holomorphic function $g : D_{2r} \to H_{\mathcal{D}}(N,s')$ such that $(f+g)|D_{r/2}$ is in $H(D_{r/2}, H_o)$ and $(f+g)(D_{r/3}) \subset U$.

This property of the space H_o (its definition is too complicated to be given here) makes it possible to apply the Mittag-Leffler construction again, in order to show the surjectivity of the mapping $R^1_{H_{\mathcal{D}}(N,s')}$:

Let $T \in \mathcal{D}'(\mathbb{R}, H_{\mathcal{D}}(N,s'))$ be given. Choose $\varphi_n \in \mathcal{D}(\mathbb{R})$ with $\varphi|[-n,n] \equiv 1$ and put $F_n : z \to \left\langle T_x, \frac{1}{2\pi i} \frac{\varphi_n(x)}{x-z} \right\rangle$. Then $G_{n-1} := F_n - F_{n-1}$ is in $H(D_{n-1}, H_{\mathcal{D}}(N,s'))$. By the above-mentioned property of the space H_o one finds inductively functions g_n $(g_o = 0)$ such that $L_{n-1} := (G_{n-1} - g_{n-1}) + g_n$ restricted to $D_{\frac{n-1}{2}}$ is in $H(D_{\frac{n-1}{2}}, H_o)$ and such that the sequence $H_n := F_n + g_n = H_k + \sum_{j=k}^{n-1} L_j$ $(n>k)$ converges

pointwise on $\mathbb{C}\setminus\mathbb{R}$ to a function H, where $H|(D_k\setminus\mathbb{R}) = F_k + R_k$ and R_k is holomorphic on D_k. This implies $H \in H_{\mathcal{D}}(1,H_{\mathcal{D}}(N,s'))$ and $R^1_{H_{\mathcal{D}}(N,s')}(H) = T$. Since $\mathcal{D}'_b(\mathbb{R}^N)$ is an ultrabornological space, it follows from the definition of $H_{\mathcal{D}}(N)$ and from the continuity and surjectivity of the mapping R^N by the closed graph theorem of de Wilde [53] that R^N is a topological homomorphism.
The statement concerning the kernel is proved in the same way as in the proof of 2.8.

Remark. The proof shows also that R^N_E is surjective for $E = s'$, $H_{\mathcal{D}}(N)$, $\mathcal{D}'_b(\mathbb{R}^N)$ and for arbitrary (F)-spaces (see Vogt [47], 5.3).

A further analysis of his proof led Vogt to a new class of l.c. spaces with a countable fundamental system of bounded sets and the following nice characterization.

3.6 Theorem (Vogt [49]). Let E be a quasi-complete l.c. with a countable fundamental system of bounded sets. The following are equivalent:

(1) $R^1_E : H_{\mathcal{D}}(1,E) \to \mathcal{D}'_b(\mathbb{R},E)$ is surjective.

(2) $\frac{\partial}{\partial \bar{z}} : C^\infty(\mathbb{R}^2,E) \to C^\infty(\mathbb{R}^2,E)$ is surjective.

(3) E has the following property (A): There exists a closed, circled, convex and bounded set B in E, such that for every $k \in \mathbb{N}$ one can find $p \in \mathbb{N}$ and $C > 0$, such that $B_k \subset rB + \frac{C}{r}B_{k+p}$ holds for every $r > 1$.

Remark. By duality, property (A) gives also rise to a property for metrizable l.c. spaces, called (DN). Vogt [50] showed that nuclear metrizable spaces with property (DN) are exactly the subspaces of s, the space of all rapidly decreasing sequences.

4. ULTRADISTRIBUTIONS

In the year 1960 resp. 1961 Roumieu [34] resp. Beurling [4] (see also Björck [6]) proposed two different generalizations of the theory of distributions of Schwartz. The general idea consists in taking smaller spaces of test functions than $\mathcal{D}(\Omega)$ in such a way that the essential statements of the theory remain true (in a generalized form), while the dual spaces are enlarged. Beurling defines spaces of test functions by growth properties of their Fourier transforms, while Roumieu uses classes of ultradifferentiable functions

taken from classical analysis.
In 1973 Komatsu [22] gave a unified treatment of both theories in the Roumieu setting and proved deep and important structure theorems for these ultradistributions. (It should be mentioned that the proof of one structure theorem in Roumieu's article was not complete.)
Since the theory of ultradistributions is not as well known as the theory of distributions, we begin with the necessary definitions (following Komatsu [22]).

4.1 Throughout this section let $(M_p)_{p\in\mathbb{N}_o}$ denote a fixed sequence of positive real numbers with $\lim_{p\to\infty} M_p^{1/p} = \infty$, which satisfy the following conditions:

(M1) (logarithmic convexity) For every $p\in\mathbb{N}$ we have $M_p^2 \le M_{p-1}M_{p+1}$.

(M2) (stability under ultradifferential operators) There exist constants $A > 0$ and $H > 1$ such that for every $p\in\mathbb{N}_o$: $M_p \le A\cdot H^p \cdot \min_{o\le q\le p} M_q M_{p-q}$.

(M3) (strong non-quasi-analyticity) There is a constant $A > 0$ such that for every $p\in\mathbb{N}$: $\sum_{q>p} \frac{M_{q-1}}{M_q} \le A\cdot p\cdot \frac{M_{p-1}}{M_p}$.

<u>4.2 Definition.</u> Let $(M_p)_{p\in\mathbb{N}_o}$ be a sequence as in 4.1, and let Ω be an open subset of $\mathbb{R}^N$. We define

a) $\mathcal{E}^{(M_p)}(\Omega) := \{f\in C^\infty(\Omega)$: for every compact subset K in Ω and every

$$h > 0,\ \sup_{\alpha\in\mathbb{N}_o^N} \sup_{x\in K} \frac{|f^{(\alpha)}(x)|}{h^{|\alpha|}M_{|\alpha|}} := p_{K,h}(f) < \infty\}.$$

On this space, we take the topology induced by semi-norms $p_{K,h}$.

b) $\mathcal{E}^{\{M_p\}}(\Omega) := \{f\in C^\infty(\Omega)$: for every compact subset K in Ω there exists some $h > 0$ such that

$$\sup_{\alpha\in\mathbb{N}_o^N} \sup_{x\in K} \frac{|f^{(\alpha)}(x)|}{h^{|\alpha|}M_{|\alpha|}} < \infty\}.$$

Since the definition suggests a representation as a projective limit of an inductive limit, we endow $\mathcal{E}^{\{M_p\}}(\Omega)$ with this natural topology.

c) Using the symbol * for (M_p) or $\{M_p\}$, we put for every compact set K in Ω :

$\mathcal{D}^*_K(\Omega) := \{f\in \mathcal{E}^*(\mathbb{R}^N)$: $\mathrm{Supp}(f) \subset K\}$ and $\mathcal{D}^*(\Omega) := \mathrm{ind}_{K\to} \mathcal{D}^*_K(\Omega)$.

The elements of $\mathcal{D}^{(M_p)'}(\Omega)$ $(\mathcal{D}^{\{M_p\}'}(\Omega))$ are called ultradistributions of Beurling (Roumieu) type on Ω.

4.3 Definition. Let $(M_p)_{p\in\mathbb{N}_0}$ be a sequence as in 4.1.

a) We define $M^* : (0,\infty) \to \mathbb{R}$, the so-called associated function of the sequence $\left(\frac{M_p}{p!}\right)_{p\in\mathbb{N}_0}$, by $M^*(t) := \sup\limits_{p\in\mathbb{N}_0} \ln \frac{t^p M_0 p!}{M_p}$. Then M^* is increasing and continuous.

b) A sequence $(a_\alpha)_{\alpha\in\mathbb{N}_0}$ of complex numbers defines an ultradifferential operator $P(D) = \sum\limits_{\alpha\geq 0} a_\alpha D^\alpha$ of class (M_p) $(\{M_p\})$, if it satisfies the following condition:

There exist constants L and C (for every $L > 0$ there exists some C) such that for every $\alpha\in\mathbb{N}_0^N$ the estimate $|a_\alpha| \leq \frac{C\cdot L^{|\alpha|}}{M_{|\alpha|}}$ holds true.

4.4 Example. After some calculations one sees that for every $s > 1$ the Gevrey sequence $(M_p)_{p\in\mathbb{N}_0}$, defined by $M_p := p^{ps}$, satisfies the conditions (M1)-(M3). Ultradifferentiable functions in Gevrey classes appear in quite a natural way in the theory of partial differential operators with constant coefficients, because for every hypoelliptic differential operator $P(D)$ (on $\mathbb{R}^N$) with constant coefficients, there exists $s > 1$ such that every zero solution $f\in C^\infty(\Omega)$ of $P(D)$ belongs to the class $\mathcal{E}^{\{p^{ps}\}}(\Omega)$.

The associated function M^* of $\left(\frac{p^{ps}}{p!}\right)_{p\in\mathbb{N}_0}$ can be estimated in the following way:

There exist $L_j > 0$ and $C_j\in\mathbb{R}$, $j = 1,2$, such that for every $t > 0$

$$(L_1 t)^{\frac{1}{s-1}} + C_1 \leq M^*(t) \leq (L_2 t)^{\frac{1}{s-1}} + C_2.$$

4.5 Remark. Condition (M2) implies that ultradifferential operators of class * define continuous sheaf homomorphisms on $\mathcal{E}^*$ and $\mathcal{D}^{*'}$, while (M3) implies that there exist "enough" functions in $\mathcal{D}^*(\Omega)$. Many results of the theory can be derived by using weaker conditions, called (M2)' and (M3)' (see Komatsu [22]). But we shall need (M2) and (M3) eventually, hence we required them from the very beginning, in order to keep things as simple as possible.

4.6 Definition. Let $(M_p)_{p\in\mathbb{N}_0}$ be as in 4.1 and $N\in\mathbb{N}$.

a) Let $v_m : (\mathbb{C}\setminus\mathbb{R})^N \to \mathbb{R}$ be given by $v_m(z) := \prod_{j=1}^{N} \exp(-M^*(\frac{m}{|\operatorname{Im} z_j|}))$.

Using these weight functions we define $H_{\mathcal{D}(M_p)}(N)$ in the same way as $H_{\mathcal{D}}(N)$ in 3.3.

b) Let $G_n := \{z\in\mathbb{C} : |\operatorname{Re} z| < n,\ 0 < |\operatorname{Im} z| < n\}$ and $v_n : (\mathbb{C}\setminus\mathbb{R})^N \to \mathbb{R}$, $v_n(z) := \prod_{j=1}^{N} \exp(-M^*(\frac{1}{n|\operatorname{Im} z_j|}))$. Then the sequence of spaces $H_n := \{f\in H(G_n^N) : \|f\|_n := \sup_{z\in G_n^N} |f(z)|v_n(z) < \infty\}$ gives a projective system, whose limit $\operatorname{proj}_{\leftarrow n} H_n$ is denoted by $H_{\mathcal{D}\{M_p\}}(N)$.

It was proved by Petzsche [32], II, 2.3, 5.1 and 5.6, that proposition 3.4 is also true for the spaces $H_{\mathcal{D}^*}(N)$. The following analogue of theorem 3.1 for ultradistributions was shown by Komatsu [22].

4.7 Theorem (Komatsu [22]). The mapping $R^1 : H_{\mathcal{D}^*}(1) \to \mathcal{D}^{*\prime}(\mathbb{R})$ is surjective.

Proof. This is the last theorem in Komatsu's long paper [22], Thm. 11.8. The greater part of this article is used in its proof, hence we can only give a rough impression of what is needed for the proof. First of all it is not obvious that R^1 can be defined as a mapping from $H_{\mathcal{D}^*}(1)$ into $\mathcal{D}^{*\prime}(\mathbb{R})$. Komatsu [22], Thm. 11.5 shows that for a given $f\in H_{\mathcal{D}^*}(1)$ and a compact set K in $\mathbb{R}$ there exist an open neighbourhood U, an ultradifferential operator $P(D)$ and a function $g\in H((\mathbb{C}\setminus\mathbb{R})\cap U)$ such that $f = P(D)g$ and such that g is bounded. Then g has distributional boundary values, hence $f(x+io) - f(x-io) = P(D)(g(x+io)-g(x-io))$ is in $\mathcal{D}^{*\prime}$ locally. Therefore $R^1(f)\in\mathcal{D}^{*\prime}(\mathbb{R})$. If f runs through a bounded set in $H_{\mathcal{D}^*}(1)$, the corresponding functions g can be selected as a family of uniformly bounded functions on U. (It was noticed by Petzsche [32], II, 2.6 that this implies the continuity of R^1.)

In the case of $\mathbb{R}$ one can use the Cauchy-transform of ultradistributions with compact support (i.e. the function $\tilde{T} : z \to \left\langle T_x, \frac{1}{x-z}\right\rangle$, $T\in\mathcal{D}^{*\prime}(\mathbb{R})$, $\operatorname{Supp}(T)$ compact) and the Mittag-Leffler argument to show the surjectivity of R^1.

4.8 Remarks. a) Concerning the representation of ultradistributions by boundary values of holomorphic functions, Komatsu [22] derives also deep necessary and sufficient results of local nature. These

are not given here because, for the representation of ultradistributions on $\mathbb{R}^N$, it is sufficient to use 4.7, as we will see below.
b) In the case of ultradistributions of Roumieu type Komatsu [22], Thm. 11.8, assumes that the sequence $(M_p)_{p\in\mathbb{N}_0}$ satisfies a further condition. Wild [52] showed that this condition is superfluous.
Since the structures of $\mathcal{D}^{(M_p)}$ and $\mathcal{D}$ are quite similar, it was reasonable to conjecture that a suitable modification of Vogt's proof of theorem 3.5 should apply also to yield the representation of ultradistributions of Beurling type. It was shown by Petzsche [32] that this is indeed true. As a first step into this direction the kernel theorem for ultradistributions had to be proved. This is a consequence of nuclearity results of Komatsu [22], Thm. 2.6, and of theorems of Grothendieck [15], and was proved independently by Komatsu [23] and Petzsche [33], I.

<u>4.9 Theorem</u> (Komatsu [23], Petzsche [33]). Let Ω_1 resp. Ω_2 be open subsets of $\mathbb{R}^N$ resp. $\mathbb{R}^M$. Then $\mathcal{D}^{*\prime}_b(\Omega_1)\,\hat{\otimes}_\pi\mathcal{D}^{*\prime}_b(\Omega_2) = \mathcal{D}^{*\prime}_b(\Omega_1\times\Omega_2)$.

<u>4.10 Theorem</u> (Petzsche [32]). For every $N\in\mathbb{N}$ the mapping $R^N : H_{\mathcal{D}*}(N) \to \mathcal{D}^{*\prime}_b(\mathbb{R}^N)$ is a surjective topological homomorphism with $\ker R^N = \prod_{j=1}^{N} H_{\mathcal{D}*}(j-1)\,\hat{\otimes}_\pi\, H(\mathbb{C})\,\hat{\otimes}_\pi\, H_{\mathcal{D}*}(N-j)$.

<u>Proof.</u> a) Ultradistributions of Roumieu type: This case is now rather easy, because $H_{\mathcal{D}\{M_p\}}(N)$ and $\mathcal{D}^{\{M_p\}\prime}(\mathbb{R}^N)$ are (FN)-spaces. As we already mentioned in the proof of 4.8, R^1 is a continuous surjection, hence an open mapping by the classical open mapping theorem. So the surjectivity of R^N is - by induction - a consequence of a theorem of Grothendieck on the tensor product of topological homomorphisms between (F)-spaces (see e.g. Treves [44], Prop. 4.39).

b) Ultradistributions of Beurling type: Following the lines of Vogt's proof one has to look for suitable spaces $S(M_p,N)$ with the property that $\mathcal{D}^{(M_p)\prime}(\mathbb{R}^N)$ is a continuously projected subspace of $((S(M_p,N)'_b)^{\mathbb{N}}$. In the case $N = 1$ one can choose the space of periodic functions in $\mathcal{E}^{(M_p)}(\mathbb{R})$ which has a representation as a sequence space $S(M_p,1)$. Unfortunately $S(M_p,1)$ is not stable under the formation of π-tensor products, so $S(M_p,N) := \hat{\bigotimes}_{\pi,\,j=1}^{N} S(M_p,1)$ is

different from $S(M_p,1)$. Therefore the surjectivity of R^N is reduced to the surjectivity of R^1_E for all the spaces $E = H_{\mathcal{D}(M_p)}(N, S(N_p,M)'_b)$. In these cases one can apply Vogt's idea to construct a continuously embedded (F)-space H_o which has the same properties as described in the proof of 3.5 (see Petzsche [32], II. 5.8-5.10).
Concerning the kernel, Petzsche [32] gave a new proof avoiding hyperfunctions. He generalized a result of Vogt [46] for distributions and gave an explicit representation for $(H_{\mathcal{D}*}(N))'_b$ and then used duality theory to show $\ker R^1 = H(\mathbb{C})$.

Remarks. a) As in the case of distributions also for ultradistributions of Beurling type R^N_E is surjective for all spaces $E = H_{\mathcal{D}(M_p)}(N)$, $\mathcal{D}^{(M_p)}{}'_b(\mathbb{R}^N)$ and for arbitrary (F)-spaces.

b) The spaces $S(*,1)$ are also useful in other situations, because, in some sense, they generate all the spaces occuring in the theory. E.g. Petzsche [33], II could give sharp estimates on the type of λ-nuclearity of all these spaces and could show that certain conditions known to be sufficient are even necessary.

4.11 Remark. In the case of ultradistributions of Roumieu type Körner [19] had proved another version of 4.10 (before Petzsche) in the more general setting of the spaces $\mathcal{D}^{\{M_\alpha\}'}(\mathbb{R}^N)$, defined by multi-indexed $(M_\alpha)_{\alpha\in\mathbb{N}_o^N}$ (which satisfy conditions slightly different from 4.1; also the spaces $H_{\mathcal{D}*}(N)$ have to be changed a little bit.).
These spaces had been introduced by Roumieu [35] (see also Chou [12]).
Körner used the following structure theorem for ultradistributions of Roumieu type:
For every $T \in \mathcal{D}^{\{M_\alpha\}'}(\mathbb{R}^N)$ there exist continuous functions T_α on $\mathbb{R}^N$ such that $T = \sum_{\alpha\in\mathbb{N}_o^N} D^\alpha T_\alpha$. Then he applied a trick of Sato and constructed an entire function g which has the property that $f_\alpha := T_\alpha/g \in L^1(\mathbb{R}^N)$ for every $\alpha\in\mathbb{N}_o^N$. Hence f_α has a holomorphic representation $\tilde{f}_\alpha$. Putting $g_\alpha := g\cdot\tilde{f}_\alpha$ and $f := \sum_{\alpha\geq o} D_z^\alpha g_\alpha$, one gets a holomorphic representation of T. As in Komatsu's article [22], it is more difficult to prove that a function which satisfies the required growth conditions has boundary values in $\mathcal{D}^{\{M_\alpha\}'}(\mathbb{R}^N)$. Körner used a conformal mapping and showed that such a function f can be decom-

posed as $f = \sum_{\alpha \geq 0} f_\alpha$ with suitable holomorphic functions f_α. Estimates for the f_α imply $R^N(f) \in \mathcal{D}^{\{M_\alpha\}'}(\mathbb{R}^N)$.

Eventually we want to draw the attention of the reader to the thesis of Langenbruch [25]. He uses boundary values of zero solutions on $(\mathbb{R}^{N+1} \setminus \mathbb{R}^N)$ of certain hypoelliptic partial differential operators with constant coefficients on $\mathbb{R}^{N+1}$ to get a representation of distributions as well as of ultradistributions of Roumieu type in certain Gevrey classes (on $\mathbb{R}^N$).

REFERENCES

1. Beltrami, E. J., Wohlers, M. R. (1966) Distributions and the Boundary Values of Analytic Functions (Academic Press, New York).
2. Bengel, G. (1963) Distributionen aus $\mathcal{D}'_{L^p}$ und Randwertverteilungen analytischer Funktionen (Diplomarbeit Heidelberg).
3. Bengel, G. (1974) Darstellung skalarer und vektorwertiger Distributionen aus $\mathcal{D}'_{L^p}$ durch Randwerte holomorpher Funktionen, manuscripta math. 13, 15-25.
4. Beurling, A. (1961) Quasi-analyticity and general distributions, AMS Summer institute, Lectures 4. and 5. (Stanford, mimeographed).
5. Bierstedt, K.-D., Meise R. (1973) Distributionen mit Werten in topologischen Vektorräumen II, manuscripta math. 10, 313-357.
6. Björck, G. (1966) Linear partial differential operators and generalized distributions, Ark. Mat. 6, 351-407.
7. Bremermann, H. (1965) Distributions, Complex Variables and Fourier Transforms (Addison Wesley, Reading Mass.).
8. Carmichael, R. D. (1971) Distributional boundary values of functions analytic in tubular radial domains, Indiana U. Math. J. 20, 843-853.
9. Carmichael, R. D. (1974) Representation of distributions with compact support, manuscripta math. 11, 305-338.
10. Carmichael, R. D. (1975) Distributional boundary values in the dual spaces of spaces of type $\mathcal{S}$, Pacific J. Math. 56, 385-422.
11. Colojoară, I., Foiaş, C. (1968) Theory of Generalized Spectral Operators (Gordon and Breach, New York).
12. Chou, C.-C. (1973) La transformation de Fourier complexe et l'équation de convolution (Lecture Notes Math. 325, Springer-Verlag, Berlin).
13. Friedman, A. (1963) Generalized Functions and Partial Differential Equations, (Prentice-Hall, Englewood Cliffs, New Yersey)
14. Gelfand, I. M., Schilow, E. G. (1962) Verallgemeinerte Funktionen (Distributionen) II (Deutscher Verlag der Wissen-

schaften, Berlin).

15. Grothendieck, A. (1966) Produits tensoriels topologiques et espaces nucléaires (Memoirs of the AMS).

16. Hasumi, M. (1961) Note on the n-dimensional tempered ultra-distributions, Tohoku Math. J. 13, 94-104.

17. Hörmander, L. (1966) Linear Partial Differential Operators (Springer-Verlag, Berlin).

18. Itano, M. (1968) On the distributional boundary values of vector valued holomorphic functions, J. Sci. Hiroshima Ser. A 32, 397-440.

19. Körner, J. (1975) Roumieusche Ultradistributionen als Randverteilungen holomorpher Funktionen (Dissertation Kiel).

20. Köthe, G. (1952) Die Randverteilungen analytischer Funktionen, Math. Z. 57, 13-33.

21. Köthe, G. (1953) Dualität in der Funktionentheorie, J. reine angew. Math. 191, 30-49.

22. Komatsu, H. (1973) Ultradistributions, I Structure theorems and a characterization, J. Fac. Sci. Univ. Tokyo, Sec. I A, 20, 25-105.

23. Komatsu, H. (to appear) Ultradistributions, II The kernel theorem and ultradistributions with support in a submanifold.

24. Konder, P. P. (1971) Funktionentheoretische Charakterisierung der Topologie in Distributionenräumen, Math. Z. 123, 241-263.

25. Langenbruch, M. (1976) Randwerte von Nullösungen hypoelliptischer Differentialoperatoren (Dissertation Mainz).

26. Luszczki, Z., Zielezny, Z. (1961) Distributionen der Räume $\mathcal{D}'_{L^p}$ und Randverteilungen analytischer Funktionen, Colloq. Math. 8, 125-131.

27. Martineau, A. (1964) Distributions et valeurs au bord des fonctions holomorphes, in Theory of Distributions, Proc. Intern. Summer Inst., Inst. Gulbenkian de Ciència, Lisboa, 193-326.

28. Meise, R. (1968) Darstellung von Distributionen durch holomorphe Funktionen (Diplomarbeit Mainz).

29. Meise, R. (1972) Räume holomorpher Vektorfunktionen mit Wachstumsbedingungen und topologische Tensorprodukte, Math. Ann. 199, 293-312.

30. Meise, R. (1972) Darstellung temperierter vektorwertiger Distributionen durch holomorphe Funktionen I, Math. Ann. 198, 147-159.

31. Meise, R. (1972) Darstellung temperierter vektorwertiger Distributionen durch holomorphe Funktionen II, Math. Ann. 198, 161-178.

32. Petzsche, H. J. (1976) Darstellung der Ultradistributionen vom Beurlingschen und Roumieuschen Typ durch Randwerte holomorpher Funktionen (Dissertation Düsseldorf).

33. Petzsche, H. J. (to appear) Die Nuklearität der Ultradistributionsräume und der Satz vom Kern I, II.

34. Roumieu, C. (1960) Sur quelques extensions de la notion de distribution, Ann. Ec. Norm. Sup. 77, 41-121.

35. Roumieu, C. (1962/63) Ultra-distributions définies sur $\mathbb{R}^n$ et sur certaines classes de variétés differentiables, J. d'Analyse Math. 10, 153-192.

36. Schmidt, E. (1969) Funktionentheoretische Charakterisierung der Topologie im Raum der gemäßigten Distributionen (Dissertation Mainz).

37. Schwartz, L. (1966) Théorie des distributions (Hermann, Paris).

38. Sebastião e Silva, J. (1958) Les fonctions analytiques comme ultradistributions dans le calcul opérationel, Math. Ann. 136, 58-96.

39. Swartz, C. (1970) The nuclearity of $K\{M_p\}$ spaces, Math. Nachr. 44, 193-197.

40. Tillmann, H. G. (1953) Randverteilungen analytischer Funktionen und Distributionen, Math. Z. 59, 61-83.

41. Tillmann, H. G. (1961) Distributionen als Randverteilungen analytischer Funktionen II, Math. Z. 76, 5-21.

42. Tillmann, H. G. (1961) Darstellung der Schwartzschen Distributionen durch analytische Funktionen, Math. Z. 77, 106-124.

43. Tillmann, H. G. (1963) Darstellung vektorwertiger Distributionen durch holomorphe Funktionen, Math. Ann. 151, 286-295.

44. Treves, F. (1967) Topological Vector Spaces, Distributions and Kernels (Academic Press, New York).

45. Vladimirov, V. S. (1966) Methods of the theory of functions of several complex variables (M. I. T. Press, Cambridge, Mass.).

46. Vogt, D. (1972) Randverteilungen holomorpher Funktionen und die Topologie von $\mathcal{D}'$, Math. Ann. 196, 281-292.

47. Vogt, D. (1973) Distributionen auf dem $\mathbb{R}^N$ als Randverteilungen holomorpher Funktionen, J. reine angew. Math. 261, 134-145.

48. Vogt, D. (1973) Temperierte vektorwertige Distributionen und langsam wachsende holomorphe Funktionen, Math. Z. 132, 227-237.

49. Vogt, D. (1975) Vektorwertige Distributionen als Randverteilungen holomorpher Funktionen, manuscripta math. 17, 267-290.

50. Vogt, D. (to appear) Charakterisierung der Unterräume von s.

51. Wiegner, M. (1971) Distributionen als Randverteilungen analytischer Funktionen (Diplomarbeit Bonn).

52. Wild, W. (1975) Ultradistributionen mit kompaktem Träger und Randverteilungen holomorpher Funktionen (Staatsexamensarbeit Mainz).

53. de Wilde, M. (1967) Théorème du graphe fermé et espaces à réseau absorbant, Bull. Math. Soc. Sci. Math. Roumaine 11, 225-238.

54. Wloka, J. (1966) Reproduzierende Kerne und nukleare Räume I, Math. Ann. 163, 167-188.

55. Zielezny, Z. (1968) On the space of convolution operators in $\mathcal{K}_1'$, Studia Math. 31, 111-124.

K.-D. Bierstedt, B. Fuchssteiner (eds.)
Functional Analysis: Surveys and Recent Results

RICHNESS OF THE CLASS OF HOLOMORPHIC FUNCTIONS ON AN INFINITE DIMENSIONAL SPACE

Martin Schottenloher

In this note a survey on some results and recent developments in the field of Infinite Dimensional Holomorphy is given guided by the following simple and general question: *How many holomorphic functions are there on a given open subset* Ω *of a locally convex Hausdorff space over* $\mathbb{C}$ *?* This question will be considered under various aspects:
1. Bounding Sets 2. Prescribed Radius of Convergence
3. Domains of Holomorphy and Analytic Continuation
4. Holomorphic Completion 5. Compact Holomorphic Mappings and the Approximation Property.

1. BOUNDING SETS

It is natural to compare the holomorphic functions on a given locally convex Hausdorff space E over $\mathbb{C}$ with the linear continuous forms $\nu \in E'$ on E. Let us recall that a function $f: \Omega \to \mathbb{C}$ on an open subset $\Omega \subset E$ is holomorphic if it is continuous and holomorphic on each complex line, i.e. for all $x \in \Omega$ and $a \in E$, the function $\lambda \longmapsto f(x+\lambda a)$ is holomorphic in a neighborhood of $0 \in \mathbb{C}$. (General reference for holomorphic functions is [23] and [25].) A finite product of linear forms $\nu \in E'$ and a locally uniformly convergent sequence of such products is holomorphic on E, hence the space $H(E)$ of holomorphic functions on E contains E' properly. In order to know more about the difference between $H(E)$ and E' it is of interest to study those subsets of E which remain bounded under all functions $f \in H(E)$:

DEFINITION. A subset $B \subset E$ is called *bounding* if

$$\|f\|_B := \sup\{|f(b)| \mid b \in B\} < \infty \text{ for all } f \in H(E) .$$

Since a subset $B \subset E$ is bounded if and only if it is weakly bounded, i.e. if $\|\nu\|_B < \infty$ for all $\nu \in E'$, a bounding subset of E is always bounded. The following results show that the bounding subsets are in many cases much smaller than the bounded sets which implies that there must be substantially more holomorphic functions on E than linear forms $\nu \in E'$:

PROPOSITION 1: Every closed bounding subset of a separable, quasicomplete locally convex space E over $\mathbb{C}$ is compact.

Proof. It suffices to show that a subset B of E which is not precompact cannot be a bounding set. If B is not precompact one can find a sequence (x_n) contained in B and a continuous seminorm $p: E \to \mathbb{R}$ on E with $p(x_n - x_m) \geq 2s$ for $n \neq m$, where $s > 0$ is suitably chosen. Let (z_n) be dense in E. Put

$$V_n = \tilde{co}\{z_1,\dots,z_n\} + \{x \in E \mid p(x) \leq s\},$$

where $\tilde{co}\, A$ denotes the absolutely convex hull of a set A. Since $\tilde{co}\{z_1,\dots,z_n\}$ is compact, each V_n contains at most finitely many x_m. Hence there is an increasing sequence n_k with $x_{n_k} \in V_{n_{k+1}} \setminus V_{n_k}$. Without loss of generality we can assume that $x_k \in V_{k+1} \setminus V_k$. Since V_k is absolutely convex there is a continuous linear form $\nu_k \in E'$ with

$$\|\nu_k\|_{V_k} < 1 < |\nu_k(x_k)| .$$

By induction one can find $p(k) \in \mathbb{N}$ such that the polynomial $f_k = \nu_k^{p(k)} \in H(E)$ satisfies

$$\|f_k\|_{V_k} < 2^{-k} \quad \text{and} \quad |f_k(x_k)| \geq k + 1 + \sum_{j=1}^{k-1} |f_j(x_k)| .$$

Since $(\mathring{V}_n)$ is an increasing open covering of E, $f = \sum f_k$ is holomorphic on E and

$$|f(x_k)| \geq |f_k(x_k)| - \sum_{j=1}^{k-1} |f_j(x_k)| - \sum_{j=k+1}^{\infty} \|f_j\|_{V_j} \geq k .$$

COROLLARY 1: Let E be a Banach space which is isomorphic to a subspace of a space $C(T)$, where T is a sequentially compact Hausdorff space and where the space $C(T)$ of continuous functions on T is endowed with the sup norm. Then every bounding subset of E is relatively compact.

Proof. We show that a sequence (x_n) of $C(T)$ with $x_n \neq x_m$ for $n \neq m$ and without convergent subsequence is not bounding in $C(T)$. There are $t_{n,m} \in T$ with $|x_n(t_{n,m}) - x_m(t_{n,m})| = \|x_n - x_m\| \geq \varepsilon$ for $n \neq m$ and a suitable $\varepsilon > 0$.

Passing to subsequences (y_k) of (x_n) and (t_k) of $(t_{n,m})$, we can assume that (y_k) converges pointwise on $\{t_{n,m} \mid n \neq m\}$ to a function y with $|y_k(t_k) - y(t_k)| \geq \delta > 0$ and that (t_k) is a discrete and convergent sequence in T. Now consider $S = \{t_k \mid k \in \mathbb{N}\} \cup \{\lim t_k\}$ with the topology induced from T. $C(S)$ is separable (isomorphic to c_0) and the restrictions $y_{k|S}$ have no convergent subsequence in $C(S)$. Hence, according to the proposition, there exists $h \in H(C(S))$ with $\sup|h(y_{k|S})| = \infty$. Since the restriction map $r\colon C(T) \to C(S)$, $x \longmapsto x_{|S}$, is linear and continuous the function $f = h \circ r$ satisfies $f \in H(C(T))$ and $\sup|f(x_n)| = \infty$.

The class of Banach spaces which can be embedded into $C(T)$, T sequentially compact and Hausdorff, includes the *reflexive spaces* and, more generally, the *weakly compactly generated* (WCG) spaces (cf. [1]), since for a WCG space E the closed unit ball U^0 of the dual E' is weak-star sequentially compact and E is canonically embedded into $C(U^0)$. Not every closed subspace of a WCG space is again WCG, hence the class of spaces for which the corollary applies is strictly larger than the class of WCG spaces.

The corollary was proven by HIRSCHOWITZ [13] in a different way. A similar result can be found in DINEEN [5] where the above assertion is shown for Banach spaces for which the closed unit ball of the dual is weak-star sequentially compact.

The corollary has some consequences regarding an attempt to define a natural topology on $H(E)$, or $H(\Omega)$, $\Omega \subset E$ open. If E is an infinite dimensional Banach space the compact open topology on the space $H(\Omega)$ of holomorphic functions on Ω, which is denoted by τ_0, is neither metrizable nor barrelled nor bornological. Now the corollary implies that for many Banach spaces the situation does not change by considering the topology of uniform convergence on the bounding subsets of U. A topology on $H(\Omega)$ with some good properties is the bornological topology τ_b associated with τ_0. Besides being barrelled this topology is adequate for the study of analytic continuation and also for approximation problems (See sections 2 - 5).

Corollary 1 has some obvious generalizations to locally convex spaces. But instead of describing more spaces for which the closed bounding subsets agree with the sequentially compact subsets we present the example of DINEEN [6]:

PROPOSITION 2: The set $D = \{e_j \mid j \in \mathbb{N}\}$ of unit vectors $e_j \in \ell_\infty$, $e_j = (\delta_{ij})_{i\in\mathbb{N}}$, is bounding in ℓ_∞.

We omit the rather technical proof. A combination of the two propositions yields:

COROLLARY 2: Not every holomorphic function $f \in H(c_0)$ can be continued analytically to ℓ_∞.

Proof. According to proposition 1 there exists $f \in H(c_0)$ with $\sup|f(e_j)| = \infty$. But $\{e_j \mid j \in \mathbb{N}\}$ is bounding in ℓ_∞, hence there cannot exist a holomorphic $F \in H(\ell_\infty)$ with $F|_{c_0} = f$.

As a consequence, c_0 cannot be a complemented subspace of ℓ_∞. Similarly, it can be shown that ℓ_∞ has no infinite dimensional and separable complemented subspace. Thus one can obtain a result on the geometry of Banach spaces [22] with the aid of holomorphic functions.

Proposition 2 and Corollary 2 imply that the richness of $H(\ell_\infty)$ is restricted in comparison to separable or reflexive Banach spaces. The bounding subsets of ℓ_∞ are studied in detail by JOSEFSON [18]:

PROPOSITION 3: For a bounded subset B of ℓ_∞ the following properties are equivalent:

1^0 B is a bounding subset of ℓ_∞.

2^0 Every sequence (ν_n) in ℓ_∞' converging pointwise to 0 converges to 0 uniformly on B.

3^0 There is no linear continuous $g: \ell_1 \to \ell_\infty$ with continuous inverse $g(\ell_1) \to \ell_1$ such that $g(U) \subset \widetilde{co}(B)$, where U is the unit ball of ℓ_1.

4^0 There is no sequence (a_n) in B which is equivalent to the usual ℓ_1-basis.

$1^0 \Rightarrow 2^0$ follows easily from the fact that $\sum \nu_n^n$ converges uniformly on compact subsets of ℓ_∞, whenever $\nu_n \to 0$ pointwise, hence $\sum \nu_n^n$ is holomorphic on ℓ_∞. The proofs of the other implications in proposition 3 are very involved. The proofs [18] give some insight in the geometry of ℓ_∞. One of the methods used in [18] is similar to a construction of ROSENTHAL [27].

COROLLARY 3: Let $B, C \subset \ell_\infty$ be bounding. Then $\widetilde{co}(B)$ and $B + C$ are bounding.

Proposition 3 connects in a fruitful way the linear and the "holomorphic" structure of ℓ_∞ . This is also true for another result of JOSEFSON [18] which generalizes proposition 2 :

PROPOSITION 4: A subset $B \subset c_0$ is bounding as a subset of ℓ_∞ if and only if B is bounded.

This result suggests that bounding sets can be rather big in some Banach spaces. However, DINEEN [5] has shown that in ℓ_∞ the bounding sets are nowhere dense. This is generalized by JOSEFSON [17]:

PROPOSITION 5: A bounding subset of an infinite dimensional locally convex Hausdorff space is nowhere dense.

In order to show proposition 5 Josefson proved the following:

For every infinite dimensional Banach space E there exists a sequence (ν_n) in E' converging to 0 pointwise but not in norm.

This result was obtained independently by NISSENZWEIG [24], and shows once more relations between the linear and the holomorphic theory. With the aid of this result, proposition 5 is easy to prove:

Proof of Proposition 5. It suffices to consider Banach spaces. Let (ν_n) be a sequence in E converging pointwise to 0 with $\inf\|\nu_n\| = \delta > 0$. $\sum \nu_n^n$ converges uniformly on the compact subsets of E , hence $f = \sum \nu_n^n \in H(E)$. Let $r > 0$ and assume f to be bounded on $B(0,r) = \{x \in E \mid \|x\| < r\}$. Then $\sum \|\nu_n\|^n s^n < \infty$ for $s < r$ and hence $\frac{1}{\liminf \|\nu_n\|} > s$. It follows $\frac{1}{\delta} \geq r$. In particular, f is not bounded on $B(0,R)$ for $R > \frac{1}{\delta}$. Consequently, no ball of E is a bounding set which implies that bounding sets are nowhere dense.

In spite of the deep results on bounding subsets of ℓ_∞ the theory of bounding sets is not yet satisfactory. For example, in view of the corollaries 1 and 3 , for many Banach spaces E the class of bounding subsets of E is a convex bornology, but it is not known whether this holds in general. Also, no linear characterization of those Banach spaces is known for which the bounding sets agree with the relatively compact subsets. The results of corollary 1 and proposition 1 suggest the following conjecture: A Banach space E has no subspace isomorphic to ℓ_∞ if and only if every bounding subset of E is relatively compact. The "if" part follows from proposition 2. The "only if" part is at least true for spaces $E = C(K)$ where K is a compact space.

2. PRESCRIBED RADIUS OF CONVERGENCE

In the preceding section we have studied the notion of a bounding set which is closely related to concepts in Functional Analysis. The next result which strengthens proposition 5 leads to problems of a more geometric nature.

PROPOSITION 6 (ARON [2]): Let E be an infinite dimensional Banach space over $\mathbb{C}$. There exists a holomorphic function $f \in H(E)$ such that to any $\varepsilon > 0$ there corresponds a point $x_\varepsilon \in B(0,1)$ such that f is unbounded on the ball $B(x_\varepsilon, \varepsilon)$.

The radius of convergence ρ_f of an entire function $f \in H(E)$ can be defined by

$$\rho_f(x) = \sup\{r > 0 \mid \|f\|_{B(x,r)} < \infty\}, \quad x \in E.$$

Now, proposition 6 claims the existence of an entire function $f \in H(E)$ with $\inf\{\rho_f(x) \mid \|x\| < 1\} = 0$. In connection with this result it is natural to ask whether the radius of convergence can be prescribed rather arbitrarily (cf. [19], [20]). Of course, an answer to the question provides another information about the richness of the class of holomorphic functions on a Banach space.

We first derive some properties which are shared by any radius of convergence ρ_f of a function $f \in H(\Omega)$, $\Omega \subset E$ open. At a given point $x \in \Omega$, f has a convergent taylor series expansion

$$f(x+h) = \sum_{n \geq 0} f_n(h), \quad \|h\| \text{ small},$$

with n-homogeneous, continuous polynomials $f_n : E \to \mathbb{C}$ depending analytically on x. The radius of convergence $\rho_f(x)$ of f at x is now defined by

$$\rho_f(x) = \sup\{r > 0 \mid \sum \|f_n\| r^n < \infty\} = \liminf (\|f_n\|)^{-\frac{1}{n}}$$

where $\|f_n\| = \sup\{|f_n(a)| \mid \|a\| \leq 1\}$. In the case $\Omega = E$ this definition agrees with the one above. In general, with $d_\Omega(x) := \operatorname{dist}(x, \partial\Omega) = \inf\{\|x-y\| \mid y \in \partial\Omega\}$ one has

$$(*) \quad \sup\{r > 0 \mid \|f\|_{B(x,r)} < \infty, B(x,r) \subset \Omega\} = \inf\{\rho_f(x), d_\Omega(x)\} .$$

Since $f_n = f_{n,x}$ depends analytically on x, the functions $x \mapsto -\log \|f_{n,x}\|$ are plurisubharmonic (Recall that a function $u: \Omega \to [-\infty,\infty[$ is plurisubharmonic if u is upper semicontinuous and if the restrictions of u to complex lines are subharmonic). Consequently, $-\log \rho_f : \Omega \longrightarrow [-\infty,\infty[$ is plurisubharmonic. Moreover, from (*) it is easy to conclude $\rho_f(x) \leq \rho_f(y) + \|x - y\|$ for $\|x - y\| \leq d_\Omega(x)$, $x,y \in \Omega$. We thus have deduced the following two conditions which a numerical function $\rho: \Omega \to]0,\infty]$ has to satisfy in order to be the radius of convergence of some holomorphic function on Ω :

(1) $-\log \rho$ is plurisubharmonic and $\rho \leq d_\Omega$,

(2) $\|\rho_f(x) - \rho_f(y)\| \leq \|x - y\|$ for all $(x,y) \in E \times E$, $\|x - y\| \leq d_\Omega(x)$.

Condition (1) implies in particular that Ω is pseudoconvex, i.e. $-\log d_\Omega$ is plurisubharmonic. Hence, the problem of finding holomorphic functions with a prescribed radius of convergence is closely related to the Levi problem which can be formulated in the following way: Given a pseudoconvex domain Ω, does there exist a holomorphic function $f \in H(\Omega)$ with $\rho_f \leq d_\Omega$, i.e. with Ω being the domain of existence of f ?

In the case of $E = \mathbb{C}^n$ every $f \in H(\Omega)$ satisfies $\rho_f \geq d_\Omega$, in particular $\rho_f = \infty$ if $\Omega = \mathbb{C}^n$. Hence, the problem of constructing a holomorphic function $f \in H(E)$ with prescribed radius of convergence $\rho \neq \infty$ is a purely infinite dimensional problem.

KISELMAN has given the following example [19]: For each $p \in]1,\infty[$, the function $\rho(x) := (\|x\| + 1)^{-1}$, $x \in \ell_p$, satisfies (1) and (2), but there is no $f \in H(\ell_p)$ with $\rho = \rho_f$. However, we can obtain a result which is close to a solution of our problem:

<u>PROPOSITION 7:</u> Let E be a Banach space with a Schauder basis and let $\rho: \Omega \to]0,\infty]$ be a function on a domain $\Omega \subset E$.

1^0 If ρ satisfies (1) there exists $f \in H(\Omega)$ with $\rho_f \leq \rho$.

2^0 If ρ satisfies (1) and (2) there exists $f \in H(\Omega)$ with $\frac{1}{3M}\rho \leq \rho_f \leq \rho$ where M is the basis constant. In the case of $E = \ell_p$, $p \in [1,\infty[$, $\frac{1}{3M}$ can be replaced by $(\frac{1}{2})^{1/q}$ where $1/p + 1/q = 1$. Thus, in the particular case $E = \ell_1$ there is $f \in H(\Omega)$ with $\rho_f = \rho$.

3^0 Under the above assumptions, the sets $\{f \in H(\Omega) \mid \rho_f \leq \rho\}$, resp.

$\{f \in H(\Omega) \mid (3M)^{-1}\rho \le \rho_f \le \rho\}$, are dense in $(H(\Omega), \tau_b)$ resp. $(H(\Omega), \tau_0)$.

Results similar to 2^0 were proven by KISELMAN [21] for the case $\Omega = E$ and by COEURE [4] for polynomially convex $\Omega \subset E$. Proposition 7 can be found in a slightly more general form (for example including non-schlicht domains) in [31]. We present here a proof of 1^0 and 2^0 of the case $\Omega = E$ which can be modified to work also for the general case. The basic idea is to construct a suitable covering (V_n) of E as in the proof of proposition 1 .

Proof of proposition 7 for $E = \Omega$. Let (e_n) be the Schauder basis of E with $\|e_n\| = 1$ and with the projections $\pi_m: E \to E$, $x \mapsto \sum_{n=1}^{m} \xi_n e_n$, $x \in E$. Let $\rho: E \to]0,\infty]$ be given with (1). Let $(z_n) \subset E$ be dense and $C > 0$ with $z_n \in E_n := \pi_n(E)$ and $\|z_n\| + \rho(z_n) \le n + C$ (The case $\rho \equiv \infty$ is trivial). Set $\varepsilon_1 = \rho(z_1)$ and $\varepsilon_{n+1} = \frac{1}{2}\inf\{\rho(z_{n+1}), \varepsilon_n, \rho(x_n)\}$ where $x_n = z_n + (\rho(z_n) - \varepsilon_n/2)e_{n+1}$. Define

$$V_n = \{x \in E \mid \|\pi_m(x)\| \le n + C \text{ and } \rho\circ\pi_k(x) \ge \varepsilon_n + \|\pi_m(x) - \pi_k(x)\| \text{ for all } m \ge k \ge n\} .$$

Then $V_n \subset V_{n+1}$, $\pi_{n+1}(V_n) = V_n \cap E_{n+1}$ and for every $x \in E$ there are $n \in \mathbb{N}$ and $r > 0$ with $B(x,r) \subset V_n$. By construction of x_n and V_n we have $x_n \in V_{n+1} \setminus V_n$: $\rho\circ\pi_n(x_n) = \rho(z_n) < \varepsilon_n + \rho(z_n) - \varepsilon_n/2 = \varepsilon_n + \|\pi_n(x_n) - x_n\|$, hence $x_n \notin V_n$, and $\rho(x_n) \ge \varepsilon_n > \varepsilon_{n+1}$ and $\|x_n\| \le n + C$, hence $x_n \in V_{n+1}$. Since $V_n \cap E_{n+1}$ is defined by global plurisubharmonic functions on E_{n+1} , $V_n \cap E_{n+1}$ is polynomially convex [15, p.91], i.e.

$$V_n \cap E_{n+1} = \{x \in E_{n+1} \mid |f(x)| \le \|f\|_{V_n \cap E_{n+1}} \text{ for all polynomials } f \text{ on } E_{n+1}\} .$$

Thus there is a sequence (f_j) of polynomials, $f_j \in H(E_j)$, with $f_1 = 0$,

$$|f_{n+1}(x_n)| \ge n + 1 + \sum_{j=1}^{n} |f_j\circ\pi_j(x_n)| \quad \text{and} \quad \|f_{n+1}\|_{V_n \cap E_{n+1}} < 2^{-(n+1)} ,$$

hence $\|f_{n+1}\circ\pi_{n+1}\|_{V_n} < 2^{-(n+1)}$.

As in the proof of proposition 1 , $f = \sum_{n=1}^{\infty} f_n\circ\pi_n$ is holomorphic on E with $|f(x_n)| \ge n$ and with

$$\|f\|_{V_n} \le \sum_{j=1}^{n} \|f_j\circ\pi_j\|_{V_n} + \sum_{j=n}^{\infty} \|f_{j+1}\circ\pi_{j+1}\|_{V_j} < \infty .$$

Now for $x \in E$ there is a subsequence $(z_{n(j)})$ of (z_n) with $\lim z_{n(j)} = x$. For $r > \liminf \|x_{n(j)} - z_{n(j)}\|$, f is unbounded on $B(x,r)$. Therefore,

$$\rho_f(x) \leq \liminf \|x_{n(j)} - z_{n(j)}\| \leq \liminf (\rho(z_{n(j)}) - \frac{\varepsilon_{n(j)}}{2}) \leq \rho(x) .$$

The last inequality follows from the lower semicontinuity of ρ .

Assume now that ρ satisfies (1) and (2) . We have to show that for any $s < \frac{1}{3M}\rho(x)$, where $M = \sup\|\pi_m\|$, there exists $n \in \mathbb{N}$ with $B(x,s) \subset V_n$. For then f is bounded on $B(x,s)$, and it follows $\rho_f(x) \geq s$.

Certainly there is $n_o \in \mathbb{N}$ with $M(\|x\| + s) \leq n_o$. Now choose $\delta > 0$ and $n \geq n_o$ with

$$3Ms + 2\delta + \varepsilon_n < \rho(x) \qquad \text{and}$$

$$\|\pi_k(x) - \pi_m(x)\| < \delta \qquad \text{for } m,k \geq n .$$

For $y \in B(x,s)$ and $m,k \geq n$ it follows that $\|\pi_m(y)\| \leq n$ and $\|\pi_k(y) - x\| \leq \|\pi_k(y-x)\| + \|\pi_k(x) - x\| \leq Ms + \delta$, hence by (2) and $\|\pi_k(y) - \pi_m(y)\| \leq 2Ms + \delta$:

$$\rho(\pi_k(y)) \geq \rho(x) - \|\pi_k(y) - x\| \geq 3Ms + 2\delta + \varepsilon_n - Ms - \delta$$

$$\geq \varepsilon_n + 2Ms + \delta \geq \varepsilon_n + \|\pi_k(y) - \pi_m(y)\| .$$

Let us finally consider the case $E = \ell_1$ (the case $E = \ell_p$ is similar). For $s < \rho(x)$ there are $\delta > 0$ and n with $s + 2\delta + \varepsilon_n < \rho(x)$ and $\|\pi_m(x) - \pi_k(x)\| < \delta$ for $m,k > n$. Hence,

$$\rho\circ\pi_k(y) \geq \rho(x) - \|\pi_k(y) - x\| \geq s + \delta + \varepsilon_n - \|\pi_k(y) - \pi_k(x)\| .$$

Moreover, for $m \geq k \geq n$, $\|\pi_m(y) - \pi_k(y)\| + \|\pi_k(y) - \pi_k(x)\| = \|\pi_m(y) - \pi_k(x)\| \leq s + \delta$, which implies $\rho\circ\pi_k(y) \geq \varepsilon_n + \|\pi_m(y) - \pi_k(y)\|$ for all $m,k \geq n$. Thus $B(x,s) \subset V_n$, and we have shown $\rho_f(x) \geq s$ for all $s < \rho(x)$. The proof of 1° and 2° for $E = \Omega$ is complete.

Whether proposition 7 holds for general separable Banach spaces is not known. Some results similar to proposition 7 can be proven for convex domains in a separable Banach space by employing the method of proposition 1 once again. For nonseparable Banach spaces, however, proposition 7 cannot be true. This follows from the counterexamples to the Levi problem:
HIRSCHOWITZ has shown in [12] that the open unit ball B in $C(\omega_1)$ is not a domain of existence, where ω_1 is the space of all countable ordinals endowed with the order topology. This follows from the fact that every $f \in H(B)$ depends on a countable number of variables, i.e. there is a countable interval $I \subset \omega_1$ with $f(x+y) = f(x)$ whenever $x \in B$ and $y \in C(\omega_1 \setminus I) = \{y \in C(\omega_1) \mid \operatorname{supp} y \subset \omega_1 \setminus I\}$. Consequently, f has an analytic continuation $\tilde{f}$ to $B \cap C(I) + C(\omega_1 \setminus I)$ which

implies $\rho_f(y) > d_\Omega(y)$ for points $y \in C(\omega_1 \setminus I)$ near the boundary of B, although $-\log d_\Omega$ is not only plurisubharmonic but even convex.

Another example is due to JOSEFSON [16]. He shows that in $c_0(\Gamma)$, Γ uncountable, a holomorphic function depends locally on countably many variables and constructs a pseudoconvex and polynomially convex domain $\Omega \subset c_0(\Gamma)$ which even fails to be a domain of holomorphy: 0 is a boundary point of Ω and has a neighbourhood V of 0 to which all holomorphic functions on Ω can be continued analytically.

3. DOMAINS OF HOLOMORPHY AND ANALYTIC CONTINUATION

A *domain of holomorphy* Ω in a locally convex space E is by definition a domain with sufficiently many holomorphic functions in the following sense: For every $x \in \partial\Omega$, every connected neighborhood V of x and every nonempty open $U \subset \Omega \cap V$ there exists $f \in H(\Omega)$ whose restriction $f_{|U}$ cannot be continued analytically to V. If $\dim_{\mathbb{C}} E \geq 2$ not every domain in E is a domain of holomorphy and one is interested to characterize the domains of holomorphy by geometric properties of the boundary. It is a well-known and deep result of Complex Analysis that for finite dimensional spaces E the appropriate property is *pseudoconvexity*. This result has been extended to Banach spaces with a Schauder basis by GRUMAN and KISELMAN [9]: Every pseudoconvex domain $\Omega \subset E$ is a domain of existence of a holomorphic function $f \in H(\Omega)$, hence a domain of existence. A stronger form of this result is formulated in proposition 7. The method used in [9] has been applied to more general situations for example in [8], [10] and [30].
However, a pseudoconvex domain in a non-separable Banach space need not be a domain of holomorphy (cf. end of section 2).

For every domain Ω there exists a maximal domain $\tilde{\Omega}$ to which all holomorphic functions $f \in H(\Omega)$ can be continued analytically. Clearly, this is true only if we admit non-schlicht domains spread over E. $\tilde{\Omega}$ is called the *envelope of holomorphy* of Ω. Evidently, $\tilde{\Omega}$ has the same class of holomorphic functions as Ω, and $\tilde{\Omega}$ is the smallest domain of holomorphy containing Ω. Existence and uniqueness of $\tilde{\Omega}$ can be shown as in the case of $E = \mathbb{C}^n$ using sheaf theory (cf. [14]). In another approach, $\tilde{\Omega}$ can be obtained as a certain subset of the spectrum $S(H(\Omega))$ of $H(\Omega)$, i.e. the space of non-zero homomorphisms on $H(\Omega)$ (see below).

In contrast to the finite dimensional case, the restriction map $r: H(\tilde{\Omega}) \to H(\Omega)$ need not be an open map with respect to the compact open topologies (cf. [16]).

However:

PROPOSITION 8 [14]: Let $\tilde{\Omega}$ be the envelope of holomorphy of a domain Ω in a normed space E. Then the restriction map

$$r: (H(\tilde{\Omega}), \tau_b) \rightarrow (H(\Omega), \tau_b)$$

is an open map, hence a homeomorphism.

Proof. We consider only the case of schlicht domains $\Omega \subset \tilde{\Omega} \subset E$. Let $c(\Omega)$ be the class of countable covers $\mathcal{V} = (V_n)$ of Ω with bounded V_n and $V_n + B(0,\delta_n) \subset V_{n+1}$ for suitable $\delta_n > 0$. Put

$$A_{\mathcal{V}} = \{f \in H(\Omega) \mid \|f\|_{V_n} < \infty \quad \text{for all} \quad n \in \mathbb{N}\} .$$

Endowed with the topology of uniform convergence on all V_n, $n \in \mathbb{N}$, $A_{\mathcal{V}}$ is a Frêchet algebra. (Note that $A_{\mathcal{V}} \neq H(\Omega)$ when $\dim_{\mathbb{C}} E = \infty$ according to proposition 5.) Every bounded set $B \subset A_{\mathcal{V}}$ is evidently bounded in $(H(\Omega), \tau_0)$. Conversely, a bounded set B of $(H(\Omega), \tau_0)$ is equicontinuous, hence with

$$U_n := \{x \in \Omega \mid \|x\| \leq n , \quad d_{\Omega}(x) \geq 2^{-n}, \ \sup\{|f(x)| \mid f \in B\} \leq n \}$$

and $V_n := \{x \in U_n \mid B(x, 2^{-n}) \subset U_n\}$ we have defined a cover $\mathcal{V} = (V_n) \in c(\Omega)$ such that B is bounded in $A_{\mathcal{V}}$. As a result $(H(\Omega), \tau_b) = \lim\limits_{\mathcal{V} \in c(\Omega)} \operatorname{ind} A_{\mathcal{V}}$.

We now consider the points $y \in \tilde{\Omega}$ as homomorphisms on $A_{\mathcal{V}}$: $f \mapsto \tilde{f}(y)$, $f \in A_{\mathcal{V}}$, where $\tilde{f} \in H(\tilde{\Omega})$ denotes the analytic continuation of f to $\tilde{\Omega}$. Since $A_{\mathcal{V}}$ is in particular barrelled, y is continuous on $A_{\mathcal{V}}$. Now, $\tilde{V}_n :=$ $\{y \in \tilde{\Omega} \mid |f(y)| \leq \|f\|_{V_n}$ for all $f \in A_{\mathcal{V}}\}$, $n \in \mathbb{N}$, defines a cover $\tilde{\mathcal{V}} \in c(\tilde{\Omega})$ such that $A_{\mathcal{V}} \rightarrow A_{\tilde{\mathcal{V}}}$, $f \mapsto \tilde{f}$, is a homeomorphism. It follows that the composition $A_{\mathcal{V}} \longrightarrow A_{\tilde{\mathcal{V}}} \longrightarrow H(\tilde{\Omega})$ is continuous, and hence the restriction $(H(\tilde{\Omega}), \tau_b) \rightarrow (H(\Omega), \tau_b)$ is an open map.

The proof of proposition 8 given here indicates that the envelope of holomorphy of a domain Ω in a Banach space E can be constructed using the spectrum of $(H(\Omega), \tau_b)$ [28]. In fact, the set $S(A_{\mathcal{V}})$ of non-zero, continuous homomorphisms on $A_{\mathcal{V}}$, $\mathcal{V} \in c(\Omega)$, can be endowed with a canonical structure of a complex manifold spread over E, and a suitable component $\tilde{\Omega}_{\mathcal{V}}$ of $S(A_{\mathcal{V}})$ is the maximal domain to which all holomorphic $f \in A_{\mathcal{V}}$ extend analytically. Now the envelope of holomorphy $\tilde{\Omega}$ of Ω is the "intersection" of all the domains $\tilde{\Omega}_{\mathcal{V}}$ (i.e. the projective limit of the $\tilde{\Omega}_{\mathcal{V}}$, $\mathcal{V} \in c(\Omega)$, in the category of domains spread over E), or, equivalently, $\tilde{\Omega}$ is a component of the space of all τ_b-continuous homomorphisms h on $H(\Omega)$ whose restrictions $h|_{A_V}$ are in $\tilde{\Omega}_V$, for all $V \in c(\Omega)$.

A direct consequence of proposition 8 is

COROLLARY 4: Every holomorphic mapping $f: \Omega \to F$ with values in a complete locally convex space F can be continued analytically to $\tilde{\Omega}$.

It is an open question whether the statement of the corollary holds if F is replaced by a domain of holomorphy Σ in F.

4. HOLOMORPHIC COMPLETION

A notion which is analogous to the notion of the envelope of holomorphy but which cannot be defined for finite dimensional spaces is the notion of the holomorphic completion of a non-complete locally convex space E over $\mathbb{C}$. The *holomorphic completion* E_0 of E is defined to be the maximal subspace of the completion $\hat{E}$ of E to which all holomorphic functions $f \in H(E)$ extend analytically. This is also analogous to the following description of the completion $\hat{E}$ of E, which could be called the linear completion of E: $\hat{E}$ is the maximal locally convex space to which all continuous linear forms on E extend uniquely as continuous linear forms. Of course, it has to be shown that E_0 is well-defined: Any $f \in H(E)$ can be extended analytically to an open neighborhood $\Omega \subset \hat{E}$ of E as can be shown by using the taylor series expansion of f at the points of E. It follows that f has a unique maximal domain of existence $\Omega_f \subset \hat{E}$. Now $\cap \{\Omega_f \mid f \in H(E)\}$ is a linear subspace of $\hat{E}$, hence $E_0 = \cap\{\Omega_f \mid f \in H(E)\}$. A locally convex space E is called *holomorphically complete* if $E = E_0$. An example of a space which is not holomorphically complete was given by HIRSCHOWITZ [11]: Let $E \subset c_0$ be the space of summable sequences endowed with the sup norm. Then E_0 contains $\cap\{\ell_p \mid p > 1\}$. An example of a holomorphically complete, non-complete space was given by NOVERRAZ [25]:

PROPOSITION 9: Let E be a Banach space with a basis (e_n). Then the linear span F of $\{e_n \mid n \in \mathbb{N}\}$ is holomorphically complete.

Proof. As in the proof of proposition 7 let $\pi_n: E \to E$ denote the projection onto the first n basis vectors. Let $y \in D \setminus F$ and choose a sequence (ε_n),

$\varepsilon_n > 0$, so that $\sum_{n=1}^{\infty} \varepsilon_n \log\|\pi_n(y) - y\| > -\infty$. Then $\rho(x) := \sum_{n=1}^{\infty} \varepsilon_n \log\|\pi_n(x) - x\|$, $x \in E$, is a plurisubharmonic function, and $\Omega = \{x \in E \mid \rho(x) < \rho(y)\}$ is a pseudoconvex domain in E . $F \subset \Omega$ because of $\rho(x) = -\infty$ for $x \in \bigcup \pi_n(E) = F$. According to proposition 7 , Ω is the domain of existence of a holomorphic function $f \in H(\Omega)$. Hence $F_0 \subset \Omega$, in particular $y \notin F_0$. Since y was an arbitrary point $y \in E \setminus F$ we conclude $F_0 = F$.

A more general result asserts that in a Banach space E with a basis a dense subspace F is holomorphically complete if and only if F is a polar set, i.e. if there exists a plurisubharmonic $u: E \longrightarrow [-\infty,\infty[$, $u \not\equiv -\infty$, with $u(y) = -\infty$ for all $y \in F$. Also, any normed space with a countable algebraic basis is holomorphically complete.

Using corollary 4 HIRSCHOWITZ has shown the following functorial property of the holomorphic completion [14]:

PROPOSITION 10: Let E and $F = F_0$ be normed spaces. Then every analytic map $f: E \to F$ can be continued analytically to E_0 .

Similar to proposition 8 , the restriction map $r: (H(E_0), \tau_b) \to (H(E), \tau_b)$ is a homeomorphism whenever E is a metrizable locally convex space [29]. We conclude this section with a proof of another functorial property of the holomorphic completion [29]:

PROPOSITION 11: $(E \times F)_0 = E_0 \times F_0$ for metrizable E and F .

Proof. Evidently, $(E \times F)_0 \subset E_0 \times F_0$ for arbitrary locally convex spaces E and F . Therefore, it remains to show that in our situation every $f \in H(E \times F)$ extends analytically to $E_0 \times F_0$. Now f can be regarded to be a holomorphic map $f: E \to (H(F), \tau_0)$. Since proposition 10 can be generalized to metrizable E and arbitrary complete F [29] it follows that $f: E \to (H(F), \tau_0)$ can be continued analytically to E_0 . This means that our original $f \in H(E \times F)$ has an analytic continuation f_1 to $E_0 \times F$. Now $f_1: F \to (H(E_0), \tau_0)$ can be continued analytically to F_0 for the same reasons as before. We thus have found an analytic continuation $\tilde{f} \in H(E_0 \times F_0)$ of $f \in H(E \times F)$.

An application of proposition 11 to the above mentioned results of Hirschowitz and Noverraz yields an example of a normed space E with $E \subsetneqq E_0 \subsetneqq \hat{E}$. Additional results on the holomorphic completion of locally convex spaces can be found in [7] and [26].

5. COMPACT HOLOMORPHIC MAPPINGS AND THE APPROXIMATION PROPERTY

The results discussed in section 1 and 2 show that there are much more holomorphic functions on a Banach space E than continuous linear forms. However, concerning the uniform approximation on compact subsets of E by finite rank mappings the linear mappings are as good as the holomorphic mappings:
(Throughout this section, let E and F denote Banach spaces over $\mathbb{C}$.)

PROPOSITION 12 [3]: The following properties are equivalent:

1^0 E has the *approximation property*, i.e. for every compact $K \subset E$ and $\varepsilon > 0$ there is $g \in E' \otimes E$ with $\|g - id_E\|_K < \varepsilon$.

2^0 For every compact $K \subset E$ and $\varepsilon > 0$ there is $f \in H(E) \otimes E$ with $\|f - id_E\|_K < \varepsilon$.

3^0 $H(E) \otimes F$ is dense in $(H(E,F), \tau_0)$ for all F .

4^0 $(H(E), \tau_0)$ has the approximation property.

In view of the propositions 1 and 5 the above implication $2^0 \Rightarrow 1^0$ can be explained only by the fact that the compact subsets of an infinite dimensional Banach space are very small.

As in the linear case there are results on holomorphic maps which relate the approximation property with the compact maps. A holomorphic map $f: E \to F$ between Banach spaces is said to be *compact* if there exists a neighborhood U of $0 \in E$ such that $f(U)$ is relatively compact. Let $H_K(E,F)$ denote the space of compact holomorphic mappings from E to F .

PROPOSITION 13 [3]: $(H(E), \tau_b)$ has the approximation property if and only if for every Banach space F , $H(E) \otimes F$ is dense in $(H_K(E,F), \tau_b)$.

Here, τ_b again denotes the bornological topology associated with the compact open topology τ_0 on $H_K(E,F)$. The proposition remains true when τ_b is replaced by τ_∞ , the topology of uniform convergence on compact sets of the holomorphic maps and all their derivatives, or by τ_ω , the ported topology, introduced by NACHBIN [23]: τ_ω is generated by all seminorms $p: H(E,F) \to \mathbb{R}$ (resp. $p: H_K(E,F) \to \mathbb{R}$) with the following property: There exist a compact subset $N \subset E$ and for every neighborhood V of N a constant $c(V)$ with

$$p(f) \leq c(V) \|f\|_V \qquad \text{for all } f \in H(E,F) \quad (\text{resp. } f \in H_K(E,F)).$$

In general, $\tau_0 \subset \tau_\infty \subset \tau_\omega \subset \tau_b$, and all inclusions can be strict. Proposition 13 can be proven for τ_ω with the aid of the formula

$$(H_K(E,F), \tau_\omega) = (H(E), \tau_\omega) \varepsilon F ,$$

where the right hand side of the equation is the ε-product of L. SCHWARTZ. The following result is also based on this formula [3]:

PROPOSITION 14: E has the approximation property if and only if for every Banach space F , $H(F) \otimes E$ is dense in $(H_K(F,E), \tau_\omega)$.

A direct application of the taylor series expansion shows that the Banach space $P(^nE)$ of the n-homogeneous continuous polynomials $P: E \to \mathbb{C}$ with the sup norm is a complemented subspace of $(H(E), \tau)$ where τ is any of the three topologies $\tau_\infty, \tau_\omega, \tau_b$. We thus have the sufficiency of the following proposition which supplements proposition 13:

PROPOSITION 15: $(H(E), \tau)$ has the approximation property for $\tau = \tau_\infty$, $\tau = \tau_\omega$ or $\tau = \tau_b$ if and only if for all $n \in \mathbb{N}$, $P(^nE)$ has the approximation property.

Using this result it can be seen that, for instance, $(H(\ell_1), \tau_\omega)$ has the approximation property by showing $P(^n\ell_1) \cong \ell_\infty$. Whether or not $(H(\ell_2), \tau_\omega)$ has the approximation property is closely related to the unsolved problem of whether $L(\ell_2,\ell_2)$ has the approximation property, since $P(^2\ell_2)$ can be represented as a complemented subspace of $L(\ell_2,\ell_2)$. In general, $(H(E), \tau_\omega)$ need not have the approximation property even if E and all its duals E', E'', etc. have a Schauder basis. A modification of an example of W.B. JOHNSON gives the following: There exists a reflexive space E with a Schauder basis such that $P(^2E)$ does not have the approximation property. E is an ℓ_2-sum of a suitable sequence (E_n) of finite dimensional spaces: $E = (\sum_{n=1}^{\infty} E_n)_2$.

REFERENCES

1. AMIR, E., LINDENSTRAUSS, J.: The structure of weakly compact sets in Banach spaces. Ann. of Math. 88 (1968), 35-46.

2. ARON, R.: Entire functions of unbounded type on a Banach space. Boll. Un. Mat. Ital. 9 (1974), 28-31.

3. ARON, R., SCHOTTENLOHER, M.: Compact holomorphic mappings on Banach spaces and the approximation·property. Journal of Funct. Anal. 21 (1976), 7-30.

4. COEURE, G.: Sur le rayon de bornologie des fonctions holomorphes. Preprint.

5. DINEEN, S.: Unbounded holomorphic functions on a Banach space. J. London Math. Soc. 4 (1972), 461-465.

6. DINEEN, S.: Bounding subsets of a Banach space. Math. Ann. 192 (1971), 61-70.

7. DINEEN, S.: Holomorphically complete locally convex topological vector spaces. In: Sém. Lelong 71/72, Lecture Notes in Mathematics 332 (Springer 1973), 77-111.

8. DINEEN, S., NOVERRAZ, Ph., SCHOTTENLOHER, M.: Le problème de Levi dans certains espaces vectoriels topologiques localement convexes. Bull. Soc. math. France 104 (1976), 87-97.

9. GRUMAN, L., KISELMAN, C.O.: Le problème de Levi dans les espaces de Banach à base. C.R. Acad. Sci. Paris, A 274 (1972), 1296-1299.

10. HERVIER, Y.: Sur le problème de Levi pour les espaces étalés banachiques. C.R. Acad. Sci. Paris, A 275 (1972), 821-824.

11. HIRSCHOWITZ, A.: Sur les suites des fonctions analytiques. Ann. Inst. Fourier, Grenoble 20 (1970), 403-413.

12. HIRSCHOWITZ, A.: Diverses notions d'ouverts d'analyticité en dimension infinie. In: Sém. Lelong 70. Lecture Notes in Mathematics 205 (Springer 1971).

13. HIRSCHOWITZ, A.: Bornologie des espaces de fonctions analytiques en dimension infinie. Sém. Lelong 70. Lecture Notes in Mathematics 205 (Springer 1971).

14. HIRSCHOWITZ, A.: Prolongement analytique en dimension infinie. Ann. Inst. Fourier, Grenoble, 22 (1972), 255-292.

15. HÖRMANDER, L.: An Introduction to Complex Analysis in Several Variables. Princeton: Van Nostrand 1966.

16. JOSEFSON, B.: A counterexample to the Levi problem. In: Proceedings on Infinite Dimensional Holomorphy. Lecture Notes in Mathematics 364 (Springer 1974), 168-177.

17. JOSEFSON, B.: Weak sequential convergence in the dual of a Banach space does not imply norm convergence. Ark. Mat. 13 (1975), 79-89.

18. JOSEFSON, B.: Bounding Subsets of $\ell_\infty(A)$. Thesis, University of Uppsala, 1975.

19. KISELMAN, C.O.: On the radius of convergence of an entire function in a normed space. Ann. Polon. Math. 33 (1976), 39-55.

20. KISELMAN, C.O.: Geometric aspects of the theory of bounds for entire functions in normed spaces. In: Infinite Dimensional Holomorphy and Applications. Ed. M.C. Matos. To appear at North-Holland, Amsterdam.

21. KISELMAN, C.O.: Constructions de fonctions entières à rayon de convergence donné. Preprint.

22. LINDENSTRAUSS, J.: On complemented subspaces of m. Isr. J. of Math. 5 (1967), 153-156.

23. NACHBIN, L.: Topology on spaces of holomorphic functions. Erg. der Math. 47 (Springer 1969).

24. NISSENZWEIG, A.: w* sequential convergence. Israel J. Math. 22 (1975), 266-272.

25. NOVERRAZ, Ph.: Pseudoconvexité, convexité polynomiale et domaines d'holomorphie en dimension infinie. Amsterdam: North-Holland 1973.

26. NOVERRAZ, Ph.: Pseudoconvex completion of locally convex topological vector spaces. Math. Ann. 208 (1974), 59-69.

27. ROSENTHAL, H.: A characterization of Banach spaces containing ℓ^1. Proc. Nat. Acad. Sci. USA 71 (1974), 2411-2413.

28. SCHOTTENLOHER, M.: Über analytische Fortsetzung in Banachräumen. Math. Ann. 199 (1972), 313-336.

29. SCHOTTENLOHER, M.: Holomorphe Vervollständigung metrisierbarer lokalkonvexer Räume. Bayer. Akad. d. Wiss., Sitz.-ber. 1973, 57-66.

30. SCHOTTENLOHER, M.: The Leviproblem for domains spread over locally convex spaces with a finite dimensional Schauder decomposition. Ann. Inst. Fourier, Grenoble 26 (1976), 207-237.

31. SCHOTTENLOHER, M.: Holomorphe Funktionen auf Gebieten über Banachräumen zu vorgegebenen Konvergenzradien. To appear in manuscripta mathematica.

Martin Schottenloher

Mathematisches Institut
der Universität München

Theresienstr. 39

D 8000 MÜNCHEN 2

K.-D. Bierstedt, B. Fuchssteiner (eds.)
Functional Analysis: Surveys and Recent Results

ABSTRACT HARDY ALGEBRA THEORY

HEINZ KÖNIG

Fachbereich Mathematik
Universität des Saarlandes
D 6600 Saarbrücken

The present article is devoted to a portion of functional analysis which is an abstract version of a portion of classical analytic function theory as circumscribed by boundary value theory and Hardy spaces H^p. The fascination of the field comes from the fact that famous classical theorems of typical complex-analytic flavor appear as instant outflows of an abstract theory the tools of which are standard real--analytic methods such as elementary functional analysis and measure theory. The abstract theory started in the fifties (Arens-Singer, Gleason, Helson-Lowdenslager, Bishop, Wermer,...) and went through several steps of abstraction (Dirichlet algebras, logmodular algebras, ...). We present the ultimate step which has been under work for about ten years.

The central concept is the abstract Hardy algebra situation. It can be looked upon as a local section of the abstract function algebra situation. To achieve the localization is the main business of the abstract F.and M.Riesz theorem and of the resultant Gleason part decomposition procedure. The abstract Hardy algebra situation as a comprehensive as well as pure and simple concept calls for methodic plainness and adequacy. It permits to build up a coherent theory of remarkable width and depth which even in its ultimate state of abstraction radiates back and illuminates the concrete classical theory.

We start to concentrate on the abstract Hardy algebra situation and then turn to the localization procedure and to certain standard applications. The presentation follows the new Lecture Notes BARBEY-KÖNIG (1977) to which we also refer for detailed references. Compared with the actual Paderborn lecture in November 1976 some new results on the Marcel Riesz estimation for conjugate functions have been added.

The author cannot conclude the Introduction without expression of his warmest thanks to Klaus Bierstedt and Benno Fuchssteiner who did the work and created the atmosphere of an unusually intense and pleasant mathematical conference.

1. The Abstract Hardy Algebra Situation

We fix a nonzero finite positive measure space (X,Σ,m). A Hardy algebra situation (H,φ) on (X,Σ,m) is defined to consist of a complex subalgebra $H\subset L^\infty(m)$ which contains the constants and is closed in the weak$*$ topology $\sigma(L^\infty(m),L^1(m))$, and of a nonzero multiplicative linear functional $\varphi:H\to ¢$ which is weak* continuous, that is which can be represented in the form $\varphi(u)=\int uFdm\ \forall u\in H$ for some $F\in L^1(m)$. Let M consist of the representative functions $0\leqq F\in L^1(m)$. It will be seen in Section 3 that $M\neq\emptyset$. From $M\neq\emptyset$ one concludes that there are functions $F\in M$ such that $[V>0]\subset[F>0]\ \forall V\in M$, called the dominant representative functions. From Section 4 on we shall adopt the reducedness assumption $[F>0]=X$ for the dominant $F\in M$, which is no real restriction.

It will be essential to extend H into the domain of unbounded functions. We define $H^\#$ to consist of the functions $f\in L(m)$ for which there exists a sequence of functions $u_n\in H$ such that $|u_n|\leqq 1, u_n\to 1$ pointwise in the L(m) sense, and $u_nf\in H$ for all $n\geqq 1$. Then $H^\#\subset L(m)$ is an algebra and $H^\#\cap L^\infty(m)=H$.

Also we define $L^\#$ to consist of the functions $f\in L(m)$ for which there exists a sequence of functions $u_n\in H$ such that $|u_n|\leqq 1$, $u_n\to 1$ pointwise, and $u_nf\in L^\infty(m)$ for all $n\geqq 1$. Then $L^\#\subset L(m)$ is an algebra and $L^\infty(m)\subset L^\#$. Also $H^\#\subset L^\#$. And if a sequence of functions $f_n\in H^\#$ tends $\to f\in L(m)$ and satisfies $|f_n|\leqq$ some $F\in L^\#$ then $f\in H^\#$.

Note that $H^\#$ and $L^\#$ do not depend on the functional φ. One proves that $\varphi:H\to ¢$ possesses a unique extension $\varphi:H^\#\to ¢$ which is continuous in the adequate sense: If $f_n\in H^\#$ with $f_n\to f$ and $|f_n|\leqq F\in L^\#$ (so that $f\in H^\#$) then $\varphi(f_n)\to\varphi(f)$. And $\varphi:H^\#\to ¢$ continues to be a multiplicative linear functional.

If $F\in L^1(m)$ is such that $\varphi(u)=\int uFdm\ \forall u\in H$ then likewise $\varphi(u)=\int uFdm$ $\forall u\in H^\#$ with $uF\in L^1(m)$. But it is important to note that for $u\in H^\#$ there need not exist an $F\in M$ with $uF\in L^1(m)$. In fact, there are (even in the unit disk situation) examples of real-valued functions $u\in H^\#$ such that $\varphi(u)$ is not real.

The class H^+ is defined to consist of the functions $f\in L(m)$ with $\mathrm{Re}\, f\geqq 0$ such that $\frac{1}{f+s}\in H$ for all $s\in ¢^+$:= the open halfplane Re>0. Equivalent is $\frac{1}{f+s}\in H$ for some $s\in ¢^+$. Also equivalent is $e^{-tf}\in H\ \forall t\geqq 0$, as it is seen from

$$e^{-tf} = \lim_{n\to\infty}(1+\tfrac{1}{n}tf)^{-n}\ \forall t\geqq 0 \text{ and } \frac{1}{f+s} = \int_0^\infty e^{-t(f+s)}dt\ \forall s\in ¢^+.$$

We have $H^+\subset H^\#$ (since for $f\in H^+$ the definition of $H^\#$ applies with $u_n:=\frac{n}{f+n}\in H$). But as a rule H^+ does not comprise all functions $f\in H^\#$ with $\operatorname{Re} f\geqq 0$.

2. Example: The Unit Disk Situation

For $G\subset ¢$ an open subset let $\operatorname{Hol}(G)$ denote the class of holomorphic functions $G\to ¢$ and $\operatorname{Hol}^\infty(G)$ the class of bounded functions in $\operatorname{Hol}(G)$. We define $\operatorname{Hol}^\#(G)$ to consist of the functions $f:G\to ¢$ for which there exists a sequence of functions $u_n\in\operatorname{Hol}^\infty(G)$ such that $|u_n|\leqq 1, u_n\to 1$ pointwise on G (and hence uniformly on each compact subset of G), and $u_nf\in\operatorname{Hol}^\infty(G)$ for all $n\geqq 1$. Then $\operatorname{Hol}^\#(G)\subset\operatorname{Hol}(G)$ is an algebra and $\operatorname{Hol}^\infty(G)\subset\operatorname{Hol}^\#(G)$. And $\operatorname{Hol}^\#(G)$ contains the class $\operatorname{Hol}^+(G)$ of the functions $f\in\operatorname{Hol}(G)$ with $\operatorname{Re} f\geqq 0$ (since we can take $u_n:=\frac{n}{f+n}\in\operatorname{Hol}^\infty(G)$).

We turn to the unit disk $D:=\{z\in ¢:|z|<1\}$ and unit circle $S:=\{s\in ¢:|s|=1\}$ and consider the measure space (S,Baire,λ) with $\lambda:=$arc length on S normalized to $\lambda(S)=1$. For $f:D\to ¢$ and $0\leqq R<1$ we put $f_R:f_R(s)=f(Rs)\ \forall s\in S$. The basic fact is that for $f\in\operatorname{Hol}^\infty(D)$ the sections $f_R\in C(S)\subset L^\infty(\lambda)$ tend for $R\uparrow 1$ to a limit function $F\in L^\infty(\lambda)$ in the weak* topology $\sigma(L^\infty(\lambda),L^1(\lambda))$ as well as pointwise for λ - almost all $s\in S$. And the map $f\mapsto F$ is injective with inverse map

$$F\mapsto f:f(z)=\int FP(z,\cdot)d\lambda =: \langle F\lambda\rangle(z) \qquad \forall z\in D,$$

$$\text{with } P(z,s) = \operatorname{Re}\frac{s+z}{s-z} = \frac{1-|z|^2}{|s-z|^2} \quad \forall s\in S \text{ the Poisson kernel.}$$

Let $H^\infty(D)$ consist of the limit functions $F\in L^\infty(\lambda)$ thus obtained. There are several classical characterizations of $H^\infty(D)\subset L^\infty(\lambda)$ which show that it is a weak* closed subalgebra. And the Weierstraß approximation theorem implies that $\operatorname{Re}H^\infty(D)$ is weak* dense in $\operatorname{Re}L^\infty(\lambda)$.

Thus for each $z\in D$ we obtain the Hardy algebra situation $(H^\infty(D),\varphi_z)$ on (S,Baire,λ) with the point evaluation

$$\varphi_z : \varphi_z(F) = f(z) = \int FP(z,\cdot)d\lambda = \langle F\lambda\rangle(z) \qquad \forall\, F\in H^\infty(D).$$

We have the uniqueness situation $M_z=\{P(z,\cdot)\}$ since there are no non-zero real-valued functions $\in L^1(\lambda)$ which annihilate $H^\infty(D)$.

Now one proves from the definition that for $f\in \mathrm{Hol}^\#(D)$ the sections $f_R\in C(S)\subset L^\infty(\lambda)$ tend for $R\uparrow 1$ to a limit function $F\in L(\lambda)$ pointwise for λ-almost all $s\in S$. Also the map $f\mapsto F$ is injective, and the class $H^\#(D)$ of the limit functions $F\in L(\lambda)$ thus obtained is equal to the homonymous associated algebra which comes from each of the Hardy algebra situations $(H^\infty(D),\varphi_z)$ $\forall z\in D$ after Section 1. And the extension $\varphi_z:H^\#(D)\to\mathbb{C}$ after Section 1 continues to be the point evaluation $\varphi_z:\varphi_z(F)=f(z)$ $\forall F\in H^\#(D)$.

In particular the class $H^+(D)$ of the limit functions $F\in L(\lambda)$ obtained from the functions $f\in \mathrm{Hol}^+(D)$ turns out to be equal to the homonymous function class which comes from each of the $(H^\infty(D),\varphi_z)$ $\forall z\in D$ after Section 1.

As an example consider the function $f:f(z)=\frac{1+z}{1-z}$ $\forall z\in D$ with limit function $F:F(s)=i\cot\mathrm{an}\frac{t}{2}$ $\forall s=e^{it}\in S$. We have $f\in \mathrm{Hol}^+(D)$ and hence $F\in H^+(D)$. Now $\mathrm{Re}F=0$. Thus $-F\in H^\#(D)$ with $\mathrm{Re}(-F)=0$ and hence $\geqq 0$, whereas it is clear that $-F\notin H^+(D)$. Moreover the function $iF\in H^\#(D)$ is real-valued, and is such that $\varphi_z(iF)=if(z)=i\frac{1+z}{1-z}$ is not real $\forall z\in D$.

3. The Basic Inequalities

We fix a Hardy algebra situation (H,φ) on (X,Σ,m). We define the functional $\alpha:\mathrm{ReL}(m)\to[-\infty,\infty]$ to be

$$\alpha(f) = \mathrm{Inf}\{\mathrm{Sup}(f+\log|u|):u\in H \text{ with } \varphi(u)=1\}$$

$$= \mathrm{Inf}\{\mathrm{Sup}(f+\log|u|):u\in H^\# \text{ with } \varphi(u)=1\} \qquad \forall f\in \mathrm{ReL}(m).$$

It is to reflect the essentials of the (H,φ) under consideration. We list some simple properties. i) $\mathrm{Inf}\, f \leq \alpha(f) \leq \mathrm{Sup}\, f$ $\forall f\in \mathrm{ReL}(m)$. ii) α is subadditive, provided that the convention $\infty+(-\infty)=\infty$ is adopted.

But we do not claim that $\alpha(tf)=t\alpha(f)$ $\forall f\in \mathrm{ReL}(m)$ and $t>0$. iii) If $f\in \mathrm{ReL}(m)$ fulfills $e^{-f}\in L^{\#}$ then $\alpha(f)=\mathrm{Sup}(f+\log|u|)$ for some $u\in H^{\#}$ with $\varphi(u)=1$. In particular $\alpha(f)>-\infty$.

The main properties are relative to the set M of the representative functions. M enters in form of the functional $\Theta:\mathrm{ReL}(m)\to[-\infty,\infty]$ defined to be

$$\Theta(f) = \mathrm{Sup}\{\int fVdm : V\in M \text{ such that } \int fVdm \ \exists\} \qquad \forall\ f\in \mathrm{ReL}(m),$$

where $\exists$ means existence in the extended sense, that is $\int f^{+}Vdm$ and $\int f^{-}Vdm$ are not both $=\infty$, and where the convention $\mathrm{Sup}\ \emptyset=-\infty$ has to be adopted. We list some immediate properties. i) $\mathrm{Inf}\, f \leqq \Theta(f) \leqq \mathrm{Sup}\, f$ $\forall f\in \mathrm{ReL}(m)$. ii) Θ is subadditive at least on the cone of the functions $\in \mathrm{ReL}(m)$ which are bounded below. And $\Theta(tf)=t\Theta(f)$ $\forall f\in \mathrm{ReL}(m)$ and $t>0$.

<u>3.1</u> <u>THEOREM</u>: i) We have $-\Theta(-f)\leqq\alpha(f)$ $\forall f\in \mathrm{ReL}(m)$. ii) We have $\alpha(f)\leqq \Theta(f)$ $\forall f\in \mathrm{ReL}(m)$ with $e^{-f}\in L^{\#}$ (recall that $-\infty<\alpha(f)$ from the above). In particular M is nonvoid.

Here i) is an alienated version of the classical Jensen inequality while ii) is an alienated version of the classical Szegö theorem. ii) is much more powerful then i). It is desirable to transform ii) into an equation. In this respect we have the subsequent result.

<u>3.2</u> <u>THEOREM</u>: Assume that (H,φ) is reduced. If $f\in \mathrm{ReL}(m)$ satisfies $\alpha^{-}(f):=\liminf_{t\downarrow 0}\frac{1}{t}\alpha((1+tf)^{-})=0$ (which in particular is true if f is bounded below) then $e^{-f}\in L^{\#}$ and $\mathrm{Sup}_{t>0}\frac{1}{t}\,\alpha(tf)=\Theta(f)$.

A rapid application of 3.1 is the subsequent theorem which characterizes the so-called Szegö situation $M=\{F\}$. It comprises the unit disk situation and is the frame where the most prominent classical theorems remain true in the same form as in the unit disk situation.

<u>3.3</u> <u>THEOREM</u>: Assume that (H,φ) is reduced. For $F\in M$ then the subsequent properties are equivalent.

i) $M=\{F\}$.

ii) We have $L^{\#}=L^{o}(Fm):=\{h\in L(m): \int(\log|h|)^{+}Fdm<\infty\}$. And $\alpha(f)=\int fFdm$ $\forall f\in ReL(m)$ with $e^{-f}\in L^{\#}$ which here means $\int f^{-}Fdm<\infty$.

iii) $\alpha(f)\leqq\int fFdm$ $\forall f\in ReL^{\infty}(m)$.

It is natural to ask for the inequality which is opposite to iii). We define MJ to consist of the functions $0\leqq F\in L^{1}(m)$ such that $\log|\varphi(u)|\leqq\int(\log|u|)Fdm$ $\forall u\in H$, called the Jensen functions. Exponentiation shows that $MJ\subset M$. The above question is then answered as follows.

3.4 THEOREM: For $F\in M$ the subsequent properties are equivalent.

i) $F\in MJ$.

ii) For $f\in ReL(m)$ with $\alpha(f)<\infty$ we have $\int f^{+}Fdm<\infty$ and $\int fFdm\leqq\alpha(f)$.

iii) $\int fFdm\leqq\alpha(f)$ $\forall f\in ReL^{\infty}(m)$.

Therefore $M=\{F\}$ implies that $F\in MJ$ (at least when (H,φ) is reduced which however can be dispensed with). But there are worlds between $F\in MJ$ and the Szegö situation $M=\{F\}$.

4. Substitution Theorems and Real-Valued Functions in $H^{\#}$

In Sections 4-10 we shall discuss central points in the abstract Hardy algebra theory under the assumption that (H,φ) be reduced. The present section centers around the substitution theorems 4.1 which is simple and 4.3 which is more sophisticated.

4.1 SUBSTITUTION THEOREM: Let $f:\mathbb{C}\rightarrow\mathbb{C}$ be an entire function and $F:F(t)=\operatorname{Max}\{|f(z)|:|z|\leqq t\}$ $\forall t\geqq 0$. If $u\in H^{\#}$ with $F(|u|)\in L^{\#}$ then $f(u)\in H^{\#}$ and $\varphi(f(u))=f(\varphi(u))$.

4.2 SPECIAL CASE: If $u\in H^{\#}$ with $e^{|u|}\in L^{\#}$ then $e^{u}\in H^{\#}$ and $\varphi(e^{u})=e^{\varphi(u)}$.

The reducedness assumption is seen to mean that all real-valued functions in H must be constant. As an application of 4.2 we obtain an essential extension: If $u\in H^{\#}$ is real-valued with $e^{|u|}\in L^{\#}$ then $u=\text{const}$. In fact, for $t\in\mathbb{R}$ we have $e^{itu}\in H$ and hence $\sin tu\in H$ so that $\sin tu=\text{const}=c(t)$, and it follows that $u=\text{const}$. In the Szegö situation $M=\{F\}$ the above means after 3.3.ii) that the real-valued functions

$u \in H^{\#}$ with $\int |u| F dm < \infty$ must be constant. In the unit disk situation this is a classical theorem. But as a rule there do exist nonconstant real-valued functions in $H^{\#}$. In the unit disk situation we had an example at the end of Section 2. For the Szegö situation see 4.5 below. In spite of the above the functions obtained are rather small as we shall see in Section 8.

In the other substitution theorem we want to substitute functions $u \in H$ of modulus $|u| \leqq 1$ into functions $f \in Hol(D)$. It is natural that the admissible functions $f \in Hol(D)$ must have limit functions on S and as in 4.1 must fulfill some boundedness condition. Thus we restrict ourselves to $f \in Hol^{\#}(D)$. Now the functions $u \in H$ of modulus $|u| \leqq 1$ fulfill $|\varphi(u)| < 1$ except when $u = const = c$ with $|c| = 1$, and one proves that otherwise for the Baire sets $B \subset S$ one has $\lambda(B) = 0 \Rightarrow m([u \in B]) = 0$. Thus for $f \in Hol^{\#}(D)$ with the assistance of its limit function $F \in H^{\#}(D)$ the substitution function $f(u) \in L(m)$ is well-defined.

4.3 SUBSTITUTION THEOREM: Assume that $f \in Hol^{\#}(D)$. For $u \in H$ of modulus $|u| \leqq 1$ with $|\varphi(u)| < 1$ then $f(u) \in H^{\#}$ and $\varphi(f(u)) = f(\varphi(u))$. And $f \in Hol^{+}(D)$ implies that $f(u) \in H^{+}$.

4.4 CONSEQUENCE: $H^{+} = \{\frac{1-u}{1+u} : u \in H$ with $|u| \leqq 1$ and $u \neq const -1\}$. In particular the nonconstant functions $u \in H$ of modulus $|u| = 1$ (:=the inner functions) correspond to the nonconstant purely imaginary functions $f = \frac{1-u}{1+u} \in H^{+}$. Moreover we see that the functions $f \in H^{+}$ fulfill $Re\varphi(f) > 0$ except when $f = const \in i\mathbb{R}$.

4.5 THEOREM: Assume the Szegö situation $M = \{F\}$ and $H \neq \mathbb{C}$. Then there exist nonconstant inner functions and hence nonconstant purely imaginary functions in H^{+}.

The substitution theorem 4.3 can be transferred from D to the half-plane $\mathbb{C}^{+}$ via the fractional-linear map $z \mapsto \frac{1-z}{1+z}$. The result is as follows.

4.6 SUBSTITUTION THEOREM: Assume that $f \in Hol^{\#}(\mathbb{C}^{+})$. For $u \in H^{+}$ with $Re\varphi(u) > 0$ then $f(u) \in H^{\#}$ and $\varphi(f(u)) = f(\varphi(u))$. And $f \in Hol^{+}(\mathbb{C}^{+})$ implies that $f(u) \in H^{+}$.

4.7 SPECIAL CASE: Let $u \in H^+$ be $\neq 0$. For each $\tau \in \mathbb{R}$ then $u^\tau \in H^\#$ and $\varphi(u^\tau) = (\varphi(u))^\tau$ (with the main branch of the power function). In particular $u^\tau \in H^+$ if $|\tau| \leqq 1$.

5. The Moduli of the Invertible Elements of $H^\#$

5.1 THEOREM: Assume that $0 < F \in L(m)$. Then there exist functions $f \in (H^\#)^\times$ (:= the class of invertible elements of $H^\#$) of modulus $|f| = F$ iff $\alpha(\log F) + \alpha(-\log F) = 0$. In this case $f \in (H^\#)^\times$ with $|f| = F$ is unique up to a constant factor of modulus one.

5.2 THEOREM: For $f \in H^\#$ with $m([f=0]) = 0$ we have

$$\log|\varphi(f)| \leqq -\alpha(-\log|f|) \leqq \alpha(\log|f|),$$

$$-\infty < \log|\varphi(f)| = \alpha(\log|f|) \Leftrightarrow f \in (H^\#)^\times.$$

Let us specialize to the Szegö situation $M = \{V\}$. For $0 < F \in L(m)$ here $F, \frac{1}{F} \in L^\#$ is equivalent to $\int |\log F| V dm < \infty$, and in this case we have $\alpha(\log F) + \alpha(-\log F) = 0$ after 3.3.ii). Thus we obtain: For $0 < F \in L(m)$ there exist functions $f \in (H^\#)^\times$ of modulus $|f| = F$ iff $\int |\log F| V dm < \infty$. And after 5.2 a function $f \in H^\#$ is in $(H^\#)^\times$ iff $-\infty < \log|\varphi(f)| = \int (\log|f|) V dm$, which is the traditional definition of the outer functions. An obvious consequence is the factorization theorem which in the unit disk situation is a classical theorem: Each function $h \in H^\#$ with $\varphi(h) \neq 0$ can be factored $h = uf$ with $u \in H$ an inner function and $f \in (H^\#)^\times$ an outer function.

6. The Conjugation Operation

In the unit disk situation the classical conjugation is the operation which associates with each harmonic function $p: D \to \mathbb{R}$ the unique harmonic function $q: D \to \mathbb{R}$ with $q(0) = 0$ such that $p + iq \in Hol(D)$. In order to extend the conjugation to the abstract Hardy algebra situation we have to redefine it as an operation which takes place on the unit circle S: that is which associates with each function P from a certain subclass E of $\operatorname{ReL}(\lambda)$ a unique function $Q\in\operatorname{ReL}(\lambda)$. The immediate idea to define it via $P+iQ\in H^{\#}(D)$ plus some normalization of Q is bound to fail since there are lots of nonconstant real-valued functions in $H^{\#}(D)$. So let us seek to transplant the initial definition form D to S.

For $P\in\operatorname{ReL}^1(\lambda)$ we can form the harmonic function $p:p(z)=\int PP(z,\cdot)d\lambda=:\langle P\lambda\rangle(z)$ $\forall z\in D$. Its classical conjugate function $q:D\to\mathbb{R}$ is obtained from

$$(p+iq)(z) = \int P(s)\frac{s+z}{s-z}\,d\lambda(s) \quad \forall\ z\in D.$$

We have $p+iq\in\operatorname{Hol}^+(D)-\operatorname{Hol}^+(D)\subset\operatorname{Hol}^{\#}(D)$ and hence a limit function $P+iQ\in H^{\#}(D)$. Of course we define $Q\in\operatorname{ReL}(\lambda)$ to be the conjugate function of $P\in\operatorname{ReL}^1(\lambda)$. But how can the transit through D be avoided? The key observation is that $e^{p+iq}\in\operatorname{Hol}^{\#}(D)$ and hence also $e^{t(p+iq)}\in\operatorname{Hol}^{\#}(D)$ $\forall t\in\mathbb{R}$. Thus $e^{t(P+iQ)}\in H^{\#}(D)$ $\forall t\in\mathbb{R}$. But the latter condition determines $Q\in\operatorname{ReL}(\lambda)$ up to an additive real constant after the uniqueness assertion in 5.1. Thus for $P\in\operatorname{ReL}^1(\lambda)$ the conjugate function can be defined to be the unique $Q\in\operatorname{ReL}(\lambda)$ such that $e^{t(P+iQ)}\in H^{\#}(D)$ $\forall t\in\mathbb{R}$ plus some normalization which can be written $\varphi_o(e^{t(P+iQ)})>0$ $\forall t\in\mathbb{R}$. Moreover Section 5 tells us that each function $P\in\operatorname{ReL}(\lambda)$ to which a conjugate function $Q\in\operatorname{ReL}(\lambda)$ can thus be associated must be in $\operatorname{ReL}^1(\lambda)$.

At this point it is clear how to proceed in the abstract Hardy algebra situation (H,φ). The basic theorem is as follows.

6.1 THEOREM: For $P\in\operatorname{ReL}(m)$ the subsequent properties are equivalent.

i) $\alpha(tP)+\alpha(-tP)=0$ $\forall t\in\mathbb{R}$.

ii) $\alpha(P)\in\mathbb{R}$ and $\alpha(tP)=t\alpha(P)$ $\forall t\in\mathbb{R}$.

iii) There exists a function $Q\in\operatorname{ReL}(m)$ such that $e^{t(P+iQ)}\in H^{\#}$ $\forall t\in\mathbb{R}$.

In this case the function $Q\in\operatorname{ReL}(m)$ is unique up to an additive real constant. Hence there exists a unique $Q\in\operatorname{ReL}(m)$ such that in addition $\varphi(e^{t(P+iQ)})=e^{t\alpha(P)}$ $\forall t\in\mathbb{R}$.

The function class E is defined to consist of the functions $P\in\operatorname{ReL}(m)$ which posses the equivalent properties i)ii)iii) above. The functions $P\in E$ are called conjugable. For $P\in E$ the unique function

$Q \in \mathrm{ReL}(m)$ such that $e^{t(P+iQ)} \in H^{\#}$ and $\varphi(e^{t(P+iQ)}) = e^{t\alpha(P)}$ $\forall t \in \mathbb{R}$ is called the conjugate function of P and written $Q =: P^*$.

We list some basic properties. i) $E \subset \mathrm{ReL}(m)$ is a linear subspace and the conjugation $E \to \mathrm{ReL}(m): P \mapsto P^*$ is a linear operator. ii) The restriction $\alpha|E: E \to \mathbb{R}$ is a positive linear functional. And if $T \subset \mathrm{ReL}(m)$ is a linear subspace such that $\alpha|T$ is finite-valued and linear then $T \subset E$. Thus E is the largest linear subspace of ReL(m) on which α is finite-valued and linear. iii) Let $h = P+iQ \in H^{\#}$ such that $e^{|h|} \in L^{\#}$. Then $P \in E$ and $Q = P^* + \mathrm{Im}\varphi(h)$ after the substitution theorem 5.1. In particular $\mathrm{Re}H \subset E$. iv) For $P \in E$ we have $e^{\pm P} \in L^{\#}$ and hence $P \in L^{\#}$. If in addition $P^* \in L^{\#}$ then $P+iP^* \in H^{\#}$ and $\varphi(P+iP^*) = \alpha(P)$. v) There is an obvious connection with H^+: If $0 \leqq P \in E$ then $P+iP^* \in H^+$. Thus if $P \in E$ is bounded below or bounded above then $P+iP^* \in H^{\#}$ and hence $\varphi(P+iP^*) = \alpha(P)$.

The most important results of the section are on the characterization of E with the means of M. In one direction we obtain from 3.1 an instant final result.

<u>6.2</u> <u>THEOREM</u>: Let $P \in \mathrm{ReL}(m)$ with $e^{\pm P} \in L^{\#}$. Assume that the integral $\int PVdm$ has the same value $c \in [-\infty, \infty]$ for all those $V \in M$ for which it exists in the extended sense. Then $P \in E$ and $\alpha(P) = c \in \mathbb{R}$.

In the other direction it would be most pleasant to deduce from $P \in E$ that the integral $\int PVdm$ exists and is $= \alpha(P)$ for all $V \in M$, or at least that $\int PVdm = \alpha(P)$ for all those $V \in M$ for which it exists in the extended sense. But we cannot prove this unless we impose an additional boundedness condition upon $P \in E$ which appears to be sharper than the implicated condition $e^{\pm P} \in L^{\#}$. From 3.2 we obtain the subsequent results.

<u>6.3</u> <u>THEOREM</u>: If $P \in E$ satisfies $\alpha^-(P) = 0$ (which in particular is true if P is bounded below) then $\alpha(P) = \Theta(P)$.

<u>6.4</u> <u>THEOREM</u>: If $P \in E$ satisfies $\alpha^-(\pm P) = 0$ (which in particular is true if P is bounded) then $\alpha(P) = \int PVdm$ for all those $V \in M$ for which $\int PVdm$ exists in the extended sense.

In 6.3 the special case that $P \in E$ is bounded below (and hence in 6.4 the special case that $P \in E$ is bounded) admits an alternative proof

which is simpler than the route via 3.2. We can of course assume that $P \geqq 0$. In this case 6.3 results from 3.1 combined with the subsequent more comprehensive fact on H^+.

6.5 PROPOSITION: For $f=P+iQ \in H^+$ we have $\int PVdm \leqq Re\varphi(f)$ $\forall V \in M$.

From 6.2 and 6.4 we obtain at least a final characterization of the subclass $E^\infty := E \cap ReL^\infty(m)$.

6.6 THEOREM: Let $P \in ReL^\infty(m)$. Then $P \in E^\infty \Leftrightarrow \int PVdm$ has the same value for all $V \in M \Leftrightarrow \int PVdm = \alpha(P)$ $\forall V \in M$.

We conclude the section with another characterization of the Szegö situation which is a simple consequence of the above results. It makes clear that the transition beyond the Szegö situation is a very serious step.

6.7 THEOREM: For $F \in M$ the subsequent properties are equivalent. i) $M=\{F\}$. ii) $E=ReL^1(Fm)$. iii) $ReL^\infty(m) \subset E$.

7. The Basic Approximation Theorem

7.1 APPROXIMATION THEOREM: Assume that $P \in E^\infty$ and hence $h:=P+iP^* \in H^\#$ with $\varphi(h)=\alpha(P)$. Then for each $\varepsilon>0$ there exists a function $h_\varepsilon \in H$ such that

i) $|h_\varepsilon| \leqq |h|$ and $|Reh_\varepsilon| \leqq |Reh| = |P|$,

ii) $|h_\varepsilon - h| \leqq \varepsilon \, Max(1,|h|^{1+\varepsilon})$,

iii) $\varphi(h_\varepsilon)$ is real and $|\varphi(h_\varepsilon)-\varphi(h)| \leqq \varepsilon$.

The above theorem is not hard to prove and has fundamental consequences. Let us introduce the real-linear span $N:=\mathbb{R}(M-M) \subset ReL^1(m)$. Then from 7.1 and 6.6 we obtain the chain of inclusions $E^\infty \subset \{P \in ReL^\infty(m): \exists P_n \in ReH$ with $|P_n| \leqq |P|$ and $P_n \to P\} \subset \overline{ReH}^{weak*} \subset N^\perp = E^\infty$, where $\perp$ is to denote the annihilator in the respective dual system. Thus we have the subsequent theorem.

7.2 THEOREM: We have

$$E^{\infty}=\{P\in ReL^{\infty}(m):\exists P_n\in ReH \text{ with } |P_n|\leqq|P| \text{ and } P_n\to P\}=\overline{ReH}^{weak*}=N^{\perp}.$$

After dualization we obtain $(ReH)^{\perp}=\overline{N}^{ReL^1(m)}$.

We pass to the complexified version of 7.2 which is of at least equal importance. Let us put $H_{\varphi}:=\{u\in H:\varphi(u)=0\}$.

7.3 THEOREM: For each dominant $F\in M$ we have

$$H = N^{\perp}\cap(H_{\varphi}F)^{\perp}:=\{f\in L^{\infty}(m):f\perp N \text{ and } \perp H_{\varphi}F\}.$$

After dualization we obtain $H^{\perp}=\overline{H_{\varphi}F+N+iN}^{L^1(m)}$.

From 7.2 we obtain still another characterization of the Szegö situation.

7.4 THEOREM: (H,φ) is Szegö iff $\overline{ReH}^{weak*}=ReL^{\infty}(m)$ (:= the weak* Dirichlet property).

8. The Extended Kolmogorov Theorems

It is natural to extend the central properties of the classical conjugation to the abstract Hardy algebra situation. We shall discuss the famous Kolmogorov and Marcel Riesz theorems. In both cases the results will be of interest and profit even in the unit disk situation.

For $P\in ReL^1(\lambda)$ the conjugate function $P^*\in ReL(\lambda)$ need not be in $ReL^1(\lambda)$, but it is always of so-called weak-$L^1(\lambda)$ type. For $F\in L(\lambda)$ this is defined to mean that $t\lambda([|F|\geqq t])$ is bounded for $t\uparrow\infty$. Note that for $F\in L^1(\lambda)$ one has

$$t\lambda([|F|\geqq t])\leqq\int_{[|F|\geqq t]}|F|d\lambda \quad \forall t\geqq 0 \text{ and hence } \to 0 \text{ for } t\uparrow\infty.$$

Kolmogorov proved the existence of a constant such that each $P\in ReL^1(\lambda)$ satisfies

$$t\lambda([|P^*|\geqq t]) \leqq \text{const } ||P||_{L^1(\lambda)} \qquad \forall\ t\geqq 0.$$

From this one deduces that $t\lambda([|P^*|\geqq t])\to 0$ for $t\uparrow\infty$. Furthermore there is a constant such that each $P\in \mathrm{Re}L^1(\lambda)$ satisfies

$$(\cos\frac{\tau\pi}{2})\int|P^*|^\tau d\lambda \leqq \text{const } ||P||^\tau_{L^1(\lambda)} \qquad \forall 0<\tau<1.$$

And from this one deduces that $(\cos\frac{\tau\pi}{2})\int|P^*|^\tau d\lambda\to 0$ for $\tau\uparrow 1$.

In order to extend these facts to the abstract Hardy algebra situation (H,φ) we take a severe but wise decision: to restrict ourselves to functions $P\geqq 0$. In the Szegö situation this can at once be removed after 6.7 but otherwise it is of course a serious restriction. But once it has been made the above results can be transferred and at the same time admit another essential extension: from the class of the functions $P+iP^*$ with $0\leqq P\in E$ to the whole of H^+, which is much more even in the unit disk situation after Section 2.

8.1 THEOREM: Let $f=P+iQ\in H^+$ with $\varphi(f)=a+ib$. For each $V\in M$ then

$$t(Vm)([|f-ib|\geqq t]) \leqq 2a \qquad \forall\ t\geqq 0,$$

$$\frac{\pi}{2}t(Vm)([|f|\geqq t]) \to a-\int PVdm \quad \text{for } t\uparrow\infty.$$

8.2 THEOREM: Let $f=P+iQ\in H^+$ be $\neq 0$ with $\varphi(f)=a+ib$. For each $V\in M$ then

$$(\cos\frac{\tau\pi}{2})\int|f|^\tau Vdm \leqq \mathrm{Re}(\varphi(f))^\tau \quad \forall \text{ real } \tau \text{ with } |\tau|<1,$$

$$(\cos\frac{\tau\pi}{2})\int|f|^\tau Vdm \to a-\int PVdm \quad \text{for } \tau\uparrow 1.$$

In both theorems the inequalities are rapid consequences of the results in Section 4. In contrast, the two limit relations are deeper results of tauberian nature. They can be deduced from the converse to the classical Fatou theorem which is due to Loomis.

8.3 REMARK: Each nonconstant purely imaginary $f\in H^+$ (see 4.4-4.5) produces a nonconstant real-valued $F:=if\in H^\#$. After 8.2 we have $\int|F|^\tau Vdm<\infty$ for $0<\tau<1$ $\forall V\in M$. But of course $\int|F|Vdm<\infty$ since otherwise $\varphi(F)=\int FVdm$ and hence $\mathrm{Re}\varphi(f)=0$ so that f must be constant.

In conclusion let us return from H^+ to the functions $P+iP^*$ with $0 \leqq P \in E$. We want to remove the restriction $P \geqq 0$. One can prove the subsequent result.

8.4 THEOREM: For $0<\tau<1$ we have

$$(\cos\frac{\tau\pi}{2})\ \Theta\ (|P+iP^*|^{\tau}) \leqq 2^{1-\tau}\ (\Theta(|P|))^{\tau} \quad \forall\ P \in E^{\infty}.$$

In the case $\dim N<\infty$ the estimation is true $\forall P \in E$.

9. The Extended Marcel Riesz Theorem

In 1924 Marcel Riesz proved for $1<p<\infty$ the existence of a constant such that

$$\|P^*\|_{L^p(\lambda)} \leqq \text{const} \|P\|_{L^p(\lambda)} \qquad \forall\ P \in \mathrm{ReL}^p(\lambda),$$

so that the situation is simpler than for $p=1$ as discussed in Section 8. But it took almost fifty years until PICHORIDES (1972) after partial results of GOKHBERG-KRUPNIK (1968) determined the smallest constant $R(p)$ to be

$$A(p) := \mathrm{Max}\ (\tan\frac{\pi}{2p}, \operatorname{cotan}\frac{\pi}{2p}) = \begin{Bmatrix} \tan\frac{\pi}{2p} & \text{for } 1 \leqq p \leqq 2 \\ \operatorname{cotan}\frac{\pi}{2p} & \text{for } 2 \leqq p < \infty \end{Bmatrix} \geqq 1.$$

In the proofs the unit disk situation permits certain reductions: Thus $R(p)=R(q)$ for conjugate exponents $1<p,q<\infty$ so that it suffices to deal with the case $1<p \leqq 2$. And except for the precise value of $R(p)$ one can restrict oneself to functions $P \geqq 0$. We see that either reduction depends on the basic fact 6.7.ii). Thus in order to obtain the results in the abstract Hardy algebra situation (H,φ) both reductions have to be dispensed with except in the Szegö situation. Recent methodic advances have been achieved in YABUTA(1977) and BARBEY-KÖNIG(1977), see also KÖNIG(1978). The presentation in BARBEY-KÖNIG(1977) is such that when specialized to the unit disk situation it provides for the first time independent proofs for the individual values $1<p<\infty$ for the Pichorides theorem and even for the Marcel Riesz theorem.

For $1\leqq p<\infty$ and $V\in M$ we use the relation

$$(O)\qquad ||\operatorname{Im} u||_{L^p(Vm)} \leqq R\, ||\operatorname{Re} u||_{L^p(Vm)} + |\operatorname{Im}\varphi(u)|,$$

in order to define in a uniform manner

$$\left.\begin{array}{r} R(p,V) \\ R^+(p,V) \\ R^*(p,V) \end{array}\right\} \begin{array}{l} \text{to be the smallest} \\ \text{constant } 0\leqq R\leqq\infty \\ \text{such that (O) is} \\ \text{fulfilled for all} \end{array} \left\{\begin{array}{l} u\in H, \\ u\in H \text{ with } \operatorname{Re} u\geqq 0, \\ u\in H \text{ with } \operatorname{Im} u\geqq 0 \text{ and } \operatorname{Re}\varphi(u)=0. \end{array}\right.$$

Then of course $\operatorname{Max}(R^+(p,V),R^*(p,V)\leqq R(p,V)$. From the approximation theorem 7.1 we see that

$$||P^*||_{L^p(Vm)} \leqq R(p,V)||P||_{L^p(Vm)} \qquad \forall\, P\in E^\infty.$$

In the Szegö situation $M=\{V\}$ the latter estimation extends at once to the functions $P\in \operatorname{Re}L^p(Vm)$. We proceed to collect the results.

<u>9.1</u> <u>REMARK</u>: Assume that H contains nonconstant inner functions. For $1\leqq p<\infty$ and $V\in M$ then

$$R^+(p,V) \geqq \tan\frac{\pi}{2p}, \quad \text{in particular } R^+(1,V)=\infty,$$

$$R^*(p,V) \geqq \operatorname{cotan}\frac{\pi}{2p},$$

and hence $R(p,V)\geqq A(p)$. This is a rather simple consequence of the results in Section 4.

<u>9.2</u> <u>THEOREM</u>: For $1\leqq p<\infty$ and $V\in MJ$ we have $R(p,V)\leqq A(p)$.

<u>9.3</u> <u>CONSEQUENCE</u> (Extended Pichorides Theorem): Assume the Szegö situation $M=\{V\}$ and $H\neq ¢$. For $1\leqq p<\infty$ then $R(p,V)=A(p)$.

<u>9.4</u> <u>THEOREM</u>: For $p=2n\,(n=1,2,\ldots)$ and $V\in M$ we have $R(p,V)\leqq A(p)= =\operatorname{cotan}\frac{\pi}{2p}$.

<u>9.5</u> <u>THEOREM</u>: For $V\in M$ we have

$$R^+(p,V) \leqq \tan\frac{\pi}{2p} = A(p) \quad \text{for } 1\leqq p\leqq 2,$$

$$R^*(p,V) \leqq \operatorname{cotan}\frac{\pi}{2p} = A(p) \quad \text{for } 2\leqq p<\infty.$$

9.6 THEOREM: i) For each $1<p<\infty$ with $p \neq 2n+1 (n=1,2,...)$ there exists a universal constant $0<A^{+}(p)<\infty$ such that

$$R^{+}(p,V) \leqq A^{+}(p) \qquad \forall V \in M.$$

ii) For each $1 \leqq p<\infty$ there exists a universal constant $0<A^{*}(p)<\infty$ such that

$$R^{*}(p,V) \leqq A^{*}(p) \qquad \forall V \in M.$$

Here 9.6.i) is the direct transfer of the basic step $P \geqq 0$ in the traditional proofs of the Marcel Riesz theorem. These proofs had in common to work in $1<p<\infty$ except for the values $p=2n+1 (n=1,2,...)$. Of course these exceptional values could be circumvented, for example via duality. But they appeared as a curious phenomenon which has defied explanation ever since.

Now the extension to the abstract Hardy algebra situation (beyond the Szegö situation!) permits to obtain such an explanation. The obvious question whether 9.6.i) remains true, or at least whether $R^{+}(p,V)<\infty \ \forall V \in M$, for the exceptional values $p=2n+1 (n=1,2,...)$ has the answer no: In KÖNIG (1978) an example (H,φ) is constructed where there are functions $V_n \in M$ such that $R^{+}(2n+1,V_n)=\infty \ (n=1,2,...)$.

As to the bound $R(p,V)$ the situation is quite different. By now one knows several examples where it is $=\infty$. It seems natural to conjecture that for each $1<p<\infty$ except $p=2n (n=1,2,...)$ there exist examples (H,φ) and $V \in M$ such that $R(p,V)=\infty$. But there remains a piece of a sane world as our final result reveals.

9.7 THEOREM: Assume that $\dim N<\infty$ and that $V \in M$ is an internal point of the convex set $M \subset \mathrm{Re}L^{1}(m)$. For $1<p<\infty$ then $R(p,V)<\infty$ and

$$||P^{*}||_{L^{p}(Vm)} \leqq R(p,V)||P||_{L^{p}(Vm)} \qquad \forall P \in E.$$

10. The Maximality Theorem

Our final point in the abstract Hardy algebra theory is a basic maximality theorem. In essence it is due to Gamelin and Lumer, see

GAMELIN (1969). An extension of (H,φ) is defined to be a Hardy algebra situation $(\tilde{H},\tilde{\varphi})$ on the same measure space such that $H\subset\tilde{H}$ and $\varphi=\tilde{\varphi}|H$. The extension $(\tilde{H},\tilde{\varphi})$ is defined to be a small extension iff

$$\mathrm{Re}\tilde{H} \subset (\log|H^{\times}|)^{\perp\perp} = \overline{\text{real-linear span}(\log|H^{\times}|)}^{\text{weak}*} .$$

In this connection observe that $\mathrm{Re}H=\log|e^{H}|\subset\log|H^{\times}|$. In the subsequent theorem we do not assume a priori that $(\tilde{H},\tilde{\varphi})$ be reduced.

<u>10.1</u> <u>MAXIMALITY THEOREM</u>: Assume that $\dim N<\infty$. If $(\tilde{H},\tilde{\varphi})$ is a small extension of (H,φ) then $\tilde{H}=H$.

11. Application to the Abstract Function Algebra Situation: Abstract Mergelyan Theorems

So far our review of the abstract Hardy algebra theory. The present section introduces the abstract function algebra situation and describes the localization step to the abstract Hardy algebra situation via the abstract F.and M.Riesz theorem. The previous theory will then be applied to obtain two abstract Mergelyan type theorems.

We fix a measurable space (X,Σ). Let $B(X,\Sigma)$ consist of the bounded measurable complex-valued functions on X and $ca(X,\Sigma)$ of the complex-valued measures on Σ, in particular $Pos(X,\Sigma)\subset ca(X,\Sigma)$ of the measures with values $\geqq 0$ and $Prob(X,\Sigma)\subset Pos(X,\Sigma)$ of those with total mass=1.

We fix a complex subalgebra $A\subset B(X,\Sigma)$ which contains the constants. With A we associate the two classes of measures

$$A^{\perp}:=\{\sigma\in ca(X,\Sigma):\int u\,d\sigma = 0 \quad \forall u\in A\},$$

$$M(A):=\{\eta\in Pos(X,\Sigma):\int uv\,d\eta = \int u\,d\eta \int v\,d\eta \quad \forall\, u\in A\}\subset Prob(X,\Sigma),$$

called the annihilating and the multiplicative measures for A. Let $A^{\vee}$ consist of those $\sigma\in A^{\perp}$ which are absolutely continuous (as usual$\ll$) with respect to some $\eta\in M(A)$, and $A^{\wedge}$ consist of those $\sigma\in A^{\perp}$ which are singular to all $\eta\in M(A)$. Then we can formulate the decisive implication of the abstract F.and M.Riesz theorem: $A^{\perp}$ is the $ca(X,\Sigma)$-norm closed linear span of $A^{\vee}\cup A^{\wedge}$. Let us reformulate the result in terms of certain closures of A.

We form the closure $\overline{A}^{\omega} \subset B(X,\Sigma)$ in the weak topology $\omega := \sigma(B(X,\Sigma), ca(X,\Sigma))$, and for nonzero $m \in Pos(X,\Sigma)$ the closure $A^m := \overline{A \bmod m}^{weak*} \subset L^{\infty}(m)$ in the weak* topology $\sigma(L^{\infty}(m), L^1(m))$. These are complex subalgebras as well. For $f \in B(X,\Sigma)$ the bipolar theorem tells us that $f \in \overline{A}^{\omega} \Leftrightarrow f \bmod m \in A^m\ \forall 0 \neq m \in Pos(X,\Sigma)$. Likewise the above result can be formulated as follows.

<u>11.1</u> <u>LOCALIZATION PRINCIPLE</u>: The function $f \in B(X,\Sigma)$ is in $\overline{A}^{\omega}$ iff 0) $\int f d\sigma = 0\ \forall \sigma \in A^{\wedge}$, and i) for each $\eta \in M(A)$ there is an $m \in Pos(X,\Sigma)$ with $\eta << m$ such that $f \bmod m \in A^m$.

Now the abstract Hardy algebra situation comes in as follows. Define the spectrum $\Sigma(A)$ of A to consist of those nonzero multiplicative linear functionals $\varphi : A \rightarrow ¢$ which are continuous in ω, that is which can be represented in the form $\varphi(u) = \int u d\sigma\ \forall u \in H$ for some $\sigma \in ca(X,\Sigma)$. For $\varphi \in \Sigma(A)$ let $M(A,\varphi)$ consist of the representative measures $\sigma \in Pos(X,\Sigma)$ and hence $\in Prob(X,\Sigma)$. It will at once become clear that $M(A,\varphi) \neq \emptyset$. Of course $M(A) =$ the union of the $M(A,\varphi)\ \forall \varphi \in \Sigma(A)$.

Consider now $\varphi \in \Sigma(A)$ and $m \in Pos(X,\Sigma)$ such that φ has representative measures $\sigma \in ca(X,\Sigma)$ which are $<<m$. Then there is a unique weak* continuous linear functional $\varphi^m : A^m \rightarrow ¢$ with $\varphi^m(u \bmod m) = \varphi(u)\ \forall u \in A$. And φ^m is $\neq 0$ and multiplicative. Thus (A^m, φ^m) is a Hardy algebra situation for which we see that $M = \{0 \leqq V \in L^1(m) : Vm \in M(A,\varphi)\}$. Hence $M(A,\varphi) \neq \emptyset$ after 3.1. If in particular m itself is $\in M(A,\varphi)$ then (A^m, φ^m) will be reduced with $1 \in M$ so that all of the previous theory can be applied. And 11.1.i) above makes clear that it will often be possible to end up with the case $m \in M(A,\varphi)$. Thus we can combine 11.1 with the fundamental 7.3 to obtain the subsequent characterization of the functions in $\overline{A}^{\omega}$.

<u>11.2</u> <u>THEOREM</u>: For $f \in B(X,\Sigma)$ condition i) in 11.1 is equivalent to each of the conditions

ii) $\int f d\sigma = \int f d\tau \quad \forall \varphi \in \Sigma(A)$ and $\sigma, \tau \in M(A,\varphi)$, and
$\int f u dm = \int f dm \int u dm \quad \forall m \in M(A)$ and $u \in A$,

iii) $\int (f+u)^2 dm = (\int (f+u) dm)^2 \quad \forall m \in M(A)$ and $u \in A$.

Thus $f \in B(X,\Sigma)$ is in $\overline{A}^{\omega}$ iff it satisfies 0) in 11.1 and some of the equivalent conditions i) ii) iii).

The most prominent consequence is the subsequent maximality theorem.

11.3 THEOREM: Assume that $B \subset B(X,\Sigma)$ is a complex subalgebra with $A \subset B$. Then $B \subset \overline{A}^{\omega}$ iff $A^{\wedge} \subset B^{\perp}$ and $M(A) \subset M(B)$ (which means that $M(A)=M(B)$).

11.4 CONSEQUENCE (First Abstract Mergelyan Theorem): Assume that $B \subset B(X,\Sigma)$ is a complex subalgebra with $A \subset B$ such that

o) $A^{\wedge} \subset B^{\perp}$,

1) each $\varphi \in \Sigma(A)$ has an extension $\psi \in \Sigma(B)$,

2) for each $\varphi \in \Sigma(A)$ we have $N(A,\varphi) := \mathbb{R}(M(A,\varphi) - M(A,\varphi)) \subset B^{\perp}$.

Then $B \subset \overline{A}^{\omega}$.

Condition 2) is satisfied in particular if each $\varphi \in \Sigma(A)$ has a unique representative measure $\in M(A,\varphi)$. Also it is sufficient to assume the overall condition $\mathrm{Re}\, B \subset \overline{\mathrm{Re}\, A}^{\omega}$ which resembles the Dirichlet algebra condition of the earlier theory. Condition 2) is often difficult to be verified. A much more powerful theorem is obtained when the abstract Hardy algebra theory contributes the maximality theorem 10.1. For each $\varphi \in \Sigma(A)$ we introduce

$$ML(A,\varphi) := \{\sigma \in \mathrm{Pos}(X,\Sigma) : \log|\varphi(u)| = \int (\log|u|)\, d\sigma \ \ \forall u \in (\overline{A}^{\omega})^{\times}\},$$

the class of logmodular measures for φ (with $\varphi : \overline{A}^{\omega} \to \mathbb{C}$ the obvious unique extension $\in \Sigma(\overline{A}^{\omega})$). It is seen that $ML(A,\varphi) \subset M(A,\varphi)$ and that $ML(A,\varphi) \neq \emptyset$ in the case $\dim N(A,\varphi) < \infty$.

11.5 THEOREM (Second Abstract Mergelyan Theorem): Assume that $B \subset B(X,\Sigma)$ is a complex subalgebra with $A \subset B$ such that

o) $A^{\wedge} \subset B^{\perp}$,

1) each $\varphi \in \Sigma(A)$ has an extension $\psi \in \Sigma(B)$,

*) $\dim N(A,\varphi) < \infty$ for each $\varphi \in \Sigma(A)$,

2) for each $\varphi \in \Sigma(A)$ we have $NL(A,\varphi) := \mathbb{R}(ML(A,\varphi) - ML(A,\varphi)) \subset B^{\perp}$.

Then $B \subset \overline{A}^{\omega}$.

Theorem 11.5 has the obvious weak point of condition *) which is neither necessary nor adequate but which we are unable to eliminate. Condition 2) is satisfied in particular if each $\varphi \in \Sigma(A)$ has a unique logmodular measure $\in ML(A,\varphi)$. Also it is sufficient to assume the overall condition $\mathrm{Re}\, B \subset \overline{\text{real-linear span } (\log|(\bar{A}^{\omega})^{\times}|)}^{\omega}$ which corresponds to the definition of smallness in Section 10.

In conclusion let us mention an important specialization of the above abstract function algebra situation: In the so-called compact-continuous situation X is a compact Hausdorff space and Σ consists of the Baire sets of X, and $A \subset C(X) \subset B(X,\Sigma)$ is supnorm closed. In view of $\omega|C(X)=\sigma(C(X),ca(X,\Sigma))=\sigma(C(X),C(X)')$ after the F.Riesz theorem we have $\bar{A}^{\omega} \cap C(X)=\bar{A}^{\text{supnorm}}=A$. Moreover $\Sigma(A)$ coincides with the spectrum of A in the usual Banach algebra sense.

A famous application is on the polynomial and rational approximation problems in the complex plane. For a nonvoid compact subset $K \subset ¢$ define $P(K) \subset R(K) \subset C(K)$ to be the subnorm closures of the polynomials and of the rational functions with poles off K. It is simple to see that $P(K)=R(K)$ iff K has no holes (:=bounded components of $¢-K$). The above theorems 11.4 and 11.5 applied to $X=\partial K$ and $A=R(K)|X$ combined with the classical Walsh approximation theorem and with the remarkable fact that $A^{\wedge}=\{0\}$ lead to the celebrated Mergelyan polynomial and rational approximation theorems in the subsequent versions: If $B \subset C(K)$ is a complex subalgebra with $R(K) \subset B$ and $\|f\|=\|f|\partial K\|$ $\forall f \in B$ then $B=R(K)$. This results from 11.4 for compact sets K with no holes and from 11.5 for K with a finite number of holes.

References

BARBEY Klaus and Heinz KÖNIG (1977): Abstract Analytic Function Theory and Hardy Algebras. Lecture Notes in Math. Vol.593, Springer, Berlin 1977.

GAMELIN Theodore W. (1969): Uniform Algebras. Prentice Hall, Englewood Cliffs 1969.

GOKHBERG I.Ts. and N.Ya.KRUPNIK (1968): Norm of the Hilbert transformation in the L^p spaces. Functional Analysis and its Applications 2 (1968) 180-181 (translated from the Russian).

KÖNIG Heinz (1978): On the Marcel Riesz estimation for conjugate functions in the abstract Hardy algebra situation. Commentationes Math. (to appear).

PICHORIDES S.K. (1972): On the best values of the constants in the theorems of M.Riesz, Zygmund and Kolmogorov. Studia Math. 44 (1972) 165-179.

YABUTA Kôzô (1977): M.Riesz's theorem in the abstract Hardy space theory. Arch.Math. (to appear).

K.-D. Bierstedt, B. Fuchssteiner (eds.)
Functional Analysis: Surveys and Recent Results

SINGULARITY AND ABSOLUTE CONTINUITY OF MEASURES

S.D. Chatterji, Dépt. Math. EPFL, Switzerland

and

V. Mandrekar, Dept. Stats. and Prob.,
Michigan State Univ., E. Lansing, Mich., U.S.A.

§1. INTRODUCTION :

The purpose of the present paper is to show in detail how certain very simple and straight-forward considerations starting from an intuitively obvious (but technically non-trivial although well-known) supermartingale theorem lead effortlessly to a generalisation of an important theorem of Kakutani [8] and to the dichotomy theorem for Gaussian processes of Feldman [5] and Hajek [6]. All the theorems of the paper are known; our purpose has been to present their proofs in a stream-lined form. As regards Gaussian processes, we could go further and obtain all the usual criteria for equivalence in terms of reproducing kernel Hilbert spaces, J-divergence or suitable Hilbert-Schmidt operators which appear in the literature by simply pursuing our method further. This will be done in another publication.

The paper is written in such a way that no specialized knowledge beyond standard measure and integration theory (e.g. [7]) is needed for its reading.

To understand the proof of theorem 1, one needs to know the conditional expectation operator and the convergence of supermartingales. The fact that the latter is used for arbitrary directed sets is unimportant since its use later could be limited to the standard (denumerable) case as indicated in lemma 7 of §5. The only knowledge of probability theory required is the 0-1 law for the proof of Kakutani's theorem (theorem 2, §3) and some familiarity with Gaussian measures in $\mathbb{R}^n$ (for §4). The 0-1 law is treated in Halmos [7] p. 201 and in Doob [4] p. 102; the latter along with [1] are references for the supermartingale convergence theorem invoked in theorem 1.

Kakutani's theorem is basic in many considerations. Although it seems to us that the Gaussian dichotomy theorem is really contained in Kakutani's theorem (say via Loève-Karhunen representation of Gaussian processes by series of independant random variables), no direct proof which deduces the dichotomy as a simple corollary of Kakutani's theorem is known to us. In another publication [3] (cf. also [2]) we have studied a particular case of Kakutani's theorem in great detail. The problem there is to study the set E of vectors $a \in \mathbb{R}^\infty$ such that $P \sim P_a$ where $P = \otimes \mu$, μ a probability measure on $\mathbb{R}$ and P_a is P translated by a. We show there, completing a result of Shepp [11], that if $\mu(dx) = p(x)\,dx$, $p>o$, then E is, in general, a Mazur-Orlicz space ℓ^ψ where ψ is a function which can be calculated explicitly from p. However E need not be a subspace; if it is (and we give in [3] sufficient conditions for this to be the case) then $E = \ell^\psi$.

§2. PRELIMINARIES :

Let P and Q be two probability measures on σ-algebra Σ of a space Ω. Let

$$Q(A) = \int_A f\, dP + Q(A \cap N)$$

be the Lebesgue decomposition of Q with respect to P where $P(N) = 0$ and $0 \leq f(\omega) < \infty$ a.e.(P). We note that (cf. [9])

$$P \perp Q \quad \text{iff} \quad P\{f = 0\} = 1 \tag{1}$$

and

$$P \ll Q \quad \text{iff} \quad P\{f > 0\} = 1 \tag{2}$$

where $\perp$ denotes singularity, $\ll$ denotes absolute continuity. Actually (1) is immediate from the uniqueness of the Lebesgue decomposition; for (2) note that $f>0$ a.e.(P) obviously implies that $P \ll Q$ (i.e. $Q(A) = 0 \Rightarrow P(A) = 0$). Conversely, if $P \ll Q$ and $M = \{f = 0\}$, we have that

$$P(M) = P(M \cap N) + P(M \setminus N)$$
$$\leq P(N) + P(M \setminus N) = 0$$

since $Q(M) = Q(M \cap N)$ implies that $Q(M \setminus N) = 0$ whence $P(M \setminus N) = 0$ by the absolute continuity of P with respect to Q.

From (1) and (2) we deduce immediately that

$$P \perp Q \quad \text{iff} \quad \int f^\beta \, dP = 0 \tag{3}$$

for some β in the open interval $]0, 1[$ (e.g. $\beta = \frac{1}{2}$)

and

$$P \ll Q \quad \text{iff} \quad \lim_{\beta \downarrow 0} \int f^\beta \, dP = 1. \tag{4}$$

To deduce (4), observe that

$$f^\beta(\omega) \leq \max(f(\omega), 1) \text{ for } 0 < \beta \leq 1$$

and

$$\lim_{\beta \downarrow 0} f^\beta(\omega) = \begin{cases} 1 & \text{if } f(\omega) > 0 \\ 0 & \text{if } f(\omega) = 0 \end{cases}.$$

Hence from the dominated convergence theorem,

$$\lim_{\beta \downarrow 0} \int f^\beta \, dP = P\{f > 0\}$$

which proves (4). The relationship (4) has been observed by Nemetz [10].

From the above considerations and martingale theory we have the following basic theorem for proving singularity or absolute continuity of two measures.

Theorem 1 :

Let $\{\Sigma_\alpha\}_{\alpha \in I}$ be an increasing family of σ-algebras in the space Ω where I is an arbitrary directed set. Let Σ be the σ-algebra generated by the Σ_α, $\alpha \in I$, and let P and Q be two probability measures on Σ. Let f_α be the Radon-Nikodym derivative of the absolutely continuous part of $Q_\alpha = Q \mid \Sigma_\alpha$ with respect to $P_\alpha = P \mid \Sigma_\alpha$.

Then

(i) $\{f_\alpha, \Sigma_\alpha\}_{\alpha \in I}$ is a non-negative super-martingale and $f_\alpha \overset{P}{\to} f$ (in P-measure) where f = the Radon-Nikodym derivative of the absolutely continuous part of Q with respect to P;

(ii) $\lim_\alpha \int | f_\alpha - f |^\beta \, dP = 0$ for any $\beta \in]0, 1[$;

(iii) $P \perp Q$ iff
$\lim_\alpha \int f_\alpha^\beta \, dP = \inf_\alpha \int f_\alpha^\beta \, dP = 0$
for some β in $]0, 1[$ (e.g. $\beta = \frac{1}{2}$);

(iv) $P \ll Q$ iff $\forall \varepsilon > 0 \ \exists \beta(\varepsilon) \in]0, 1[$ such that
$\int f_\alpha^\beta \, dP > (1-\varepsilon)$
for all $\alpha \in I$ and $0 < \beta < \beta(\varepsilon)$.

Proof :

(i) Let $\alpha \leqslant \delta$ in I; then for all $A \in \Sigma_\delta \supset \Sigma_\alpha$,

$Q(A) = \int_A f_\delta \, dP + Q'(A)$

where Q' is defined on Σ_δ and $Q' \perp P_\delta$. If now A belongs to Σ_α then

$Q(A) = \int_A E_\alpha f_\delta \, dP + \int_A g \, dP + Q''(A)$

where E_α denotes P-conditional expectation given the σ-algebra Σ_α and $g \cdot dP + Q''$ is the Lebesgue decomposition of $Q' \mid \Sigma_\alpha$ with respect to P_α. By the uniqueness of the Lebesgue decomposition,

$$f_\alpha = E_\alpha f_\delta + g \geqslant E_\alpha f_\delta$$

which proves that $\{f_\alpha; \Sigma_\alpha\}_{\alpha \in I}$ is a supermartingale.

The general theory of supermartingales (cf. [1]) then completes the proof.

(ii) Since $\int f_\alpha \, dP \leqslant Q(\Omega) \leqslant 1$, we have, for $0 < \beta < 1$,

$$\int f_\alpha^\beta \, dP \leqslant \left(\int f_\alpha \, dP \right)^\beta \leqslant 1,$$

$$\int (f_\alpha^\beta)^{\frac{1}{\beta}} \, dP \leqslant \int f_\alpha \, dP \leqslant 1$$

which gives the uniform integrability of $\{f_\alpha^\beta\}_{\alpha \in I}$ for fixed β in $]0, 1[$. Since $x \to x^\beta$ is concave for $0 < \beta < 1$, the general theory of supermartingales completes the proof.

(iii) and (iv) follow from the preceding discussion since

$$\int f^\beta \, dP = \inf_\alpha \int f_\alpha^\beta \, dP$$

for $0 < \beta < 1$.

§3. KAKUTANI'S THEOREM :

Let $\{\Omega_j, \sigma_j\}_{j \geqslant 1}$ be a sequence of Borel spaces (where σ_j is a σ-algebra of sets in Ω_j) and let p_j, q_j be two probability measures on σ_j. Let

$$\Omega = \Pi\, \Omega_j,\ P = \otimes\, p_j,\ Q = \otimes\, q_j,$$

$$\Pi_n : \Omega \to \prod_{j=1}^{n} \Omega_j,\ \Pi_n\,(\omega_1, \omega_2, \ldots) = (\omega_1, \omega_2, \ldots, \omega_n),$$

$$\Sigma_n = \Pi_n^{-1}\,(\bigotimes_{j=1}^{n} \sigma_j),\ P_n = P \mid \Sigma_n,\ Q_n = Q \mid \Sigma_n.$$

Let g_j be the derivative of the absolutely continuous part of q_j with respect to p_j.

<u>Theorem 2</u> (Kakutani [8]) :

(i) $P \perp Q$ iff $\prod_{j=1}^{\infty} \int g_j^{\frac{1}{2}}\, dp_j = 0$.

(ii) If $p_j \ll q_j$ for all j, then either $P \perp Q$ or $P \ll Q$. Further $P \ll Q$ iff the product $\prod_{j=1}^{\infty} \int (dp_j/dq_j)^{\frac{1}{2}}\, dq_j > 0$; $P \perp Q$ iff the product is 0.

<u>Remark</u> : Let P and Q be any two probability measures on (Ω, Σ) and $P \ll Q$; if $f = \frac{dP}{dQ}$ and $Q(A) = \int_A g\, dP + Q(A \cap N)$, $P(N) = 0$ is the Lebesgue decomposition of Q with respect to P, then $g = \frac{1}{f}$ a.e.(P). To see this, note that

$$\begin{aligned} P(A) &= P(A \setminus N) \\ &= \int_{A\setminus N} f\, dQ \\ &= \int_{A\setminus N} f\, g\, dP \\ &= \int_A f\, g\, dP \end{aligned}$$

whence $f\,g = 1$ a.e.(P). Since $P\{f = 0\} = 0$ we get $g = \frac{1}{f}$ a.e.(P). Note also that $g > 0$ a.e.(P).

Hence in the situation of theorem 2 (ii), if $h_j = dp_j/dq_j$ then $g_j = \frac{1}{h_j}$ a.e.(p_j) and

$$\begin{aligned} \int g_j^{\frac{1}{2}}\, dp_j &= \int h_j^{-\frac{1}{2}}\, dp_j \\ &= \int h_j^{\frac{1}{2}}\, dq_j \end{aligned}$$

so that the criterion in (ii) could also be given in terms of the product appearing in (i).

<u>Proof</u> :

It is easy to verify that

$$f_n(\omega) = \prod_{j=1}^{n} g_j(\omega_j) \quad , \quad \omega = (\omega_j)_{j\geq 1}$$

defines the derivative of the absolutely continuous part of Q_n with respect to P_n.

Since

$$\int f_n^{\frac{1}{2}} \, dP = \prod_{j=1}^{n} \int g_j^{\frac{1}{2}} \, dp_j$$

the assertion of (i) follows immediately from theorem 1 (iii). Now let $p_j \ll q_j$.

Since each $g_j > 0$ a.e.(p_j) we know that $f_n(\omega) > 0$ a.e.(P). Also f_n is the product of the independent random variables defined by g_j; hence by the 0-1 law (see [4]) either $f = 0$ a.e.(P) or $f > 0$ a.e.(P). From §1 we deduce then that either $P \perp Q$ or $P \ll Q$. The necessary and sufficient condition for $P \perp Q$ is, according to (i), that

$$\prod_{j=1}^{\infty} \int g_j^{\frac{1}{2}} \, dp_j = 0$$

which according to the remarks above is equivalent to

$$\prod_{j=1}^{\infty} \int (dp_j \,/\, dq_j)^{\frac{1}{2}} \, dq_j = 0.$$

Since

$$\int (dp_j \,/\, dq_j)^{\frac{1}{2}} \, dq_j \leq 1$$

if the product above does not diverge to 0, it must converge to a finite number (≤ 1) which is the necessary and sufficient condition for $P \ll Q$.

§4. GAUSSIAN MEASURES :

Let (Ω, Σ) be a Borel space equipped with two probability measures P and Q such that a real-valued stochastic process $\{X_t\}_{t \in T}$ (T an arbitrary parameter set) is Gaussian for both measures P and Q. This means that any finite linear combination of the X_t's has a Gaussian distribution under P as well as Q. We suppose further that $\Sigma = \sigma\{X_t, t \in T\}$ = the smallest σ-algebra with respect to which all the X_t's are measurable. With this notation we have the following theorem :

Theorem 3 ([5], [6]) :

Either $P \perp Q$ or $P \sim Q$ i.e. $P \ll Q$ and $Q \ll P$.

Before proving the theorem, we introduce further notation. For any probability measure P and random variables Y, Z (suitably integrable), we write,

$$E_P \, Y = \int Y \, dP$$

$$\mathrm{Cov}_P \, (Y, Z) = E_P \, (YZ) - E_P(Y) \, E_P \, (Z).$$

Let

$$E_P \, X_t = p(t), \quad \mathrm{Cov}_P \, (X_s, X_t) = C \, (s, t)$$

$$E_Q \, X_t = q(t), \quad \mathrm{Cov}_Q \, (X_x, X_t) = D \, (s, t)$$

$$I = \{\alpha \mid \alpha \subset T, \text{ card } \alpha \text{ finite}\}$$

$$\Sigma_\alpha = \sigma\{X_t, t \in \alpha\}$$

$$P_\alpha = P \mid \Sigma_\alpha , \quad Q_\alpha = Q \mid \Sigma_\alpha .$$

Proof of theorem 3 :

We shall suppose known the elementary fact that thetheorem is true if T is a finite set (cf §5). If for some $\alpha \in I$, $P_\alpha \perp Q_\alpha$ then $P \perp Q$. Suppose that $P_\alpha \sim Q_\alpha$ for all $\alpha \in I$. If P and Q are not mutually singular, then by theorem 1 (iii),

$$\inf_\alpha \int \left(\frac{dQ_\alpha}{dP_\alpha} \right)^{\frac{1}{2}} dP_\alpha > 0 \tag{5}$$

For each $\alpha \in I$, let $\{X_{t_j} , 1 \leq j \leq n(\alpha)\}$ be a basis for the linear manifold generated by $\{X_t , t \in \alpha\}$. From lemmas 2 and 3 of §5, we know that (5) is equivalent to

$$\inf_\alpha \prod_{j=1}^{n(\alpha)} \left(\frac{2 \lambda_{\alpha j}^{\frac{1}{2}}}{\lambda_{\alpha j} + 1} \right)^{\frac{1}{2}} \cdot \exp \left(- \frac{1}{4} \sum_{j=1}^{n} \frac{\lambda_{\alpha j} m_{\alpha j}^2}{\lambda_{\alpha j} + 1} \right) > 0 \tag{6}$$

where $\lambda_{\alpha j}$, $m_{\alpha j}$, $1 \leq j \leq n(\alpha)$, correspond to the λ's and m's of lemma 3 of §5.

Note that the product term in (6) is positive and less than or equal to 1 by the arithmetic-geometric mean inequality. Also the exponential term is between 0 and 1. Hence (6) is equivalent to

$$\sup_\alpha \prod_{j=1}^{n(\alpha)} \left(\frac{\lambda_{\alpha j} + 1}{2 \lambda_{\alpha j}^{\frac{1}{2}}} \right) < \infty \tag{7}$$

and

$$\sup_\alpha \sum_{j=1}^{n(\alpha)} \left(\frac{\lambda_{\alpha j} m_{\alpha j}^2}{\lambda_{\alpha j} + 1} \right) < \infty \tag{8}$$

By lemma 4 of §5, (7) is equivalent to

$$\sup_\alpha \sum_{j=1}^{n(\alpha)} \left\{ \frac{1 + \lambda_{\alpha j}}{2 \lambda_{\alpha j}^{\frac{1}{2}}} - 1 \right\} < \infty \tag{9}$$

From lemma 5 of §5, we deduce that (9) is equivalent to

$$\begin{cases} \sup_\alpha \sum_{j=1}^{n(\alpha)} (\lambda_{\alpha j} - 1)^2 < \infty \\ 0 < c_1 \leq \lambda_{\alpha j} \leq c_2 < \infty \text{ for all } \alpha \text{ and } j. \end{cases} \tag{10}$$

Thus we see that (7) and (8) are equivalent to (10) and

$$\sup_\alpha \sum_{j=1}^{n(\alpha)} m_{\alpha j}^2 < \infty \tag{11}$$

Hence P and Q are not mutually singular iff (10) and (11) hold. We shall now show that under (10) and (11),

$$\lim_{\beta \downarrow 0} \inf_\alpha \int \left(\frac{dQ_\alpha}{dP_\alpha} \right)^\beta dP_\alpha = 1 \tag{12}$$

whence by theorem 1 (iv), $P \ll Q$. The roles of P and Q being perfectly symmetric, we deduce that $P \sim Q$.

From lemma 3 of §5, (12) is equivalent to

$$\lim_{\beta \downarrow 0} \inf_{\alpha} \prod_{j=1}^{n(\alpha)} \left[\frac{\lambda_{\alpha j}^{\beta}}{\beta\lambda_{\alpha j} + (1-\beta)} \right]^{\frac{1}{2}} \cdot \exp \left[- \frac{\beta\,(1-\beta)}{2} \sum_{j=1}^{n(\alpha)} \frac{\lambda_{\alpha j}\, m_{\alpha j}^2}{\beta\lambda_{\alpha j} + (1-\beta)} \right] = 1 \tag{13}$$

Because of (10) and (11)

$$\sum_{j=1}^{n(\alpha)} \frac{\lambda_{\alpha j}\, m_{\alpha j}^2}{\beta\lambda_{\alpha j} + (1-\beta)} \leqslant M$$

where M is independent of α, β; hence the exponential term tends to 1 as $\beta \to 0$. To establish (13) it suffices then to show (by taking logarithm of the product term) that

$$\sum_{j=1}^{n(\alpha)} [\ \log \{(1-\beta) + \beta\lambda_{\alpha j}\} - \beta \log \lambda_{\alpha j}\ ] \leqslant \varepsilon \tag{14}$$

for all $\alpha \in I$ if only $0 < \beta \leqslant \beta_o\ (\varepsilon)$ where $\varepsilon > 0$ is arbitrary. Since $-1 < d_1 \leqslant \lambda_{\alpha j} - 1 \leqslant d_2 < \infty$ for all α, j, we have from lemma 6 §5 that

$$\begin{aligned} 0 &\leqslant \log \{(1-\beta) + \beta\,\lambda_{\alpha j}\} - \beta \log \lambda_{\alpha j} \\ &= \log \{1 + \beta(\lambda_{\alpha j} - 1)\} - \beta \log \{1 + (\lambda_{\alpha j} - 1)\} \\ &\leqslant \text{const.}\ \beta\ (\lambda_{\alpha j} - 1)^2 \end{aligned}$$

Hence the sum in (14) is less then or equal to const. β where because of (10) the constant is independent of α and β. This is enough to deduce (14) for $\beta \leqslant \beta_o(\varepsilon)$ and completes the proof of the theorem.

§5. MISCELLANEOUS LEMMAS :

In this section we gather together various lemmas that have been needed in §4.

Lemma 1 :

Theorem 3 is true for a finite set T.

Proof :

If P and Q are not mutually singular, $Y = a + \sum_{t \in T} a_t X_t = 0$ a.e.(P) iff the same is true a.e.(Q). This follows from the Gaussianness of the law of Y under P and Q. Let $1, X_{t_1}, \ldots\ldots, X_{t_n}$ be a maximal linearly independent set (both in $L^2(P)$ and in $L^2(Q)$. From

$$0 < V_P\ (\Sigma\, a_j\, X_{t_j}) = \Sigma\, a_j a_k\ \ Cov_P\ (X_{t_j}, X_{t_k})$$

and the same with P replaced by Q (where $V_P(Y)$ = variance of Y under P) we deduce that the variance-covariance matrices of the random vector $\xi = (X_{t_j})$ is non-singular both under P and under Q. Since the law of ξ is Gaussian in $\mathbb{R}^n$ under both P and Q, these are equivalent in $\mathbb{R}^n$. The following trivial lemma (which has been implicitly used elsewhere) then establishes the equivalence of P and Q.

Lemma 2 :

Let (Ω, Σ) , (Ω', Σ') be two Borel spaces and $\xi : \Omega \to \Omega'$ be a measurable map such that $\xi^{-1}(\Sigma') = \Sigma$. If P and Q are two probability measures in (Ω, Σ) and P' and Q' their respective images in (Ω', Σ') via ξ then

$$P \perp Q \text{ iff } P' \perp Q'$$

$$P \sim Q \text{ iff } P' \sim Q'.$$

If $P \ll Q$ then (and only then) $P' \ll Q'$ and $\frac{dP}{dQ} = \frac{dP'}{dQ'} \circ \xi$; hence

$$\int \left(\frac{dP}{dQ} \right)^\beta dQ = \int \left(\frac{dP'}{dQ'} \right)^\beta dQ'.$$

We omit the proof of this simple lemma. The calculation of the following lemma 3 must be widely known. We give it in detail for lack of any suitable reference.

We recall that a non-singular Gaussian measure in $\mathbb{R}^n$ with mean vector p and variance covariance matrix C (symmetric and positive definite) has a density given by

$$f(x) = (2\pi)^{-\frac{n}{2}} (\det C)^{-\frac{1}{2}} \exp\left[-\tfrac{1}{2} C^{-1}(x-p)\cdot(x-p) \right]$$

where $x \in \mathbb{R}^n$ and $x\cdot y$ denotes the usual scalar product in $\mathbb{R}^n$. The characteristic function of the distribution is given by

$$\int f(x) \exp(i\, t\cdot x)dx = \exp\left[-\tfrac{1}{2} C(x-p)\cdot(x-p) \right].$$

Lemma 3 :

Let P and Q be two non-singular Gaussian measures in $\mathbb{R}^n$ with mean vectors p, q and non-singular variance-covariance matrices C, D respectively. If $0 < \beta < 1$, we have the formula :

$$\int \left(\frac{dQ}{dP} \right)^\beta dP = \prod_{j=1}^{n} \left(\frac{\lambda_j^\beta}{\beta\lambda_j + (1-\beta)} \right)^{\frac{1}{2}} \cdot \exp\left[-\frac{\beta(1-\beta)}{2} \sum_{j=1}^{n} \frac{\lambda_j m_j^2}{\beta\lambda_j + 1-\beta} \right]$$

where $\lambda_1, \ldots, \lambda_n$ are the eigen-values of the positive definite matrix $C^{\frac{1}{2}} D^{-1} C^{\frac{1}{2}}$ and $m = (m_j)$ is such that $m = V^{-1} C^{-\frac{1}{2}} (q-p)$ for a certain orthogonal matrix V.

Proof :

The proof could be given by a suitable choice of basis in $\mathbb{R}^n$. We prefer to proceed by a direct (and equivalent) calculation. Let f, g be the densities of P, Q respectively. Then

$$\int \left(\frac{dQ}{dP} \right)^\beta dP = \int g^\beta(x)\, f^{1-\beta}(x)\, dx$$

$$= \frac{\{(\det A)^{1-\beta} (\det B)^{\beta}\}^{\frac{1}{2}}}{(2\pi)^{\frac{n}{2}}} \int \exp - \tfrac{1}{2} \{(1-\beta)Ax\cdot x + \beta B(x-r)\cdot(x-r)\} dx$$

where $r = q-p$, $A = C^{-1}$, $B = D^{-1}$. To calculate the integral we find a non-singular W such that $W'AW = I$ and $W'BW = \Lambda$ where I is the n x n identity matrix and Λ is a diagonal matrix. This can be done by choosing an orthogonal matrix V such that

$$V' A^{-\frac{1}{2}} B A^{-\frac{1}{2}} V = \Lambda$$

and then putting $W = A^{-\frac{1}{2}} V$. Let $\lambda_1, \ldots, \lambda_n$ the diagonal elements of Λ; all $\lambda_j > 0$. Note that $|\det W| = (\det A)^{-\frac{1}{2}}$. Substituting $x = Wy$, $dx = |\det A|^{-\frac{1}{2}} dy$ and letting $m = (m_j)$ be such that $Wm = r$ we see that the integral above becomes

$$|\det A|^{-\frac{1}{2}} \prod_{j=1}^{n} \int_{-\infty}^{\infty} \exp [- \tfrac{1}{2} \{(1-\beta)y_j^2 + \beta\lambda_j (y_j - m_j)^2\}] dy_j .$$

An easy calculation gives that

$$\int_{-\infty}^{\infty} \exp [- \tfrac{1}{2}\{(1-\beta)y^2 + \beta\lambda(y-t)^2\}]dy$$

$$= \left(\frac{2\pi}{\beta\lambda + 1-\beta} \right)^{\frac{1}{2}} \exp - \frac{\beta(1-\beta)}{2} \cdot \frac{\lambda t^2}{\beta\lambda + 1-\beta} .$$

Using the fact that

$$\frac{\det B}{\det A} = \det \Lambda = (\lambda_1 \cdot \lambda_2 \cdot \ldots \lambda_n)$$

we obtain the formula of lemma 3.

We now give some lemmas of elementary analysis which have been used in §4.

<u>Lemma 4</u> :

Let $a_{\alpha j} \geq 1$ for $\alpha \in I$ and $1 \leq j \leq n(\alpha)$. Then

$$\sup_{\alpha} \prod_{j=1}^{n(\alpha)} a_{\alpha j} < \infty \quad \text{iff} \quad \sup_{\alpha} \sum_{j=1}^{n(\alpha)} \{a_{\alpha j} - 1\} < \infty$$

<u>Proof</u> :

This follows from the inequalities

$$\prod_{j=1}^{N} (1 + b_j) \geq 1 + \sum_{j=1}^{N} b_j \quad \text{for } b_j \geq 0$$

and $\log x \leq x - 1$ for $x > 0$.

<u>Lemma 5</u> :

Let $f(x) = (1 + x) / 2\sqrt{x}$, $x > 0$; then for $M > 1$,

$$\{x \,|1 \leq f(x) \leq M\} = [x_1, x_2], \ 0 < x_1 < x_2 < \infty ,$$

and there exist constants d_1, d_2 ($0 < d_1 < d_2 < \infty$) such that for $x_1 \leq x \leq x_2$,

$$d_1(1-x)^2 \leqslant f(x)-1 \leqslant d_2(1-x)^2$$

Proof :

f attains its infimum at x = 1 and f decreases between 0 and 1 and increases from 1 onward. Also $\lim_{x \downarrow 0} f(x) = \lim_{x \to \infty} f(x) = \infty$. This gives the first assertion.

From

$$f(x) - 1 = \frac{(1-x)^2}{2\sqrt{x}\,(1+\sqrt{x})^2}$$

we deduce the second.

Lemma 6 :

If $-1 < x_1 \leqslant x \leqslant x_2$ and $0 < \beta < 1$ then there exists a constant $d > 0$ (depending on x_1 and x_2 but independent of β) such that

$$\log(1+\beta x) - \beta \log(1+x) \leqslant d\beta x^2$$

Proof :

We first note that for $-1 < x_1 \leqslant x \leqslant x_2 < \infty$, $x - d_2 x^2 \leqslant \log(1+x) < x - d_1 x^2$ for some constants d_1, d_2 such that $0 < d_1 < d_2 < \infty$.
This is immediate from consideration of the function

$$f(x) = \frac{x - \log(1+x)}{x^2}$$

which is continuous in $]-1, \infty[$ with $f(o) = \frac{1}{2}$, $f(x) > 0$; taking

$$d_1 = \sup\{f(x) \mid x_1 \leqslant x \leqslant x_2\}$$

$$d_2 = \inf\{f(x) \mid x_1 \leqslant x \leqslant x_2\}$$

we have the inequality for $\log(1+x)$. Without loss of generality, suppose that $x_1 < 0 < x_2$ (otherwise we just enlarge the interval); then $x_1 < \beta x < x_2$ if $0 < \beta < 1$ and $x_1 \leqslant x \leqslant x_2$ and we get the desired inequality with $d = d_2$.

We add a final lemma which clarifies the insignificance of the parameter set T in theorem 3.

Lemma 7 :

Theorem 3 for general parameter set T follows from the case where T is denumerable.

Proof :

We note first that $\Sigma = \cup \{\Sigma_\delta \mid \delta \subset T,\ \delta \text{ denumerable}\}$
Suppose that theorem 3 has been proved for denumerable parameter sets. Than each $P_\delta = P|\Sigma_\delta$ is either equivalent to the corresponding $Q_\delta = Q|\Sigma_\delta$ or $P_\delta \perp Q_\delta$. If for some δ, $P_\delta \perp Q_\delta$ then $P \perp Q$. If $P_\delta \sim Q_\delta$ for all denumerable δ contained in T we prove that $P \sim Q$. Indeed if $P(A) = 0$ the since $A \in \Sigma_\delta$ for some δ, $Q(A) = 0$. Thus $P \sim Q$.

REFERENCES

[1] Chatterji, S.D.
Les martingales et leurs applications analytiques.
Ecole d'été de probabilités : Processus Stochastiques.
Lecture Notes in Mathematics 307, Springer-Verlag (1973).

[2] Chatterji, S.D. and Mandrekar, V.
Sur la quasi-invariance des mesures sous les translations. C.R. Acad. Sc. Paris 281, 581-583 (1975).

[3] Chatterji, S.D. and Mandrekar, V.
Quasi-invariance of measures under translation.
To be published in Math. Zeit. 1977.

[4] Doob, J.L.
Stochastic processes. John Wiley and Sons, N.Y. (1953).

[5] Feldman, J.
Equivalence and perpendicularity of Gaussian processes. Pacific J. Math. 8, 699-708 (1958); ibid. 9, 1295-1296 (1959).

[6] Hájek, J.
On a property of normal distributions of an arbitrary stochastic process (in Russian).
Czechosl. Math. J. 8, 610-618 (1958). (Also Select. Transl. math. Statist. Probab. 1, 245-253.)

[7] Halmos, P.
Measure theory. Van Nostrand, N.Y. (1950).

[8] Kakutani, S.
On equivalence of infinite product measures. Ann. Math. 49, 214-224 (1948).

[9] Lepage, R.D. and Mandrekar, V.
Equivalence-singularity dichotomies from zero-one laws. Proc. Amer. Math. Soc. 31, 251-254 (1972).

[10] Nemetz, T.
Equivalence-orthogonality dichotomies of probability measures. Colloques de la Société Mathématique János Bolyai, Vol. 11. North Holland, Amsterdam (1975).

[11] Shepp, L.A.
Distinguishing a sequence of random variables from a translate of itself. Ann. Math. Stats. 36, 1107-1112 (1965).

K.-D. Bierstedt, B. Fuchssteiner (eds.)
Functional Analysis: Surveys and Recent Results

GENERALIZED SPECTRAL OPERATORS

E. Albrecht
Fachbereich Mathematik
Universität des Saarlandes
Saarbrücken, Germany

In this survey we present some recent results in the theory of generalized spectral operators, which have been obtained after 1968, when the monograph [12] of I. Colojoară and C. Foiaş was published. The relations between the various types of decomposable operators are discussed and the decomposability theorem and the support theorem for $\mathfrak{A}$-scalar N-tuples of commuting operators are proved. An example of a decomposable operator is constructed which is not $\mathfrak{A}$-decomposable for any inverse closed admissible algebra $\mathfrak{A}$.

1. INTRODUCTION AND NOTATIONS

There are two possibilities how to generalize the notion of a spectral operator (in the sense of N. Dunford, see [13]):

(A) The resolution of the identity of the operator can be generalized. This leads to the various types of decomposable operators and to the notion of a spectral capacity.

(B) If one admits more general functional calculi for the scalar part of the operator (i.e. functional calculi operating on algebras of functions which are possibly smaller than the algebra of bounded Borel functions) one obtains the notions: $\mathfrak{A}$-scalar operator, $\mathfrak{A}$-spectral operator, and $\mathfrak{A}$-decomposable operator.

In their monograph [12] I. Colojoară and C. Foiaş extensively treated the theory of generalized spectral operators along these lines. Inspired by this book a number of simplifications and new results have been obtained and a big part of the theory has been carried over to N-tuples of commuting operators. In this paper parts of these new developments are surveyed. For the sake of simplicity we restrict ourselves to the case of bounded linear operators on a Banach space (in spite of the fact that the most interesting examples in the theory are unbounded operators, see for instance [28],

[29]). There are also some interesting results in the case of N-tuples of non-commuting operators ([1],[10],[11],[25],[32]) which could not be treated here. First, in the following section we consider the various types of decomposable operators. Section 3 is devoted to the study of $\mathfrak{A}$-scalar N-tuples. In the fourth part we investigate the spaces $\mathcal{B}^j(U)$ and $\mathcal{B}^j(\overline{U})$ which have been of some use in spectral theory (see [2],[3],[6],[31]) and which are needed for one of the examples in the last section.

First, we need some notations. In the following, X is a complex Banach space, L(X) the Banach algebra of all continuous linear operators on X, and $\mathfrak{S}(X)$ is the family of all closed linear subspaces of X. If $T = (T_1,\dots,T_N)$ is an N-tuple of commuting operators in L(X) we have the following notions of joint spectra for T:

(a) $\sigma(T,X)$, the joint spectrum of T with respect to X in the sense of J.L. Taylor [30].

(b) If $\mathfrak{R}$ is a closed subalgebra of L(X) containing $T_1,\dots,T_N$ in its center, then as usual

$$\sigma_{\mathfrak{R}}(T) := \{z \in \mathbb{C}^N : \sum_{j=1}^{N} (z_j I - T_j)\mathfrak{R} \neq \mathfrak{R}\}$$

is the joint spectrum of T in $\mathfrak{R}$.

(c) $$\sigma_R(T) := \{z \in \mathbb{C}^N : \sum_{j=1}^{N} (z_j I - T_j)L(X) \neq L(X)\},$$

$$\sigma_L(T) := \{z \in \mathbb{C}^N : \sum_{j=1}^{N} L(X)(z_j I - T_j) \neq L(X)\}$$

are the joint right, resp. left spectra for T.

(d) For $x \in X$, we denote by $\sigma(x;T,X)$ the local spectrum of T at x in X in the sense of [1],[4].

For $A \in L(X)$ we denote by A_L (resp. A_R) the operator on L(X) which is defined by $A_L B := AB$ (resp. $A_R B := BA$) for $B \in L(X)$, and write $T_L = (T_{1,L},\dots,T_{N,L})$ (resp. $T_R = (T_{1,R},\dots,T_{N,R})$). We say that $Y \in \mathfrak{S}(X)$ is invariant for T if Y is invariant for $T_1,\dots,T_N$ and write $T|Y = (T_1|Y,\dots,T_N|Y)$. Two operators $A,B \in L(X)$ are said to be quasinilpotent equivalent (notation: $A \overset{g}{\sim} B$) if

$$\lim_{n\to\infty} \left\| \sum_{k=0}^{n} \binom{n}{k} A^k(-B)^{n-k} \right\|^{1/n} = 0 = \lim_{n\to\infty} \left\| \sum_{k=0}^{n} \binom{n}{k} B^k(-A)^{n-k} \right\|^{1/n}.$$

This defines an equivalence relation on L(X) (see [12], p. 11).

2. DECOMPOSABLE OPERATORS

2.1.DEF. $Y \in \mathfrak{S}(X)$ is called a spectral maximal subspace for the N-tuple $T = (T_1,\dots,T_N)$ of commuting operators in L(X) if it is invariant for T and if every $Z \in \mathfrak{S}(X)$, which is invariant for T and which satisfies $\sigma(T,Z) \subset \sigma(T,Y)$, is contained in Y.

For a single linear operator $T \in L(X)$ we define:

2.2.DEF. (a) T is m-decomposable ([26]) if for every finite open covering $\{U_1,\ldots,U_m\}$ of $\sigma(T,X)$ (consisting of m elements) there exist spectral maximal spaces $Y_1,\ldots,Y_m$ for T such that

(i) $\sigma(T,Y_j) \subset U_j$ for $j=1,\ldots,m$ and

(ii) $X = \sum_{j=1}^{m} Y_j$.

(b) T is decomposable ([14]) if it is m-decomposable for every $m \in \mathbb{N}$.

(c) T is strongly decomposable ([37]) if $T|Y$ is decomposable for every spectral maximal space Y of T.

(d) T is weakly decomposable ([12],[23],[24]) if for every finite open covering $\{U_1,\ldots,U_m\}$ of $\sigma(T,X)$ there are spectral maximal spaces $Y_1,\ldots,Y_m$ for T such that

(i) $\sigma(T,Y_j) \subset U_j$ for $j=1,\ldots,m$ and

(ii)' $X = \overline{\sum_{j=1}^{m} Y_j}$.

Which are the relations between these notions? Obviously, strongly decomposable implies decomposable and this implies 2-decomposable. The question, whether the converse implications are true, is still open. However, the following has been shown:

If $\dim \sigma(T,X) \leq m$ and if T is $(m+1)$-decomposable then T is decomposable (see [8], where this result is proved for N-tuples of operators). Hence, every 3-decomposable operator is decomposable, and if T is 2-decomposable and if $\sigma(T,X)$ has dimension < 2 then T is decomposable. The last statement has also been obtained by A. A. Jafarian in [21].

If X is reflexive then 2-decomposable implies decomposable by a result of Şt. Frunză [18] who has shown that for every 2-decomposable operator T the transpose T' is decomposable.

It is known that every 2-decomposable operator is weakly decomposable (cf. [16], Th. 3) but the converse is not true (by [3]). For the theory of weakly decomposable operators see [21],[22].

There is another, more direct generalization of the notion of the spectral measure, first introduced in [37] in the case $N = 1$ and in [17] for arbitrary $N \in \mathbb{N}$, which gives the possibility to generalize the theory of decomposable operators to N-tuples of commuting operators. Denote by $\mathfrak{F}(\mathbb{C}^N)$ the family of all closed subsets of $\mathbb{C}^N$.

2.3.DEF. (a) A map $\mathcal{E}:\mathfrak{F}(\mathbb{C}^N) \to \mathfrak{S}(X)$ is called an m-spectral capacity if it satisfies the following three conditions:

(i) $\mathcal{E}(\emptyset) = \{0\}$, $\mathcal{E}(\mathbb{C}^N) = X$.

(ii) If $(F_n)_{n=1}^{\infty}$ is a sequence in $\mathfrak{F}(\mathbb{C}^N)$ then $\mathcal{E}(\bigcap_{n=1}^{\infty} F_n) = \bigcap_{n=1}^{\infty} \mathcal{E}(F_n)$.

(iii) For every finite open covering $\{U_1,\dots,U_m\}$ of $\mathbb{C}^N$ (consisting of m elements): $X = \sum_{j=1}^{m} \mathcal{E}(\overline{U_j})$.

(b) If $\mathcal{E}:\mathfrak{F}(\mathbb{C}^N) \to \mathfrak{S}(X)$ is an m-spectral capacity for every $m \in \mathbb{N}$ then $\mathcal{E}$ is called a spectral capacity.

(c) $\mathcal{E}:\mathfrak{F}(\mathbb{C}^N) \to \mathfrak{S}(X)$ is called a strong spectral capacity if it satisfies (i) and (ii) in (a) and the following condition

(iii)' For every finite open covering $\{U_1,\dots,U_m\}$ of $\mathbb{C}^N$ one has $\mathcal{E}(F) = \sum_{j=1}^{m} \mathcal{E}(F \cap \overline{U_j})$ for every closed $F \subset \mathbb{C}^N$.

For N=1 one has the following theorem, due to C. Foiaş [15]:

2.4.THEOREM. *For* $T \in L(X)$ *the following statements are equivalent:*

(a) T *is* 2-*decomposable (resp. decomposable, resp. strongly decomposable).*

(b) *There exists a* 2-*spectral (resp. spectral, resp. strong spectral)capacity* $\mathcal{E}:\mathfrak{F}(\mathbb{C}) \to \mathfrak{S}(X)$ *such that the spaces* $\mathcal{E}(F)$ *are invariant for* T *and satisfy* $\sigma(T,\mathcal{E}(F)) \subset F$ *for every* $F \in \mathfrak{F}(\mathbb{C})$.

Moreover, the 2-*spectral capacity* $\mathcal{E}$ *in* (b) *is given by*

(1) $\mathcal{E}(F) = X_T(F) := \{x \in X: \sigma(x;T,X) \subset F\}$ *for* $F \in \mathfrak{F}(\mathbb{C})$.

"(a) => (b)" follows easily by choosing $\mathcal{E}$ as in (1) and by using the structure of the spectral maximal spaces of T (see [12], Th. 2.1.5). The proof of the converse implication will be given in the proof of Th. 2.6, below.

Because of Th. 2.4,it is natural to define for an N-tuple $T = (T_1,\dots,T_N)$ of commuting operators in $L(X)$:

2.5.DEF. $T = (T_1,\dots,T_N)$ is (m-) decomposable, if there exists a (m-) spectral capacity $\mathcal{E}:\mathfrak{F}(\mathbb{C}^N) \to \mathfrak{S}(X)$ such that the spaces $\mathcal{E}(F)$ are invariant for T and satisfy $\sigma(T,\mathcal{E}(F)) \subset F$ for every $F \in \mathfrak{F}(\mathbb{C}^N)$.

Decomposable N-tuples of operators have been introduced and investigated by Şt. Frunză in [17]. In particular, he proved

2.6.THEOREM. (a) *If* $T = (T_1,\dots,T_N)$ *is* (2-) *decomposable, then there is only one* (2-) *spectral capacity* $\mathcal{E}:\mathfrak{F}(\mathbb{C}^N) \to \mathfrak{S}(X)$ *which*

satisfies the conditions in Def. 2.5.
(b) $\mathcal{E}(F) = \{x \in X: \sigma(x;T,X) \subset F\}$ for all closed $F \subset \mathbb{C}^N$, if $\mathcal{E}$ satisfies the conditions in 2.5.

Of course, (b) implies (a). The proof of (b) is very deep and cannot be given here. However, there is a short and elegant proof of (a), due to F.-H. Vasilescu [33], which we reproduce now.

Proof of (a). It is sufficient to show that for every (2-)spectral capacity $\mathcal{E}$ of T, the spaces $\mathcal{E}(F)$ satisfy the condition:

$$(2) \quad \begin{cases} \text{If } Y \in \mathfrak{S}(X) \text{ is invariant for } T \text{ with the property that} \\ \sigma(T,Y) \subset F, \text{ then } Y \subset \mathcal{E}(F). \end{cases}$$

Notice that this implies that the spaces $\mathcal{E}(F)$ ($F \in \mathfrak{F}(\mathbb{C}^N)$) are spectral maximal for T, so that Vasilescu's proof gives a new proof of the implication "(b) => (a)" in Th. 2.4.

Hence, let Y be a closed invariant subspace for T with $\sigma(T,Y) \subset F$. Let $\{U,V\}$ be an open covering of $\mathbb{C}^N$ such that $F \subset U$ and $\overline{V} \cap F = \emptyset$. Then (by (iii) in 2.3) $X = \mathcal{E}(\overline{U}) + \mathcal{E}(\overline{V})$, so that (by (ii) in 2.3): $X/\mathcal{E}(\overline{U}) \cong \mathcal{E}(\overline{V})/(\mathcal{E}(\overline{U}) \cap \mathcal{E}(\overline{V})) = \mathcal{E}(\overline{V})/\mathcal{E}(\overline{U} \cap \overline{V})$. Hence (by Lemma 1.2 in [30]), $\sigma(\tilde{T},X/\mathcal{E}(\overline{U})) = \sigma(\tilde{\tilde{T}},\mathcal{E}(\overline{V})/\mathcal{E}(\overline{U} \cap \overline{V})) \subset \overline{V} \cup (\overline{V} \cap \overline{U}) = \overline{V}$, where $\tilde{T}$ (resp. $\tilde{\tilde{T}}$) denotes the N-tuple induced by T on $X/\mathcal{E}(\overline{U})$ (resp. on $\mathcal{E}(\overline{V})/\mathcal{E}(\overline{U} \cap \overline{V})$). Denote by $\varphi: X \to X/\mathcal{E}(\overline{U})$ the canonical homomorphism and put $\psi := \varphi | Y$. Then $\psi \circ (T|Y) = \tilde{T} \circ \psi$. By Prop. 4.5 in [31] one has for every function f which is analytic in a neighbourhood of $\sigma(T,Y) \cup \sigma(\tilde{T},X/\mathcal{E}(U))$: $\psi \circ f(T|Y) = f(\tilde{T}) \circ \psi$. (Here $f \to f(T)$ denotes the analytic functional calculus in the sense of J.L. Taylor [31]). If f is a function identical to 1 in a neighbourhood of F (which contains $\sigma(T,Y)$) and vanishing in a neighbourhood of the set $\overline{V}$ ($\supset \sigma(\tilde{T},X/\mathcal{E}(\overline{U}))$), we obtain $f(T|Y) = I|Y$ and $f(\tilde{T}) = 0$, i.e. $\psi \circ (I|Y) = 0 \circ \psi = 0$, and hence $Y \subset \mathcal{E}(\overline{U})$. As U was an arbitrary neighbourhood of F, we obtain (by means of (ii) in 2.3) that $Y \subset \mathcal{E}(F)$.

3. NON-ANALYTIC FUNCTIONAL CALCULI

The results of this section are due to Şt. Frunză and the author (see [7]). First, we have to define, which algebras of functions we want to admit for the functional calculus.

3.1.DEF. An algebra $\mathfrak{A}$ of $\mathbb{C}$-valued functions on a set $\Omega \subset \mathbb{C}^N$ is called admissible, if it satisfies the following three conditions:
(a) $1 \in \mathfrak{A}$, $\pi_1,\dots,\pi_N \in \mathfrak{A}$, where $\pi_j: \Omega \to \mathbb{C}$ are the coordinate functions

defined by $\pi_j(z) := z_j$ for $z = (z_1,\dots,z_N)\in\Omega$ $(j=1,\dots,N)$.

(b) $\mathfrak{A}$ is <u>normal</u>, i.e. for every finite open covering $\{U_1,\dots,U_n\}$ of $\overline{\Omega}$ there are functions $f_1,\dots,f_n\in\mathfrak{A}$ with $0 \leq f_j \leq 1$, $\operatorname{supp}(f_j)\subset U_j$ for $j=1,\dots,n$, and $f_1+ \dots + f_n = 1$ on Ω. (Here $\operatorname{supp}(f)$ is the closure in $\mathbb{C}^N$ of the set $\{z\in\Omega\colon f(z) \neq 0\}$).

(c) For every $f\in\mathfrak{A}$ and every $w\in \mathbb{C}^N\setminus\operatorname{supp}(f)$ there are $f_1,\dots,f_N\in\mathfrak{A}$, such that $\sum_{j=1}^{N} (w_j - z_j)f_j(z) = f(z)$ for all $z\in\Omega$.

In the case N=1 this definition coincides with that in [12],p.59.

<u>3.2.DEF.</u> Let $\mathfrak{A}$ be an admissible algebra on $\Omega \subset \mathbb{C}^N$. We say that $T = (T_1,\dots,T_N)$ is an $\mathfrak{A}$-scalar N-tuple if there exists a homomorphism $\Phi\colon\mathfrak{A} \to L(X)$ with $\Phi(1) = I$ and $\Phi(\pi_j) = T_j$ for $j=1,\dots,N$. Φ is then called an $\mathfrak{A}$-<u>functional calculus</u> ($\mathfrak{A}$-FC) <u>for</u> T.

In the case N=1, this definition seems to be more general than the original one in [12], p.59. However, we will see in Prop. 3.6 that both definitions are equivalent.

<u>3.3.THEOREM.</u> <u>If</u> $T = (T_1,\dots,T_N)$ <u>is an</u> $\mathfrak{A}$-<u>scalar</u> N-<u>tuple for an admissible algebra</u> $\mathfrak{A}$, <u>then</u> T <u>is decomposable and its spectral capacity is given by</u>

(3) $$\mathcal{E}(F) := \bigcap\{\ker\Phi(f)\colon f\in\mathfrak{A},\ \operatorname{supp}(f) \cap F = \emptyset\} \text{ for } F\in\mathfrak{F}(\mathbb{C}^N),$$

<u>where</u> Φ <u>is an arbitrary</u> $\mathfrak{A}$-FC <u>for</u> T.

<u>Proof.</u> Obviously, the spaces $\mathcal{E}(F)$ (as defined in (3)) are closed and satisfy (i) and (by the normality of $\mathfrak{A}$ also) (iii) in Def. 2.3 ($F\in\mathfrak{F}(\mathbb{C}^N)$). Let us now prove, that for every closed $F \subset \mathbb{C}^N$

(4) $$\sigma(T,\mathcal{E}(F)) \subset F \cap \sigma_{\mathfrak{G}}(T) ,$$

where $\mathfrak{G}$ is the closed commutative subalgebra of L(X) generated by $\{\Phi(f)\colon f\in\mathfrak{A}\}$. Indeed, if $w\in\mathbb{C}^N\setminus F$ then (by (b) and (c) in 3.1) there are $f,f_1,\dots,f_N\in\mathfrak{A}$ such that $w\notin\operatorname{supp}(f)$, $f \equiv 1$ on $\Omega\setminus V$, where V is a compact neighbourhood of w with $V \cap F = \emptyset$, and such that

$$\sum_{j=1}^{N} (w_j - z_j)f_j(z) = f(z) \quad \text{for all } z\in\Omega.$$

Hence,

(5) $$\Phi(f) = \sum_{j=1}^{N} (w_jI - T_j)\Phi(f_j) .$$

As $\mathcal{E}(F)$ is invariant for all $A\in\mathfrak{G}$, we may take in (5) the restrictions to $\mathcal{E}(F)$ and obtain $w\notin\sigma_{(T|\mathcal{E}(F))'}(T|\mathcal{E}(F))$. As (by [30], Lemma 1.1) $\sigma(T,\mathcal{E}(F)) \subset \sigma_{(T|\mathcal{E}(F))'}(T|\mathcal{E}(F))$, it follows that $\sigma(T,\mathcal{E}(F))\subset F$. On the other hand, if $w\notin\sigma_{\mathfrak{G}}(T)$, then there are $A_1,\dots,A_N\in\mathfrak{G}$ such that

$$(w_1 I - T_1)A_1 + \ldots + (w_N I - T_N)A_N = I \ .$$

Taking again the restrictions to $\mathcal{E}(F)$, we obtain (as before) that $w \notin \sigma(T,\mathcal{E}(F))$. Hence, (4) is proved.

From (4) we conclude now that Φ vanishes on $\mathbb{C}^N \setminus \sigma_{\mathfrak{A}}(T)$. Indeed, if $f \in \mathfrak{A}$ with $\mathrm{supp}(f) \cap \sigma_{\mathfrak{A}}(T) = \emptyset$, then by (4)

$$\sigma(T,\mathcal{E}(\mathrm{supp}(f))) \subset \mathrm{supp}(f) \cap \sigma_{\mathfrak{A}}(T) = \emptyset \ ,$$

so that $\mathcal{E}(\mathrm{supp}(f)) = \{0\}$ (cf. [30], Cor. to Th. 3.2). As the range of $\Phi(f)$ is contained in $\mathcal{E}(\mathrm{supp}(f))$, we obtain $\Phi(f) = 0$.

We are now able to show that $\mathcal{E}$ fulfills (ii) in 2.3. Let $(F_n)_{n=1}^{\infty}$ be a sequence in $\mathfrak{F}(\mathbb{C}^N)$. By the definition of $\mathcal{E}$ we have (with $F := \bigcap_{n=1}^{\infty} F_n$): $\mathcal{E}(F) \subset \bigcap_{n=1}^{\infty} \mathcal{E}(F_n)$. Let now x be in $\bigcap_{n=1}^{\infty} \mathcal{E}(F_n)$ and let $f \in \mathfrak{A}$ be a function with $\mathrm{supp}(f) \cap F = \emptyset$. $\{\mathbb{C}^N \setminus F_n : n \in \mathbb{N}\}$ is an open covering of the compact set $\mathrm{supp}(f) \cap \sigma_{\mathfrak{A}}(T)$, which contains a finite subcovering $\{\mathbb{C}^N \setminus F_1, \ldots, \mathbb{C}^N \setminus F_k\}$. Since $\mathfrak{A}$ is normal, there are $u_1, \ldots, u_k, u, v \in \mathfrak{A}$ such that $\mathrm{supp}(u_n) \cap F_n = \emptyset$ $(n=1,\ldots,k)$, $\mathrm{supp}(u) \cap \mathrm{supp}(f) = \emptyset$, $\mathrm{supp}(v) \cap \sigma_{\mathfrak{A}}(T) = \emptyset$, and $u_1 + \ldots + u_k + u + v = 1$ on Ω. Using $uf = 0$ and the fact that Φ vanishes on $\mathbb{C}^N \setminus \sigma_{\mathfrak{A}}(T)$ we obtain:

$$\Phi(f)x = \sum_{n=1}^{k} \Phi(u_n f)x + \Phi(uf)x + \Phi(vf)x = \sum_{n=1}^{k} \Phi(u_n f)x = 0,$$

because of $\mathrm{supp}(u_n f) \cap F_n = \emptyset$ and $x \in \mathcal{E}(F_n)$ for $n=1,\ldots,k$. Hence, we have shown that $\mathcal{E}$ is a spectral capacity, which satisfies (by (4)) the conditions in 2.5, so that T is decomposable.

By means of Th. 3.3 we can now give a characterization of the support of the $\mathfrak{A}$-functional calculi for T.

<u>3.4. THEOREM.</u> <u>Let</u> $T = (T_1,\ldots,T_N)$ <u>be an</u> $\mathfrak{A}$-<u>scalar</u> N-<u>tuple for an admissible algebra</u> $\mathfrak{A}$ <u>and let</u> $\Phi : \mathfrak{A} \to L(X)$ <u>be an</u> $\mathfrak{A}$-FC <u>for</u> T. <u>Then for all</u> $x \in X$ <u>the set of all closed</u> $F \subset \mathbb{C}^N$, <u>such that the map</u> $f \to \Phi(f)x$ <u>vanishes on</u> $\mathbb{C}^N \setminus F$, <u>has a unique minimal element, denoted by</u> $\mathrm{supp}\,\Phi(.)x$, <u>and</u> $\mathrm{supp}\,\Phi(.)x = \sigma(x;T,X)$.

<u>Proof.</u> By Th. 3.3, T is decomposable and by Th. 2.6.(b)

$$\mathcal{E}(F) = \{\, y \in X : \sigma(y;T,X) \subset F \} \quad \text{for all closed } F \subset \mathbb{C}^N, \tag{6}$$

so that $x \in \mathcal{E}(\sigma(x;T,X))$. Hence, if $\mathrm{supp}(f) \cap \sigma(x;T,X) = \emptyset$, then $\Phi(f)x = 0$ by (3). Conversely, let $F \subset \mathbb{C}^N$ be a closed set such that Φ vanishes on $\mathbb{C}^N \setminus F$. Then $x \in \mathcal{E}(F)$ by (3), so that by (6) $\sigma(x;T,X) \subset F$.

<u>3.5. THEOREM.</u> <u>Let</u> $T = (T_1,\ldots,T_N)$ <u>be an</u> $\mathfrak{A}$-<u>scalar</u> N-<u>tuple for an admissible algebra</u> $\mathfrak{A}$ <u>and let</u> $\Phi : \mathfrak{A} \to L(X)$ <u>be an</u> $\mathfrak{A}$-FC <u>for</u> T. <u>Then the set of all</u> $F \in \mathfrak{F}(\mathbb{C}^N)$, <u>such that</u> Φ <u>vanishes on</u> $\mathbb{C}^N \setminus F$, <u>has a unique minimal element, denoted by</u> $\mathrm{supp}\,\Phi$, <u>and</u>

$$\operatorname{supp}\Phi = \sigma(T,X) = \sigma_{(T)'}(T) = \sigma_{\mathfrak{B}}(T) = \sigma_R(T) = \sigma_L(T) .$$

Proof. T_R and T_L are $\mathfrak{A}$-scalar N-tuples in $L(L(X))$ and corresponding $\mathfrak{A}$-functional calculi are given by $\Phi_R(f) := \Phi(f)_R$, resp. $\Phi_L(f) := \Phi(f)_L$, $(f \in \mathfrak{A})$. By Th.3.4 we have

$$\operatorname{supp}\Phi = \operatorname{supp}\Phi_L(.)I = \sigma(I;T_L,L(X)) = \sigma_R(T) ,$$

where the last equation follows by Kor. 1.4.(b) in [4]. In the same way we obtain by means of Th.3.4 and Kor.1.4 in [4]

$$\operatorname{supp}\Phi = \sigma_L(T) = \sigma_{\mathfrak{B}}(T) = \sigma_{(T)'}(T).$$

Now, we have $\sigma(x;T,X) \subset \sigma(T,X)$ for all $x \in X$ (as a consequence of Th. 2.2 in [30]). Hence by Th.3.4, Φ vanishes on $\mathbb{C}^N \setminus \sigma(T,X)$. On the other hand $\sigma(T,X) \subset \sigma_{(T)'}(T) = \operatorname{supp}\Phi$, so that $\operatorname{supp}\Phi = \sigma(T,X)$.

Let us now consider the case N=1. If $\mathfrak{A}$ is an admissible algebra on $\Omega \subset \mathbb{C}$ and $f \in \mathfrak{A}$, then we denote for $w \in \mathbb{C} \setminus \operatorname{supp}(f)$ by f_w the function defined by

$$f_w(z) := \begin{cases} f(z)/(w-z) & \text{for } z \in \Omega \setminus \{w\} \\ 0 & \text{for } z \in \Omega \cap \{w\} \end{cases} .$$

By (c) in Def.3.1 f_w is an element of $\mathfrak{A}$.

3.6. PROPOSITION. *Let $\mathfrak{A}$ be an admissible algebra on $\Omega \subset \mathbb{C}$ and let $\Phi : \mathfrak{A} \to L(X)$ be a homomorphism with $\Phi(1) = I$. Then for all $f \in \mathfrak{A}$ the $L(X)$-valued function $w \to \Phi(f_w)$ is analytic in $\mathbb{C} \setminus \operatorname{supp}(f)$; hence Φ is an $\mathfrak{A}$-spectral function in the sense of [12],Def.3.1.3.*

Proof. For $w \in \mathbb{C} \setminus \operatorname{supp}(f)$ we have

$$(7) \qquad (wI - \Phi(id))\Phi(f_w) = \Phi((w - id)f_w) = \Phi(f).$$

On the other hand, $w \notin \sigma(\Phi(id), \mathcal{E}(\operatorname{supp}(f)))$ ($\mathcal{E}$ as in Th.3.3, with $T = \Phi(id)$), so that $((wI - \Phi(id))|\mathcal{E}(\operatorname{supp}(f)))^{-1}$ exists. When we apply this operator to (7) we obtain

$$\Phi(f_w) = ((wI - \Phi(id))|\mathcal{E}(\operatorname{supp}(f)))^{-1}\Phi(f)$$

(notice that the ranges of $\Phi(f)$ and of $\Phi(f_w)$ are contained in $\mathcal{E}(\operatorname{supp}(f))$), and the assertion follows from the fact that the resolvent of $\Phi(id)|\mathcal{E}(\operatorname{supp}(f))$ is analytic on $\mathbb{C} \setminus \sigma(\Phi(id), \mathcal{E}(\operatorname{supp}(f))) \supset \supset \mathbb{C} \setminus \operatorname{supp}(f)$.

For a further characterization of the local spectra and the spectral maximal spaces in the case of $C^\infty(\mathbb{C}^N)$-scalar N-tuples with continuous $C^\infty(\mathbb{C}^N)$-functional calculi see [36] (in the case N=1) and [9] (for arbitrary $N \in \mathbb{N}$).

3.7.DEF. An algebra $\mathfrak{A}$ of complex valued functions on $\Omega \subset \mathbb{C}^N$ is called *inverse closed*, if Ω is closed and if $\mathfrak{A}$ is a full subalgebra of $C(\Omega)$ (i.e. $\mathfrak{A} \subset C(\Omega)$ and $f \in \mathfrak{A}$, $1/f \in C(\Omega)$ implies $1/f \in \mathfrak{A}$).

3.8.THEOREM. *Let* $T = (T_1, \dots, T_N)$ *be an* $\mathfrak{A}$*-scalar* N*-tuple for an inverse closed admissible algebra* $\mathfrak{A} \subset C(\Omega)$ *and let* Φ *be an* $\mathfrak{A}$*-FC for* T.

(a) *Denote by* $\mathfrak{G}$ *the closed commutative subalgebra of* $L(X)$ *generated by* $\{\Phi(f)\colon f \in \mathfrak{A}\}$ *and let* $\mathfrak{M}$ *be the maximal ideal space of* $\mathfrak{G}$. *Then* $\mathfrak{M}$ *is homeomorphic to* $\sigma(T,X)$ *and may therefore be identified with* $\sigma(T,X)$. *With this identification the Gelfand homomorphism* $\hat{}\colon \mathfrak{G} \to C(\sigma(T,X))$ *has the property that* $\Phi(f)^\wedge = f|\sigma(T,X)$ *for all* $f \in \mathfrak{A}$.

(b) *The spectral mapping theorem holds*, i.e. $\sigma(\Phi(f),X) = f(\sigma(T,X))$ *for all* $f \in \mathfrak{A}$.

(c) *For every* $f \in \mathfrak{A}$ *the operator* $\Phi(f)$ *is decomposable and its spectral capacity* $\mathcal{E}_f\colon \mathfrak{F}(\mathbb{C}) \to \mathfrak{S}(X)$ *is given by* $\mathcal{E}_f(F) = \mathcal{E}(f^{-1}(F \cap \sigma(T,X)))$ *for* $F \in \mathfrak{F}(\mathbb{C})$, *where* $\mathcal{E}$ *is the spectral capacity of* T *(given by* Th.3.3).

(d) *The* $\mathfrak{A}$*-FC is unique up to quasinilpotent equivalence, i.e. for every* $\mathfrak{A}$*-FC* $\Psi\colon \mathfrak{A} \to L(X)$ *one has* $\Phi(f) \overset{q}{\sim} \Psi(f)$ *for all* $f \in \mathfrak{A}$.

The proof of this theorem is the same as in the case of continuous $\mathfrak{A}$-functional calculi (see [5]).

Some of the results of this section can be proved for cases, when the normality condition for $\mathfrak{A}$ is weakened and when $\mathfrak{A}$ carries a topology, such that the $\mathfrak{A}$-FC is continuous (see [4]).

4. THE SPACES $\mathcal{B}(U)$ AND $\mathcal{B}(\overline{U})$

Let $U \subset \mathbb{C}^N$ be open and let X be a locally convex Hausdorff space. We denote by $\mathcal{B}^0(U,X)$ the space of all continuous X-valued functions on U. We put $\bar{\partial}_j := \frac{\partial}{\partial \bar{z}_j} := \frac{1}{2}\left(\frac{\partial}{\partial x_j} + i\frac{\partial}{\partial y_j}\right)$ and $\partial_j := \frac{\partial}{\partial z_j} := \frac{1}{2}\left(\frac{\partial}{\partial x_j} - i\frac{\partial}{\partial y_j}\right)$ for $j=1,\dots,N$. As in [31] let now $\mathcal{B}^k(U,X)$ be the space of all those functions $f \in \mathcal{B}^0(U,X)$ such that $\bar{\partial}^\alpha f := \bar{\partial}_1^{\alpha_1} \dots \bar{\partial}_N^{\alpha_N} f$ (in the sense of distributions) is continuous for all $\alpha = (\alpha_1, \dots, \alpha_N)$ with $|\alpha| := \alpha_1 + \dots + \alpha_N \leq k$. We endow $\mathcal{B}^k(U,X)$ with the topology of uniform convergence of the functions and their $\bar{\partial}$-derivatives up to the order k on all compact subsets of U. Hence, if X is a (F)-space then $\mathcal{B}^k(U,X)$ is a (F)-space, too. We put $\mathcal{B}(U,X) := \bigcap_{j=1}^{\infty} \mathcal{B}^j(U,X)$. These spaces have been introduced in [31] for the construction of the analytic functional calculus and have also been studied in [34] and [35]. In [31], p.14, J.L. Taylor asks if for every analytic parametrized cochain complex of Banach spaces (of finite length), which is pointwise exact, the cochain complex $C^{\infty}(U,Y)$ is again exact. As the corresponding result is true for $\mathcal{B}(U,Y)$ (cf. [31], Th.2.16), the answer to this question is given by the following

4.1.THEOREM. *For every complete locally convex Hausdorff space* X *the spaces* $\mathcal{B}(U,X)$ *and* $C^{\infty}(U,X)$ *coincide*.

Proof. As $C^{\infty}(U,X) \subset \mathcal{B}(U,X)$, we have only to show that every $f \in \mathcal{B}(U,X)$ belongs to $C^{\infty}(U,X)$. To prove this, it is sufficient to show that for every $x' \in X'$ the scalar function $z \to \langle f(z),x' \rangle$ is in $C^{\infty}(U)$ (cf. [19], Chap.II, p.80, Th.13), i.e. we have only to prove that $\mathcal{B}(U) \subset C^{\infty}(U)$. Obviously it is sufficient to show that every f in $\mathcal{B}(U)$ with compact support contained in U belongs to $C^{\infty}(U)$. But then for all multi-indices $\alpha=(\alpha_1,\ldots,\alpha_N)$ the functions $\bar{\partial}^{\alpha} f$ are continuous functions with compact support, so that the theorem is an immediate consequence of the Sobolev lemma and the following lemma which follows directly from Lemma 4.2.4 in [20]:

4.2.LEMMA. *Let* f *be in* $L^2(\mathbb{C}^N)$ *with compact support such that* $\bar{\partial}^{\alpha} f$ *is in* $L^2(\mathbb{C}^N)$ *for all* $\alpha=(\alpha_1,\ldots,\alpha_N)$ *with* $|\alpha| \leq k$ $(k \in \mathbb{N})$. *Then* f *is in the Sobolev space* $W^k(\mathbb{C}^N)$.

For the second example in the following section we need a Banach space variant of the spaces $\mathcal{B}^k(U)$, which has also been used in [2], [3], and [6]. For the sake of simplicity we restrict ourselves to the case $N=1$. We put $\mathcal{B}^0(\overline{U}) := C(\overline{U})$. For $k \in \mathbb{N}$ let $\mathcal{B}^k(\overline{U})$ be the space of all $f \in \mathcal{B}^0(\overline{U})$ with $f|U \in \mathcal{B}^k(U)$ such that $\bar{\partial}^j f := \frac{\partial^j f}{\partial \bar{z}^j}$ $(j=1,\ldots,k)$ have continuous extensions to $\overline{U}$ (again denoted by $\bar{\partial}^j f$). Endowed with the norm $\|\cdot\|_{k,\overline{U}}$ given by

$$\|f\|_{k,\overline{U}} := \sum_{j=0}^{k} \frac{1}{j!} \sup_{z \in \overline{U}} |(\bar{\partial}^j f)(z)| \qquad (f \in \mathcal{B}^k(\overline{U})),$$

$\mathcal{B}^k(\overline{U})$ is a Banach space.

4.3.LEMMA. (a) *For every* $f \in \mathcal{B}^k(\overline{U})$ *and every* $\varphi \in C^k(\mathbb{C})$ *the function* φf *belongs to* $\mathcal{B}^k(\overline{U})$ *and the map* $(f,\varphi) \to \varphi f$ *from* $\mathcal{B}^k(\overline{U}) \times C^k(\mathbb{C})$ *to* $\mathcal{B}^k(\overline{U})$ *is continuous*.

(b) *If* $f \in \mathcal{B}^k(\overline{U})$ *and if* g *is locally analytic in a neighbourhood of* $\operatorname{supp}(f)$, *then* $fg \in \mathcal{B}^k(\overline{U})$ *and* $\|fg\|_{k,\overline{U}} \leq \|f\|_{k,\overline{U}} \sup\{|g(z)| : z \in \operatorname{supp}(f)\}$.

(c) *There is a* $r > 0$ (*not depending on* k) *such that for all* $w \in U$ *and all* $f \in \mathcal{B}^k(\overline{U})$ *with* $\operatorname{supp}(f) \subset \{z \in U : |z-w| < r\}$

$$\|f\|_{j,\overline{U}} \geq 2^j \|f\|_{0,\overline{U}} \quad \text{for} \quad j=1,\ldots,k.$$

Proof. (a) is obvious, (b) follows from the fact that the operator $\bar{\partial}$ commutes with multiplication by analytic functions, and (c) is a consequence of Cor. 2.6 in [3].

4.4.LEMMA. Let $U_1, U_2 \subset \mathbb{C}$ be open and bounded and let A be an operator in $L(\mathcal{B}^n(\overline{U_1}), \mathcal{B}^m(\overline{U_2}))$ with the property

(8) $\qquad \operatorname{supp}(Af) \subset \operatorname{supp}(f) \quad$ for all $\ f \in \mathcal{B}^n(\overline{U_1})$.

(a) In the case $n=m$ there exists a function $g \in \mathcal{B}^n(\overline{U_2})$ with $\operatorname{supp}(g) \subset \overline{U_1} \cap \overline{U_2}$, such that for all $f \in \mathcal{B}^n(\overline{U_1})$

(9) $\qquad (Af)(z) = g(z)f(z) \quad (z \in \overline{U_2})$.

(b) If $U_1 \cap U_2 = \emptyset$ or if $m > n$ then $A = 0$.

Proof. (cf. also the proofs of Th.6.1.15 in [12] and of Prop.2.4 in [2]). If $U_1 \cap U_2 = \emptyset$ then for all $f \in \mathcal{B}^n(\overline{U_1})$:

$$\operatorname{supp}(Af) \subset \operatorname{supp}(f) \cap \overline{U_2} \subset \partial U_2,$$

hence (by continuity) $Af = 0$.

Let now w be a point in $\overline{U_2}$. The linear functional on $C^n(\mathbb{C})$ given by $\varphi \to (A(\varphi|\overline{U_1}))(w) \quad (\varphi \in C^n(\mathbb{C}))$ is continuous with support contained in $\{w\} \cap \overline{U_1}$ (by (8)). Hence, there are $a_{i,j}(w) \in \mathbb{C}$ for $i,j \in \mathbb{N}_0$ $(= \mathbb{N} \cup \{0\})$, $i+j \leq n$ such that for all $\varphi \in C^n(\mathbb{C})$

$$(A(\varphi|U_1))(w) = \sum_{i+j \leq n} a_{i,j}(w)(\bar{\partial}^i \partial^j \varphi)(w) .$$

By applying A successively to the polynomials $1, z, \bar{z}, \ldots, z^n, z^{n-1}\bar{z}, \ldots, z\bar{z}^{n-1}, \bar{z}^n$ one concludes easily, that the functions $a_{i,j}$ are in $\mathcal{B}^m(\overline{U_2})$ and that $\operatorname{supp}(a_{i,j}) \subset \overline{U_1} \cap \overline{U_2}$. We put now

$$p := \min\{k > -2 \colon a_{i,j} \equiv 0 \text{ for all } i,j \in \mathbb{N}_0 \text{ with } i + j > k\}$$

and show that $p \leq \max\{-1, n-m\}$. Assume that $p > n - m$ and $p \geq 0$. Then there is a $k \in \{0, \ldots, p\}$ and a point $z_0 \in U_1 \cap U_2$ such that $a_{k,p-k}(z_0) \neq 0$. Let $\varphi \in C^\infty(\mathbb{C})$ be a function such that the diameter of $\operatorname{supp}(\varphi)$ is less than 1, $\operatorname{supp}(\varphi) \subset U_1 \cap U_2$, and such that $\varphi \equiv 1$ in a neighbourhood V of z_0. Then, for the function h with

$$h(z) := \varphi(z)\overline{(z - z_0)}^{m+k}(z - z_0)^{p-k} \ln|\ln|z - z_0||$$

we have $h|\mathbb{C}\setminus\{z_0\} \in C^\infty(\mathbb{C}\setminus\{z_0\})$ and $\bar{\partial}^i \partial^j h$ is continuous on $\mathbb{C}$ for all $i,j \in \mathbb{N}_0$ with $i + j < m + p$ and with $(i,j) \neq (m+k, p-k)$. For $z \in V\setminus\{z_0\}$

(10) $\qquad (\bar{\partial}^{m+k} \partial^{p-k} h)(z) = (m+k)!(p-k)! \ln|\ln|z-z_0|| + g(z) ,$

where $g \in C(V)$ with $\lim_{z \to z_0} g(z) = 0$. Hence $h \in \mathcal{B}^n(\overline{U_1})$ and thus $Ah \in \mathcal{B}^m(\overline{U_2})$.
By regularization we find a sequence $(h_\mu)_{\mu=1}^\infty$ in $C^\infty(\mathbb{C})$ of functions with support contained in $U_1 \cap U_2$ such that $\bar{\partial}^i \partial^j h_\mu$ converges uniformly to $\bar{\partial}^i \partial^j h$ for all $i,j \in \mathbb{N}_0$ with $i+j \leq m+p$ and $(i,j) \neq (m+k, p-k)$.

By the continuity of A and by the definition of the norm $\|.\|_{m,\overline{U}_2}$, $\overline{\partial}^m(Ah_\mu)$ has to converge uniformly on $U_1 \cap U_2$ to $\overline{\partial}^m(Ah)$. But

$$\overline{\partial}^m(Ah_\mu) = \partial^m(a_{k,p-k}\overline{\partial}^k\partial^{p-k}h_\mu) + \sum_{\substack{i+j\leq n\\(i,j)\neq(k,p-k)}} \overline{\partial}^m(a_{i,j}\overline{\partial}^i\partial^j h_\mu) =$$

$$= a_{k,p-k}\overline{\partial}^{m+k}\partial^{p-k}h_\mu + R_\mu ,$$

where R_μ converges uniformly to a continuous function on $U_1 \cap U_2$. For $z\in U_1\setminus\{z_o\}$ hence $a_{k,p-k}(z)\overline{\partial}^{m+k}\partial^{p-k}h_\mu(z)$ converges to $a_{k,p-k}(z)\overline{\partial}^{m+k}\partial^{p-k}h(z)$. As $a_{k,p-k}(z_o) \neq 0$ and because of (10) the sequence $(\overline{\partial}^m(Ah_\mu))_{\mu=1}^\infty$ cannot converge to a continuous function, which is a contradiction to $Ah\in\mathcal{B}^m(\overline{U_2})$. Thus, our assumption was false, and we obtain that $a_{i,j}\equiv 0$ for all $i,j\in\mathbb{N}_o$ with $i+j > n-m$.

Let now f be an arbitrary function in $\mathcal{B}^n(\overline{U_1})$ and let z_o be a point in $U_1 \cap \overline{U_2}$. Let $h\in C^\infty(\mathbb{C})$ be a function with $\mathrm{supp}(h) \subset U_1$ and $h \equiv 1$ in a neighbourhood of z_o. Then

$$(Af)(z_o) = (A(hf))(z_o) + (A((1-h)f))(z_o) = (A(hf))(z_o) .$$

By regularization we find a sequence $(g_k)_{k=1}^\infty$ of C^∞-functions, such that their restrictions to $\overline{U_1}$ converge to hf in the norm of $\mathcal{B}^n(\overline{U_1})$. Hence, we obtain by the continuity of A:

(a) In the case n=m:

$$(Af)(z_o) = \lim_{k\to\infty} (Ag_k)(z_o) = \lim_{k\to\infty} a_{o,o}(z_o)g_k(z_o) =$$

$$= a_{o,o}(z_o)f(z_o)$$

and thus (by the continuity of f,Af, and $a_{o,o}$) $Af = a_{o,o}f$ on $\overline{U_1} \cap \overline{U_2}$.

(b) In the case n<m:

$$(Af)(z_o) = \lim_{k\to\infty} (Ag_k)(z_o) = 0 ,$$

i.e. A = 0.

<u>4.5. COROLLARY. Let</u> $T\in L(\mathcal{B}^n(\overline{U}))$ <u>be the operator defined by</u> $(Tf)(z) := zf(z)$ <u>for</u> $f\in\mathcal{B}^n(\overline{U})$, $z\in\overline{U}$. <u>Then, for every</u> $A\in(T)'$ <u>there is a function</u> $a\in\mathcal{B}^n(\overline{U})$ <u>such that</u> $Af = af$ <u>for all</u> $f\in\mathcal{B}^n(\overline{U})$.

<u>Proof.</u> Obviously $\mathrm{supp}(Af) \subset \mathrm{supp}(f)$ for all $f\in\mathcal{B}^n(U)$, so that we can apply 4.4 with $U_1 = U_2 = U$ and m=n.

<u>4.6. COROLLARY. Let</u> U_1, U_2 <u>be bounded open sets in</u> $\mathbb{C}$ <u>and</u> $k_1 < k_2$. <u>Define</u> $T_j\in L(\mathcal{B}^{k_j}(\overline{U_j}))$ <u>by</u> $(T_jf)(z) := zf(z)$ <u>for</u> $f\in\mathcal{B}^{k_j}(\overline{U_j})$ (j=1,2). <u>Let</u> $\mathfrak{A}$ <u>be an inverse closed admissible algebra and suppose that there are</u> $\mathfrak{A}$<u>-functional calculi</u> Ψ_j <u>for</u> T_j (j=1,2) <u>with</u> $\Psi_j(id) = T_j$. <u>For every</u> $g\in\mathfrak{A}$, $\Psi_j(g)$ <u>is a multiplication operator</u>, $\Psi_j(g)f = a_{j,g}f$

$(f\in\mathcal{B}^{k_j}(\overline{U_j}))$, with $a_{j,g}\in\mathcal{B}^{k_j}(\overline{U_j})$ for $j=1,2$ (by 4.5). Then for all $z\in\overline{U_1\cap U_2}$ we have $a_{1,g}(z) = a_{2,g}(z)$. Moreover, $\text{supp}(a_{j,g})\subset \text{supp}(g)$.

Proof. Consider $T\in L(\mathcal{B}^{k_1}(\overline{U_1\cap U_2}))$ with $(Tf)(z):= zf(z)$ $(z\in\overline{U_1\cap U_2}$, $f\in\mathcal{B}^{k_1}(\overline{U_1\cap U_2}))$. We have $a_{j,g}|\overline{U_1\cap U_2}\in\mathcal{B}^{k_1}(\overline{U_1\cap U_2})$ for all $g\in\mathfrak{A}$ and $j=1,2$, so that $\tilde{\Psi}_1$ and $\tilde{\Psi}_2$, defined by $\tilde{\Psi}_j(g)f:= a_{j,g}f$ $(g\in\mathfrak{A}$, $f\in\mathcal{B}^{k_1}(\overline{U_1\cap U_2}))$, are two $\mathfrak{A}$-functional calculi for T. Hence, for all $g\in\mathfrak{A}$ we have $\tilde{\Psi}_1(g) \overset{q}{\sim} \tilde{\Psi}_2(g)$ (by 3.8.(d)). As the operators $\tilde{\Psi}_j(g)$ are the operators of multiplication with $a_{j,g}$, this implies $a_{1,g} \equiv a_{2,g}$ on $\overline{U_1\cap U_2}$. The final assertion follows by the fact that the spectral capacity for T_j is given by $\mathcal{E}_j(F):=\{f\in\mathcal{B}^k(U_j)\colon \text{supp}(f)\subset F\}$ $(F=\overline{F}\subset\mathbb{C})$ and that $\Psi_j(g)$ has range in $\mathcal{E}_j(\text{supp}(g))$ by 3.3.

5. EXAMPLES

We first construct a nonnormal, subnormal operator which is decomposable (thus answering a question raised in [27], p. 388). In the following, we denote by D the open unit disc in $\mathbb{C}$ and by λ the planar Lebesgue measure. Let H be the completion of the space $C_0^\infty(D)$ (of all functions in $C^\infty(D)$ having compact support contained in D) in the norm $\|\varphi\| := (\int_D |\varphi(z)|^2 d\lambda(z) + \int_D |(\partial\varphi)(z)|^2 d\lambda(z))^{1/2}$. Notice, that by 4.2, H coincides (as a vector space) with the closure of $C_0^\infty(D)$ in the Sobolev space $W^1(D)$.

5.1.PROPOSITION. The operator $S\in L(H)$, defined by $(Sf)(z):=zf(z)$ $(f\in H,\ z\in D)$, is a nonnormal, subnormal operator which is $C^1(\mathbb{C})$-scalar (and hence decomposable by Th.3.3).

Proof. Let K be the Hilbert space $L^2(D)\oplus L^2(D)$. H is isometric isomorphic to the closed subspace $H_o:=\{(f,\bar{\partial}f)\colon f\in H\}$ of K, and S is unitary equivalent to $S_o\in L(H_o)$ with $S_o(f,\bar{\partial}f):= (zf,\bar{\partial}(zf)) =$ $= (zf,z(\bar{\partial}f))$. Then $S_o= T|H_o$, where $T\in L(K)$ is the normal multiplication operator defined by $T(f,g):= (zf,zg)$ for $(f,g)\in K$. Hence, S is a subnormal operator which is obviously not normal. S is $C^1(\mathbb{C})$-scalar and a $C^1(\mathbb{C})$-FC for S is given by $\Phi(\varphi)f:= \varphi f$ for $\varphi\in C^1(\mathbb{C}), f\in H$.

Let us recall, that a continuous linear operator on a Banach space is called $\mathfrak{A}$-decomposable if it is quasinilpotent equivalent to an $\mathfrak{A}$-scalar operator. In our second example we construct a strongly decomposable operator T, such that T is not $\mathfrak{A}$-decomposable for any inverse closed admissible algebra $\mathfrak{A}$. This gives an almost complete answer to the open question (c) in [12], p.217.

We put $Q := Q_{1,1,1} := \{z\in\mathbb{C}: 0 < \operatorname{Re} z < 1,\ 0 < \operatorname{Im} z < 1\}$ and $Q_{n,j,k} := \{z\in Q: \frac{1}{n}(j - \frac{4}{3}) < \operatorname{Re} z < \frac{1}{n}(j + \frac{1}{3}),\ \frac{1}{n}(k - \frac{4}{3}) < \operatorname{Im} z < \frac{1}{n}(k + \frac{1}{3})\}$ for $j,k := 1,\dots,n$, $n\in\mathbb{N}$. Obviously $\{\overline{Q_{n,j,k}}\}_{j,k=1,\dots,n}$ is a covering of $\overline{Q}$ for all $n\in\mathbb{N}$, and $d(Q_{n,j,k}) := \sup\{|z-w|: z,w\in \overline{Q_{n,j,k}}\} \to 0$ for $n\to\infty$. Let now $X_{n,j,k}$ be the Banach space $\mathcal{B}^{p(n,j,k)}(\overline{Q_{n,j,k}})$, where $p(n,j,k) := \frac{n}{6}(n-1)(2n-1) + j + k(n-1)$ for $n\in\mathbb{N}$, $j,k=1,\dots,n$. Notice that $(n,j,k) \neq (n',j',k')$ implies $p(n,j,k) \neq p(n',j',k')$ and that $p(n,j,k) > p(n',j',k')$ if $n > n'$. Instead of $\|.\|_{p(n,j,k),\overline{Q_{n,j,k}}}$ we write $\|.\|_{n,j,k}$. Let now X be the ℓ^1-direct sum of the spaces $X_{n,j,k}$, i.e.

$$X := \{f = (f_{n,j,k}): f_{n,j,k}\in X_{n,j,k},\ \|f\| := \sum_{n=1}^{\infty}\ \sum_{j,k=1}^{n} \|f_{n,j,k}\|_{n,j,k} < \infty\}$$

and define $T\in L(X)$ by $Tf := (zf_{n,j,k})$ for $f = (f_{n,j,k})\in X$.

5.2.LEMMA. T is strongly decomposable.

Proof. Define $\mathcal{E}:\mathfrak{F}(\mathbb{C}) \to \mathfrak{S}(X)$ by

$$\mathcal{E}(F) := \{ f = (f_{n,j,k})\in X: \bigcup_{n=1}^{\infty}\ \bigcup_{j,k=1}^{n} \operatorname{supp}(f_{n,j,k}) \subset F\} \tag{11}$$

$(F\in\mathfrak{F}(\mathbb{C}))$. Obviously, $\mathcal{E}$ satisfies the conditions (i) and (ii) in 2.3. Let us now show that $\mathcal{E}$ fulfills (iii)' in 2.3:

Let $\{U_1,\dots,U_m\}$ be a finite open covering of $\mathbb{C}$ and let F be a closed subset of $\mathbb{C}$. Then there are functions $\varphi_1,\dots,\varphi_m\in C^\infty(\mathbb{C})$ with $\operatorname{supp}(\varphi_i) \subset U_i$, $0 \leq \varphi_i \leq 1$ $(i=1,\dots,m)$, and $\varphi_1 + \dots + \varphi_m \equiv 1$ on $\mathbb{C}$. For $n\in\mathbb{N}$ we put $M_{n,1} := \{(j,k): 1 \leq j,k \leq n,\ \overline{Q_{n,j,k}} \subset U_1\}$ and for $q=2,\dots,m$:

$$M_{n,q} := \{(j,k): 1 \leq j,k \leq n,\ \overline{Q_{n,j,k}} \subset U_q\} \setminus \bigcup_{p=1}^{q-1} M_{n,p}\ .$$

Because of $d(Q_{n,j,k}) \to \infty$ as $n \to \infty$, there is a $n_o\in\mathbb{N}$ such that for all $n > n_o$:

$$\bigcup_{q=1}^{n} M_{n,q} = \{(j,k): j,k=1,\dots,n\}.$$

If $f = (f_{n,j,k})\in\mathcal{E}(F)$ we define $g_q = (g^{(q)}_{n,j,k})$ by

$$g^{(q)}_{n,j,k} := \begin{cases} \varphi_q f_{n,j,k} & \text{for } n=1,\dots,n_o;\ j,k=1,\dots,n \\ f_{n,j,k} & \text{for } n > n_o \text{ and } (j,k)\in M_{n,q} \\ 0 & \text{for } n > n_o \text{ and } (j,k)\notin M_{n,q} \end{cases}$$

Then $f = \sum_{q=1}^{m} g_q$ and $g_q\in\mathcal{E}(\overline{U_q}\cap F)$. Hence, $\mathcal{E}$ fulfills (iii)' in 2.3 and is therefore a strong spectral capacity.

Obviously, the spaces $\mathcal{E}(F)$ $(F\in\mathfrak{F}(\mathbb{C}))$ are invariant for T. Let now $F \subset \mathbb{C}$ be a closed set and let z_o be a point in $\mathbb{C}\setminus F$. The function g with $g(z) := 1/(z - z_o)$ is analytic in a neighbourhood of $F \cap Q$. By 4.3.(b) the operator $R:\mathcal{E}(F) \to \mathcal{E}(F)$ defined by $R(f_{n,j,k}) := (gf_{n,j,k})$ is continuous. Moreover,

$$R(z_o I - T)|\mathcal{E}(F) = I|\mathcal{E}(F) = (z_o I - T)R ,$$

i.e. $z_o \notin \sigma(T,\mathcal{E}(F))$. Thus, T is strongly decomposable by Th. 2.4.

In the following we denote by $J_{n,j,k}:X_{n,j,k} \to X$ the canonical injections and by $P_{n,j,k}:X \to X_{n,j,k}$ the canonical projections ($n \in \mathbb{N}$, $j,k=1,\dots,n$).

<u>5.3.LEMMA.</u> <u>Let</u> $A \in L(X)$ <u>be an operator with the property</u>

$$A\mathcal{E}(F) \subset \mathcal{E}(F) \quad \text{for every closed } F \subset \mathbb{C},$$

<u>where</u> $\mathcal{E}:\mathfrak{F}(\mathbb{C}) \to \mathfrak{S}(X)$ <u>is the spectral capacity of</u> T <u>given by</u> (11). <u>Then</u>:

(a) $P_{n,j,k}AJ_{n',j',k'} = 0$ <u>if</u> $Q_{n,j,k} \cap Q_{n',j',k'} = \emptyset$ <u>or if</u> $p(n,j,k) > p(n',j',k')$.

(b) $P_{n,j,k}AJ_{n,j,k}$ <u>is a multiplication operator with a function in</u> $X_{n,j,k}$.

<u>Proof</u>. Since the operators $P_{n,j,k}AJ_{n',j',k'}$ have the property (8) in 4.4, the assertions are immediate consequences of Lemma 4.4.

<u>5.4.LEMMA.</u> <u>If</u> T <u>is</u> $\mathfrak{A}$-<u>decomposable for an admissible algebra</u> $\mathfrak{A}$, <u>then</u> T <u>is actually</u> $\mathfrak{A}$-<u>scalar</u>.

<u>Proof</u>. If T is $\mathfrak{A}$-decomposable, then there exists a homomorphism $\Phi:\mathfrak{A} \to L(X)$ such that $\Phi(1) = I$ and $\Phi(\text{id}) \overset{q}{\sim} T$. Then the operator $\Phi(\text{id})$ is decomposable and the spectral capacities of T and $\Phi(\text{id})$ coincide (this is an immediate consequence of Th. 2.4 and of Th. 2.2.1 in [12]). Hence by Th. 2.3.3 in [12], $\Phi(f)\mathcal{E}(F) \subset \mathcal{E}(F)$ for all $f \in \mathfrak{A}$ and all closed $F \subset \mathbb{C}$. By the preceding lemma the operators $P_{n,j,k}\Phi(f)J_{n,j,k}$ (and hence $\Phi(\text{id})$) are multiplication operators with functions in $X_{n,j,k}$. As $\Phi(\text{id}) \overset{q}{\sim} T$, we obtain by means of 5.3 for all $n \in \mathbb{N}$, $j,k=1,\dots,n$:

$$0 = \lim_{m\to\infty} \left\| P_{n,j,k} \sum_{q=o}^{m} \binom{m}{q} \Phi(\text{id})^q (-T)^{m-q} J_{n,j,k} \right\|^{1/m} =$$

$$= \lim_{m\to\infty} \left\| \sum_{q=o}^{m} \binom{m}{q} (P_{n,j,k}\Phi(\text{id})J_{n,j,k})^q (-P_{n,j,k}TJ_{n,j,k})^{m-q} \right\|^{\frac{1}{m}},$$

i.e. the operators $P_{n,j,k}\Phi(\text{id})J_{n,j,k}$ and $P_{n,j,k}TJ_{n,j,k}$ are quasi-nilpotent equivalent. Since these operators are multiplication operators, this is only possible if

$$P_{n,j,k}\Phi(\text{id})J_{n,j,k} = P_{n,j,k}TJ_{n,j,k} \quad (n \in \mathbb{N},\ j,k=1,\dots,n). \tag{12}$$

Define now $\Psi:\mathfrak{A} \to L(X)$ by $\Psi(f)g := (P_{n,j,k}\Phi(f)J_{n,j,k}g_{n,j,k})$ for $f\in\mathfrak{A}$ and $g = (g_{n,j,k})\in X$. Obviously $\Psi(1) = I$ and (because of (12)) $\Psi(id) = T$. As Φ is a homomorphism we obtain by means of Lemma 5.3 that Ψ is a homomorphism too. Hence, T is an $\mathfrak{A}$-scalar operator.

<u>5.5. THEOREM. The operator</u> T <u>is strongly decomposable, but</u> T <u>is not</u> $\mathfrak{A}$<u>-decomposable for any inverse closed admissible algebra</u> $\mathfrak{A}$.

<u>Proof</u>. By Lemma 5.2 T is strongly decomposable. Assume now, that T is $\mathfrak{A}$-decomposable for some inverse closed admissible algebra $\mathfrak{A}$. By the preceding lemma T is actually $\mathfrak{A}$-scalar and we can construct an $\mathfrak{A}$-FC $\Psi:\mathfrak{A} \to L(X)$ as in the proof of 5.4. The mapping $f \to P_{n,j,k}\Psi(f)J_{n,j,k}$ from $\mathfrak{A}$ to $L(X_{n,j,k})$ is then an $\mathfrak{A}$-FC for the operator $P_{n,j,k}TJ_{n,j,k}$. We apply now Cor. 4.6 to the spaces $X_{1,1,1} = \mathcal{B}^1(\overline{Q})$ and $X_{n,j,k} = \mathcal{B}^{p(n,j,k)}(\overline{Q_{n,j,k}})$. Thus, for every $f\in\mathfrak{A}$ there are functions $a_{n,j,k}(f)\in X_{n,j,k}$ with the following properties:

(13) $\quad P_{n,j,k}\Psi(f)J_{n,j,k}g = a_{n,j,k}(f)g \quad (g\in X_{n,j,k})$ and

(14) $\quad a_{n,j,k}(f) \equiv a_{1,1,1}(f)$ on $Q_{n,j,k}$

for all $n\in\mathbb{N}$, $j,k=1,\dots,n$. Hence, Ψ is of the form

$$\Psi(f)g = (a_{1,1,1}(f)g_{n,j,k}) \quad (f\in\mathfrak{A},\ g = (g_{n,j,k})\in X).$$

As $n \leq p(n,j,k)$ for all $j,k=1,\dots,n$ and as $\{\overline{Q_{n,j,k}}\}_{j,k=1,\dots,n}$ covers $\overline{Q}$ for all $n\in\mathbb{N}$, we conclude that the function $a_{1,1,1}(f)$ belongs to $\bigcap_{n=1}^{\infty} \mathcal{B}^n(\overline{Q})$ for all $f\in\mathfrak{A}$. $\mathfrak{A}$ being normal, there is a function $f\in\mathfrak{A}$ such that $f \equiv 1$ on $U \cap \Omega$ for a neighbourhood U of $z_o := (1+i)/2$ and such that $\mathrm{supp}(f) \subset K_r := \{z\in Q: |z - z_o| < r\}$, where r is chosen according to 4.3.(c). Then, by Cor. 4.6, $\mathrm{supp}(a_{1,1,1}(f)) \subset K_r$. Moreover, $a_{1,1,1}(f) \not\equiv 0$ because of $\mathrm{supp}(\Psi) = \sigma(T,X) = \overline{Q} \supset U$ (cf. Th. 3.5). By 4.3.(c):

(15) $\quad \|a_{1,1,1}(f)\|_{n,\overline{Q}} \geqq 2^n \|a_{1,1,1}(f)\|_{0,\overline{Q}}$.

Consider now $g = (g_{n,j,k})\in X$ with $g_{n,j,k} \equiv n^{-4}$ ($n\in\mathbb{N}$, $j,k=1,\dots,n$). Then we obtain for all $m\in\mathbb{N}$:

$$\|\Psi(f)g\| = \sum_{n=1}^{\infty}\sum_{j,k=1}^{n} \|n^{-4}a_{1,1,1}(f)\|_{n,j,k} \geqslant$$

$$\geqslant m^{-4}\sum_{j,k=1}^{m} \|a_{1,1,1}(f)\|_{m,j,k}.$$

Hence (because of $m \leq p(m,j,k)$):

$$\|\Psi(f)g\| \geqslant m^{-4}\sum_{j,k=1}^{m} \|a_{1,1,1}(f)\|_{m,\overline{Q_{m,j,k}}} \geqslant$$

$$\geqslant m^{-4}\|a_{1,1,1}(f)\|_{m,\overline{Q}},$$

as $\{\overline{Q_{m,j,k}}\}_{j,k=1,\ldots,m}$ covers $\overline{Q}$. By (15) we obtain

$$\|\Psi(f)g\| > m^{-4}\, 2^m \|a_{1,1,1}(f)\|_{0,\overline{Q}} \to \infty \quad \text{for } m \to \infty$$

in contradiction to $\|\Psi(f)\| < \infty$. Hence, T cannot be $\mathfrak{A}$-decomposable for any inverse closed admissible algebra $\mathfrak{A}$.

REFERENCES

[1] ALBRECHT, E.: Funktionalkalküle in mehreren Veränderlichen. Dissertation, Mainz (1972).

[2] - " - : An example of a $C^\infty(\mathbb{C})$-decomposable operator, which is not $C^\infty(\mathbb{C})$-spectral. Rev. Roum. Math. Pures et Appl. 19, p. 131-139, (1974).

[3] - " - : An example of a weakly decomposable operator which is not decomposable. Rev. Roum. Math. Pures et Appl. 20, p. 855-861, (1975).

[4] - " - : Funktionalkalküle in mehreren Veränderlichen für stetige lineare Operatoren auf Banachräumen. Manuscripta Math. 14, p. 1-40, (1974).

[5] - " - : Der spektrale Abbildungssatz für nichtanalytische Funktionalkalküle in mehreren Veränderlichen. Manuscripta Math. 14, p. 263-277, (1974).

[6] - " - : On joint spectra.To appear.

[7] ALBRECHT, E. and Şt. FRUNZĂ: Non-analytic functional calculi in several variables. Manuscripta Math. 18, p. 327-336, (1976).

[8] ALBRECHT, E. and F.-H. VASILESCU: On spectral capacities. Rev. Roum. Math. Pures et Appl. 19, p. 701-705, (1974).

[9] - " - : Non-analytic local spectral properties in several variables. Czech. Math. J. 24 (99), p. 43o-443, (1974).

[10] ANDERSON, R.F.V.: The Weyl functional calculus. J. Functional Analysis 4, p. 240-267, (1969).

[11] - " - : On the Weyl functional calculus. J. Functional Analysis 6, p. 110-115, (1970).

[12] COLOJOARĂ, I. and C. FOIAŞ: Theory of generalized spectral operators. New York-London-Paris: Gordon and Breach (1968).

[13] DUNFORD, N. and J.T. SCHWARTZ: Linear operators, Part III: Spectral operators. New York-London-Sidney-Toronto: Wiley-Interscience (1971).

[14] FOIAŞ,C.: Spectral maximal spaces and decomposable operators.

Archiv der Math. 14, p. 341-349, (1963).

[15] FOIAŞ, C.: Spectral capacities and decomposable operators. Rev. Roum. Math. Pures et Appl. 13, p. 1539-1545, (1968).

[16] FRUNZĂ, Şt.: A duality theorem for decomposable operators. Rev. Roum. Math. Pures et Appl. 16, p. 1055-1058, (1971).

[17] - " - : The Taylor spectrum and spectral decompositions. J. Functional Analysis 19, p. 390-421, (1975).

[18] - " - : Spectral decomposition and duality. Illinois J. Math. 20, p. 314-321, (1976).

[19] GROTHENDIECK, A.: Produits tensoriels topologiques et espaces nucléaires. Mem. Amer. Math. Soc. 16, Reprint (1966).

[20] HÖRMANDER, L.: An introduction to complex analysis in several variables. Princeton-Toronto-London: D. Van Nostrand (1966).

[21] JAFARIAN, A.A.: Spectral decomposition of operators on Banach spaces. Dissertation, University of Toronto (1973).

[22] - " - : Weak and quasi-decomposable operators. Rev. Roum. Math. Pures et Appl. 22, p. 195-212, (1977).

[23] LJUBIC, Ju. I. and V. I. MACAEV: On the spectral theory of linear operators in a Banach space. Doklady Akad. Nauk. SSSR 131, p. 21-23, (1960) (= Soviet Math. Dokl. 1, p. 184-186, (1960)).

[24] - " - : On operators with decomposable spectrum. Mat. Sbornik N.S. 56 (98), p. 433-468, (1962).Errata, ibid. 71 (113), (1966). (= Amer. Math. Soc. Transl. (2) 47, p.89-129).

[25] NELSON, E.: A functional calculus for non-commuting operators. Proceed. Conf. in Honour of Prof. M. Stone at Univ. of Chicago, May 1968, Ed. F.E. Browder, Berlin: Springer (1970).

[26] PLAFKER, S.: On decomposable operators. Proc. Amer. Math. Soc. 24, p. 215-216, (1970).

[27] RADJABALIPOUR, M.: On subnormal operators. Transactions Amer. Math. Soc. 211, p. 377-389, (1975).

[28] SUSSMANN, H.J.: Non-spectrality of a class of second order ordinary differential operators. Comm. Pure and Appl. Math. 23, p. 819-840, (1970).

[29] - " - : Generalized spectral theory and second order ordinary differential operators. Can. J. Math. 25, p. 178-193, (1973).

[30] TAYLOR, J.L.: A joint spectrum for several commuting operators. J. Functional Analysis 6, p. 172-191, (1970).

[31] TAYLOR, J.L.: The analytic functional calculus for several commuting operators. Acta Math. 125, p. 1-38, (1970).

[32] TAYLOR, M.E.: Functions of several self-adjoint operators. Proc. Amer. Math. Soc. 19, p. 91-98, (1968).

[33] VASILESCU, F.-H.: An application of Taylor's functional calculus. Rev. Roum. Math. Pures et Appl. 19, p. 1165-1167, (1974).

[34] - " - : Funcţii analitice şi forme diferentiale în spaţii Fréchet. St. Cerc. Mat. 26, p. 1023-1049, (1974).

[35] - " - : Asupra unei clase de funcţii vectoriale. St. Cerc. Mat. 28, p. 121-127, (1976).

[36] VRBOVÀ, P. The structure of maximal spectral spaces of generalized scalar operators. Czech. Math. J. 23 (98), p. 493-496, (1973).

[37] APOSTOL, C.: Spectral decompositions and functional calculus. Rev. Roum. Math. Pures et Appl. 13, p. 1481-1528, (1968).

K.-D. Bierstedt, B. Fuchssteiner (eds.)
Functional Analysis: Surveys and Recent Results

A RUSSO DYE THEOREM FOR JORDAN C*-ALGEBRAS

J. D. Maitland Wright,
Department of Mathematics,
University of Reading,
Reading, England.

and

M. A. Youngson,
Department of Mathematics,
University of Edinburgh,
Edinburgh, Scotland.

In [6] Russo and Dye proved that the closed unit ball of a C*-algebra is the closed convex hull of its unitary elements. In this note we answer a question posed in [9] by showing that an analogous result is true for Jordan C*-algebras (see below for definitions). The results obtained here were discovered independently by the two authors.

The method adopted is based on the elegant argument of Harris [3] who expresses each element z of the open unit ball of a C*-algebra as an integral, round the unit circle, of the generalized Möbius transform of z and shows that, on the unit circle, this Möbius transform takes unitary values. After some preliminary results on polar decompositions, we show that the generalized Möbius transform of z must lie in the norm-closed unital Jordan *-subalgebra generated by z.

For the general theory of Jordan algebras we refer to [4]. We use $a \circ b$ to denote the Jordan product. Let us recall that for any a, b, c in a Jordan algebra J the Jordan triple product $\{abc\}$ is defined to be $(a \circ b)\circ c - (c \circ a)\circ b + (b \circ c)\circ a$ which reduces to $\frac{1}{2}(abc + cba)$ when the algebra is special with $a \circ b = \frac{1}{2}(ab + ba)$.

The concept of Jordan C*-algebras is due to Kaplansky.

<u>DEFINITION</u> (Kaplansky). Let $\mathcal{A}$ be a complex Banach space and a complex unital <u>Jordan algebra</u> with an involution $*$. Then $\mathcal{A}$ is a <u>Jordan</u> C*-<u>algebra</u> when the following four conditions are satisfied.

(i) $||z \circ w|| \leq ||z||\,||w||$ for all z and w in $\mathcal{A}$,

(ii) $||z|| = ||z^*||$ for all z in $\mathcal{A}$,

(iii) $||\{zz^*z\}|| = ||z||^3$ for all z in $\mathcal{A}$.

(iv) Each norm closed associative *-subalgebra of $\mathcal{A}$ is a C*-algebra.

As observed in [9], condition (iv) is a consequence of (i) and (iii).

A Jordan C*-algebra is said to be a JC*-<u>algebra</u> if it is isometrically *-isomorphic to a norm-closed Jordan *-subalgebra of the *-algebra of all bounded operators on a complex Hilbert space. Not all Jordan C*-algebras are JC*-algebras; it is shown in [9] that there exists an exceptional Jordan C*-algebra $\mathcal{M}_3^8$, whose self-adjoint part can be identified with M_3^8 (the exceptional Jordan algebra discovered by von Neumann, Jordan and Wigner [8]). Moreover, given any Jordan C*-algebra $\mathcal{A}$ there exists a unique *-ideal $\mathcal{J}$ such that $\mathcal{A}/\mathcal{J}$ is a JC*-algebra and such that every 'factorial' representation of $\mathcal{A}$ which does not annihilate $\mathcal{J}$ is onto $\mathcal{M}_3^8$.

§1 POLAR DECOMPOSITIONS

Let z be any element of a Jordan C*-algebra $\mathcal{A}$. Let Jord(z) be the smallest norm closed Jordan *-subalgebra of $\mathcal{A}$ which contains z and 1. Then [9; Corollary 2.2], Jord(z) is a JC*-algebra. Thus we can, and shall, identify Jord(z) with a JC*-subalgebra of $\mathcal{L}(H)$, where H is some complex Hilbert space.

Let au and vb be, respectively, the left and right polar decompositions of z in $\mathcal{L}(H)$, so that $a = (zz^*)^{\frac{1}{2}}$ and $b = (z^*z)^{\frac{1}{2}}$. There is no reason to suppose, in general, that a or b is in $\mathrm{Jord}(z)$.

LEMMA 1.1. *Whenever* $w \in \mathrm{Jord}(z)$ *then* $a^2wb^2 \in \mathrm{Jord}(z)$.

We have
$$a^2wb^2 = zz^*wz^*z = \left\{z\{z^*wz^*\}z\right\}.$$

LEMMA 1.2. $\sigma(a) \cup \{0\} = \sigma(b) \cup \{0\}$.

By [7; Proposition 1.1.8], $\sigma(zz^*)\setminus\{0\} = \sigma(z^*z)\setminus\{0\}$.

PROPOSITION 1.3. *Let* ψ *be a continuous complex function on* $\sigma(a) \cup \{0\}$; *let* ϕ *be a continuous complex function on* $\sigma(a)$. *Then*

(i) $\psi(a)z = z\psi(b)$,

(ii) $\phi(a)z \in \mathrm{Jord}(z)$,

(iii) $\psi(a) + \psi(b) \in \mathrm{Jord}(z)$,

(iv) $\psi(a)\psi(b) \in \mathrm{Jord}(z)$.

We have $b^{2n} = z^*a^{2n-2}z$, for $n \geq 1$.

So $zb^{2n} = zz^*a^{2n-2}z = a^{2n}z$, for $n \geq 1$.

Hence, by the Stone-Weierstrass Theorem,

$$z\psi(b) = \psi(a)z. \tag{i}$$

In particular, $a^{4n}z = a^{2n}(a^{2n}z) = a^{2n}zb^{2n}$.

So, by Lemma 1.1, $a^{4n}z \in \mathrm{Jord}(z)$ for $n = 0, 1, 2, \ldots$.

So, by the Stone-Weierstrass Theorem, $\phi(a)z \in \mathrm{Jord}(z)$. (ii)

It follows from (ii) that $a^{2n-2}z \in \mathrm{Jord}(z)$ for $n \geq 1$. But $a^{2n} + b^{2n} = 2(a^{2n-1}z)\circ z^*$, so that $a^{2n} + b^{2n} \in \mathrm{Jord}(z)$ for $n = 0, 1, 2, \ldots$. Hence, by the Stone-Weierstrass Theorem,

$$\psi(a) + \psi(b) \in \mathrm{Jord}(z). \tag{iii}$$

Let m and n be non-negative integers and consider $a^{2m}b^{2n} + a^{2n}b^{2m}$. We can suppose, without loss of generality, that $m \geq n$ and put $n = n+r$, where $r \geq 0$. Then

$$a^{2n+2r}b^{2n} + a^{2n}b^{2n+2r} = a^{2n}(a^{2r} + b^{2r})b^{2n}.$$

So, by (iii), and repeated applications of Lemma 1.1, $a^{2m}b^{2n} + a^{2n}b^{2m} \in \mathrm{Jord}(z)$.

Let $\xi(t) = \sum_1^n \alpha_j t^j$ be any polynomial with complex coefficients. Then, since $\xi(a^2)\xi(b^2)$ is a linear combination of terms of the form $a^{2m}b^{2n} + a^{2n}b^{2m}$ it follows that $\xi(a^2)\xi(b^2) \in \mathrm{Jord}\ z$. The Stone-Weierstrass Theorem now gives (iv).

---

* $\sigma(\ )$ denotes the spectrum

§2 A RUSSO-DYE THEOREM FOR JORDAN C* ALGEBRAS

Let $||z|| < 1$ and, for $|\lambda| < \frac{1}{||z||}$, let

$$F(\lambda) = (1 - a^2)^{-\frac{1}{2}}(\lambda + z)(1 + \lambda z^*)^{-1}(1 - b^2)^{\frac{1}{2}} .$$

The following lemma of Harris [3] leads directly to his elegant and elementary proof of the Russo-Dye Theorem [6].

LEMMA 2.1 (see [2; page 210]). *When* $|\lambda| = 1$ *then* $F(\lambda)$ *is unitary and*

$$z = \frac{1}{2\pi}\int_0^{2\pi} F(e^{i\theta})d\theta .$$

The second part of the lemma follows from the observation that F is analytic on the open disc $\left\{\lambda \in \mathbb{C} : |\lambda| < \frac{1}{||z||}\right\}$.

LEMMA 2.2. *Whenever* $|\lambda| < \frac{1}{||z||}$ *then* $F(\lambda)$ *is in* Jord(z).

We have
$$(\lambda + z)(1 + \lambda z^*)^{-1} = z + \lambda(1 - a^2)(1 + \lambda z^*)^{-1} .$$

So
$$F(\lambda) = (1 - a^2)^{-\frac{1}{2}}\left(z + \lambda(1 - a^2)(1 + \lambda z^*)^{-1}\right)(1 - b^2)^{\frac{1}{2}} .$$

From Proposition 1.3 (i),

$$(1 - a^2)^{-\frac{1}{2}}z(1 - b^2)^{\frac{1}{2}} = z(1 - b^2)^{-\frac{1}{2}}(1 - b^2)^{\frac{1}{2}} = z .$$

So it only remains to show that

$$(1 - a^2)^{\frac{1}{2}}(1 + \lambda z^*)^{-1}(1 - b^2)^{\frac{1}{2}}$$

is in Jord(z). It suffices to show that its inverse is in Jord(z). But $(1 - b^2)^{-\frac{1}{2}}(1 - a^2)^{-\frac{1}{2}}$ is in Jord(z) by Proposition 1.3 (iv).

By Proposition 1.3 (i), $(1 - b^2)^{-\frac{1}{2}}z^*(1 - a^2)^{-\frac{1}{2}} = z^*(1 - a^2)^{-1}$ which is, by Proposition 1.3 (ii), in Jord z.

We recall that an element u of a Jordan C*-algebra $\mathcal{A}$ is said to be unitary [9], if $u \circ u^* = 1$ and $u^2 \circ u^* = u$, in other words u^* is the inverse of u [4; §11 Chapter 1]. In particular, when $\mathcal{A}$ is a JC*-subalgebra of $\mathcal{L}(H)$, then u is a unitary element of $\mathcal{L}(H)$.

Let $\mathcal{A}$ be a (unital) Jordan C*-algebra; let U be the set of unitaries in $\mathcal{A}$; let E be the set of all elements of $\mathcal{A}$ of the form $\exp ia$ where a is a self-adjoint element of $\mathcal{A}$.

THEOREM 2.3. *The closed unit ball of* $\mathcal{A}$ *is the closed convex hull of* U.

This is an immediate consequence of the preceding two lemmas.

COROLLARY 2.4. *The closed unit ball of* $\mathcal{A}$ *is the closed convex hull of* E .

It suffices to show that if u is unitary in $\mathcal{A}$ then $u \in \overline{\text{co}E}$. For $0 < t < 1$, replace z by tu in the definition of F . Then, see [2; page 210], $F(e^{it})$ is in E for each real θ . So $tu \in \overline{\text{co}E}$.

COROLLARY 2.5. *Let* $\mathcal{A}$ *be a Jordan* C*-*algebra and let* z_1, z_2, z_3 *be any elements of* $\mathcal{A}$. *Then*

$$||\{z_1\ z_2\ z_3\}|| \leq ||z_1||\ ||z_2||\ ||z_3|| .$$

For finite dimensional $\mathcal{A}$, this follows from [9; Lemma 2.5]. Essentially the same proof together with Theorem 2.3 gives the general result.

An element z of a complex Jordan *-algebra is said to be *normal* if the Jordan subalgebra generated by z and z^* is associative.

COROLLARY 2.6. *Let* $\mathcal{A}$ *be a complex Banach space and a complex unital Jordan algebra equipped with an involution* $*$. *Then* $\mathcal{A}$ *is a Jordan* C*-*algebra if the following conditions are satisfied.*

(i) $||z \circ w|| \leq ||z||\ ||w||$ for all z and w in $\mathcal{A}$,

(ii) $||z|| = ||z^*||$ for all z in $\mathcal{A}$,

(iii) $||w \circ w^*|| = ||w||^2$ for all normal w in $\mathcal{A}$,

(iv) $||\{zz^*z\}|| \leq ||z||^3$ for all z in $\mathcal{A}$.

First we observe that conditions (i) and (iii) imply that A , the self-adjoint part of $\mathcal{A}$, is a JB-algebra [1]. Hence, by [9], there exists a Jordan C*-norm ρ on $\mathcal{A}$ such that $\rho(x) = ||x||$ for all x in A . By [9; Lemma 1.2] the norms ρ and $||\ ||$ are equivalent. So, by [9; Lemma 1.1] $\rho(z) \geq ||z||$ for all $z \in \mathcal{A}$.

Let w be any normal element of $\mathcal{A}$, so that Jord(w) is a commutative C*-algebra. By a theorem of Kaplansky [5] (see Theorem 1.2.4 and Corollary 1.25 [7]) this implies $||w|| = \rho(w)$. So, for each unitary u in $\mathcal{A}$, $||z|| = \rho(z)$. It now follows from Theorem 2.3 that $\rho(z) \leq ||z||$ for all $z \in \mathcal{A}$.

Thus $\rho(z) = ||z||$ for all z in $\mathcal{A}$.

REFERENCES

1. Alfsen, E. M., Schultz, F. W. and Störmer, E. (to appear). A Gelfand-Naimark Theorem for Jordan Algebras.
2. Bonsall, F. F. and Duncan, J. (1973). Complete Normed Algebras. (Springer).
3. Harris, L. A. (1972). Banach algebras with involution and Möbius transformations. J. Functional Analysis 11, 1-16.
4. Jacobson, N. (1968). Structure and Representations of Jordan Algebras. Amer. Math. Soc. Colloquium Publications 39, 1-453.
5. Kaplansky, I. (1949). Normed Algebras. Duke Math. J. 16, 399-418.
6. Russo, B. and Dye, H. A. (1966). A note on unitary operators in C*-algebras. Duke Math. J. 33, 413-416.
7. Sakai, S. (1971). C*-algebras and W*-algebras. (Springer).
8. von Neumann, J., Jordan, P. and Wigner, E. (1934). On an algebraic generalization of the quantum mechanical formalism. Ann. of Math. (2) 35, 29-64.
9. Wright, J. D. M. (in press). Jordan C*-algebras. Michigan Math. J.

FUNCTIONAL ANALYSIS FOR THE PRACTICAL MAN

J. D. Maitland Wright
Department of Mathematics
University of Reading
Reading, England.

Practical analysis has its origins in geometry, so let us begin with Euclid. His famous Fifth Postulate is equivalent to 'Playfair's Axiom' which states:

> _Let AB be a line and P a point not on the line. Then there exists a unique line CD which passes through P and is parallel to AB._

For many centuries mathematicians tried to prove the Fifth Postulate from the other axioms of Euclid. Early in the nineteenth century, Gauss, Bolyai and Lobashevsky conceived the revolutionary idea that the Fifth Postulate was not a consequence of the other axioms and they introduced 'non-Euclidean' geometries in which Playfair's Axiom does not hold. But many of their contemporaries rejected this idea as nonsense, for they believed that a 'non-Euclidean' geometry must be inconsistent. Eventually, it was established in the latter half of the nineteenth century that it is impossible to deduce the Fifth Postulate from the other axioms of Euclid [5, 10].

How can you prove rigorously that the Fifth Postulate is not a consequence of the other axioms? How can you convince a Euclidean geometer that there are other geometries of equal validity? Many answers to these questions have been found but one of the earliest and simplest was discovered by Beltrami [4] over a hundred years ago. His method is as follows. First, we start with a Euclidean plane and draw an arbitrary circle K in this plane. We can construct a new geometry $\mathcal{G}$ by taking for '$\mathcal{G}$-points' all the ordinary Euclidean points inside the circle K and for '$\mathcal{G}$-lines', the restrictions to the interior of K of all ordinary Euclidean lines in the plane which cut K. Some care is needed in defining distance in $\mathcal{G}$ suitably. But this can be done in such a way that $\mathcal{G}$ satisfies all the axioms of Euclidean geometry apart from Playfair's Axiom (or, equivalently, the Fifth Postulate). [5, page 252.]

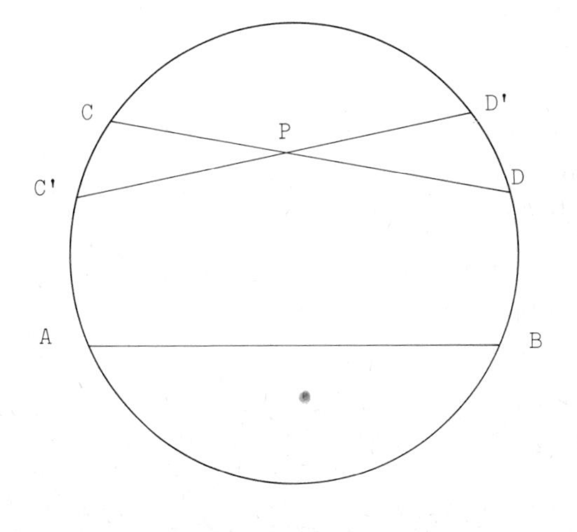

Playfair's Axiom is false in $\mathcal{G}$. For, in the diagram, CD is 'parallel' to AB because these lines do not intersect however far they are produced (in $\mathcal{G}$). Similarly, C'D' is 'parallel' to AB . But CD and C'D' intersect in P and CD ≠ C'D' .

In recent times, logicians have shown, by constructing suitable 'models' of set theory, that the Axiom of Choice can neither be proved nor disproved from the other axioms of (Zermelo-Fraenkel) set theory. So, the Axiom of Choice is merely a matter of choice! Just as we can have non-Euclidean geometry we can have non-Zornian set theory. I hope to justify my title by showing that, for certain practical purposes, it is more convenient to work in a set theory where the Axiom of Choice is replaced by a different axiom.

Our jumping off point is ZF + DC . Here 'ZF' denotes ordinary Zermelo-Fraenkel axiomatic set theory without the Axiom of Choice. To this system we have adjoined DC , the Axiom of Dependent Choice, which states that whenever X is a non-empty set and R is any binary relation on X whose domain is the whole of X (that is, for each $x \in X$ there exists $y \in X$ such that xRy holds) then there exists a sequence (x_n) $(n = 1, 2, \ldots)$ in X such that $x_n R x_{n+1}$ holds for every natural number n . This implies, in particular, a 'countable' axiom of choice.

Just as Molière's hero [12; page 19] was astonished to discover that he had been speaking prose all his life, some practical men may not yet have recognized that they work in ZF + DC . (To forestall quibbling on this point we could adopt this as our definition of 'practical man'!)

All the theorems of classical analysis and positive results of elementary measure theory are theorems of ZF + DC . Of course, in functional analysis, extensive use has been made of the Axiom of Choice. But, even in functional analysis, many theorems are derived in ZF + DC , for example, the Closed Graph Theorem and the Uniform Boundedness Theorem. Of those theorems, such as the Hahn-Banach Theorem and the Krein-Milman Theorem, whose usual proofs depend on the Axiom of Choice, most can be derived in ZF + DC , provided some mild separability conditions are imposed. For a clear account of this see the definitive treatise of Garnir, Schmets and de Wilde [6].

Solovay [16] has built a model of ZF + DC in which various marvellous things happen. For example, let LM be the proposition: Each subset of the real numbers is Lebesgue measurable. Then, when LM is interpreted in Solovay's model it becomes a true statement. So, if we adjoin LM as an axiom to ZF + DC then Solovay's model is a model for the axiomatic system ZF + DC + LM , which implies that it is impossible to prove ∿LM in ZF + DC , that is, we cannot find a non-Lebesgue measurable subset of $\mathbb{R}$ except by appealing to an 'uncountable' form of the Axiom of Choice.

Let I be the statement: There exists an inaccessible cardinal. Solovay uses the hypothesis that there exists a (transitive) model for ZFC + I when constructing * his model.

We shall adjoin an axiom BP (this will be specified later) to ZF + DC . It is known that BP becomes a true statement when interpreted in Solovay's model, that is, Solovay's model is a model for the system ZF + DC + BP . So when Q is any proposition which is a theorem in ZF + DC + BP then, provided Solovay's model exists, Q is consistent with ZF + DC . (Remark of Dr. R. O. Gandy: By standard arguments, if ZFC + I is consistent then so is ZF + DC + Q .) For example, let H be the proposition: Every linear operator mapping the whole of a Hilbert space into itself is bounded. It follows from more general results that H is a theorem in ZF + DC + BP . So H cannot be disproved in ZF + DC .

* ZFC stands for ZF + Axiom of choice

But practical men, although they do not need uncountable forms of the Axiom of Choice, do consider more general spaces than Hilbert spaces or indeed Banach spaces. Let us recall that a Frechet space is a vector space V equipped with a metric d such that the algebraic operations are continuous with respect to the metric and such that the metric space (V, d) is complete, that is, each Cauchy sequence in V is convergent, e.g. $\mathcal{E}$, the space of all infinitely differentiable functions on $\mathbb{R}$ equipped with the topology of compact convergence for all derivatives [13], is a Frechet space which is not a Banach space. Unless we explicitly state otherwise, any topological vector spaces considered are not required to be locally convex.

Let us examine the following propositions.

(A) Let X be a locally convex Frechet space, let Y be a locally convex topological vector space and let $T : X \to Y$ be a linear operator. Then T is continuous.

(B) Let X be a locally convex Frechet space and let $f : X \to \mathbb{R}$ be a convex function. Then f is continuous.

(C) Let X be a Frechet space, let Y be a Hausdorff topological vector space and let $T : X \to Y$ be a linear operator. Then T is continuous.

(D) Let X be a Frechet space. Let $\phi : X \to \mathbb{R}$ be an Archimedean subadditive functional. Then ϕ is continuous.

It would be very convenient if, say, (C) were true. For this would mean that whenever we found ourselves with a specific linear operator (for example, in connection with a differential or integral operator) then, provided it was defined on a complete metrizable space, we would know, without any checking or verification, that the operator must be continuous.

Unfortunately, if we use the Axiom of Choice, it is easy to show, by a Hamel base argument, that (A), (B), (C) and (D) are all false, even when $Y = \mathbb{R}$. At first sight this is disappointing, but all is not lost. For we can prove that (A), (B), (C) and (D) are valid theorems in ZF + DC + BP. So the practical man, who, because he is a practical man always works in ZF + DC, is entitled to suppose, without fear of contradiction, that all linear operators on Frechet spaces are continuous!

There are several possible approaches. When X is a Banach space, Ajtai [1], has shown by using deep model theoretic methods that (A) and (B) are consistent with ZF + DC. While Garnir [7] has proved that (A) and (B) are theorems in ZF + DC + LM. Garnir defines a group structure on the Cantor set and uses properties of Haar measure in an ingenious way. In [18], I showed that (A) and (B) are theorems in ZF + DC + BP. Subsequently, Dr. Fremlin (private letter) found an alternative proof of (A) which did not require local convexity. In the words of Kipling [19; page 394],

> "There are nine and sixty ways of constructing tribal lays,
> And-every-single-one-of-them-is-right!"

In this article, we shall use a different method which is essentially self-contained. First we work in ZF + DC, as practical men, and prove a theorem about subadditive functionals on Frechet spaces. The classical Uniform Boundedness Theorem of Banach-Steinhaus is an immediate corollary. Another consequence is a new proof of the Borel Graph Theorem of Schwartz [14, page 160]. Then, in §2, we apply the results of §1 in the system ZF + DC + BP and find that (D) and (C) are theorems in this system and hence are consistent with ZF + DC.

§1 SUBADDITIVE FUNCTIONALS AND LINEAR OPERATORS

In this section, all arguments are in the system ZF + DC .

Let us recall that a subset M of a metric space X is said to be meagre (or of the first category) when it is the union of countably many nowhere dense sets. Furthermore, a subset B of X is said to have the Baire Property if there exists an open set U such that the symmetric difference $(B\setminus U) \cup (U\setminus B)$ is meagre. For any metric space X let $\mathcal{BP}(X)$ be the σ-field of all subsets of X with the Baire Property. Clearly each Borel subset of X is in $\mathcal{BP}(X)$.

The following lemma is proved by Kuratowski [11; page 306]. For the convenience of the reader a proof is included here.

> LEMMA 1.1. Let X be a complete metric space and let $f : X \to \mathbb{R}$ be a function which is measurable with respect to the σ-field $\mathcal{BP}(X)$. Then there exists a meagre set M such that the restriction of f to $X\setminus M$ is continuous.

Let $B_k^n = f^{-1}\left[\frac{k}{n}, \frac{k+1}{n}\right)$. Let $h_n = \sum_{k=-\infty}^{+\infty} \frac{k}{n} \chi_{B_k^n}$. Since each B_k^n has the Baire property and since countable unions of meagre sets are meagre, we can find a meagre set M and open sets U_k^n such that $B_k^n \cap (X\setminus M) = U_k^n \cap (X\setminus M)$ for each n and each k . It follows that the restriction of each h_n to $X\setminus M$ is continuous. We observe that (h_n) $(n = 1, 2, \ldots)$ converges uniformly to f .

Let V be any (real or complex) vector space. A subadditive functional on V is a map $\rho : V \to \mathbb{R}$ such that

(i) $\rho(x + y) \leq \rho(x) + \rho(y)$, for all x and y in V ,

and

(ii) $\rho(0) = 0$.

A subadditive functional ρ on V is said to be Archimedean if, for each $a \in V$, $\limsup_{n\to\infty} \rho(\frac{1}{n}a) \leq 0$.

If a subadditive ρ is positively homogeneous, that is $\rho(\lambda a) = \lambda\rho(a)$ for each $\lambda \in \mathbb{R}^+$, then ρ is necessarily Archimedean. If there exists any Hausdorff vector topology for V such that ρ is continuous at the origin with respect to this topology then, clearly, ρ is Archimedean.

Example. Let $\delta : \mathbb{R} \to \mathbb{R}$ be defined by $\delta(t) = 1$, for $t \neq 0$, and $\delta(0) = 0$. Then δ is subadditive but not Archimedean.

We now come to the key proposition.

> THEOREM 1.2. Let X be a Frechet space. Let ρ be an Archimedean subadditive functional on X such that ρ is measurable with respect to $\mathcal{BP}(X)$. Then ρ is continuous at the origin.

Let (x_n) $(n = 1, 2, \ldots)$ be any sequence in X such that $\lim x_n = 0$. Since ρ is $\mathcal{BP}(X)$-measurable, there exists, by Lemma 1.1, a meagre set $M \subset X$ such that the restriction of ρ to $X\setminus M$ is continuous with respect to the relative topology of $X\setminus M$.

Let $L = \bigcup_{k=1}^{\infty} \bigcup_{n=1}^{\infty} k(M - x_n) \cup \bigcup_{k=1}^{\infty} kM$.

Then L is meagre and so, by the Baire Category Theorem, there exists $z \in X \setminus L$. Thus $x_n + \frac{1}{k}z \notin M$ for any k or n and $\frac{1}{k}z \notin M$ for any k .

So, for each k ,

$$\lim_{n\to\infty} \rho(x_n + \tfrac{1}{k}z) = \rho(\tfrac{1}{k}z) .$$

By subadditivity,

$$\rho(x_n) \le \rho(x_n + \tfrac{1}{k}z) + \rho(\tfrac{1}{k}(-z)) .$$

So

$$\limsup_n \rho(x_n) \le \rho(\tfrac{1}{k}z) + \rho(\tfrac{1}{k}(-z)) .$$

Thus

$$\limsup_n \rho(x_n) \le \limsup_k \rho(\tfrac{1}{k}z) + \limsup_k \rho(\tfrac{1}{k}(-z))$$
$$\le 0 .$$

On the other hand,

$$\rho(x_n) \ge \rho(x_n + \tfrac{1}{k}z) - \rho(\tfrac{1}{k}z) .$$

So

$$\liminf_n \rho(x_n) \ge \lim_n \rho(x_n + \tfrac{1}{k}z) - \rho(\tfrac{1}{k}z) = 0 .$$

It follows that $\lim \rho(x_n)$ exists and is 0 .

COROLLARY 1.3. *Let X be a Frechet space and let ρ be an Archimedean subadditive functional on X . Suppose that, for each separable closed subspace F of X , the restriction of ρ to F is $\mathcal{BP}(F)$-measurable. Then ρ is continuous at the origin.*

Let (x_n) $(n = 1, 2, \ldots)$ be any sequence in X which converges to 0 . Let F be the closed linear span of $\{x_n : n = 1, 2, \ldots\}$. Then, by Theorem 1.2 applied to F , $\lim \rho(x_n) = 0$.

The classical Uniform Boundedness Theorem follows easily.

COROLLARY 1.4. (Banach-Steinhaus). *Let X be a Banach space and Y a normed space. Let $\{T_\lambda : \lambda \in \Lambda\}$ be a family of bounded linear operators from X to Y such that, for each $x \in X$, $\{||T_\lambda x|| : \lambda \in \Lambda\}$ is a bounded set. Then $\{||T_\lambda|| : \lambda \in \Lambda\}$ is also bounded.*

For each $x \in X$, let $\rho(x) = \sup\{||T_\lambda x|| : \lambda \in \Lambda\} < +\infty$. Then ρ is subadditive, positively homogeneous and lower semicontinuous. Thus ρ is $\mathcal{BP}(X)$-measurable and hence, by Theorem 1.2, continuous at 0 .

Thus, there exists $\delta > 0$, such that $\rho(x) \le 1$ whenever $||x|| \le \delta$.

So $||T_\lambda|| \le 1/\delta$ for all $\lambda \in \Lambda$.

Let us define a quasi-norm on a (real or complex) vector space V to be a subadditive functional $\rho : V \to \mathbb{R}$ such that $\rho(x) \ge 0$ and $\rho(x) = \rho(-x)$ for all $x \in V$. We recall that a quasi-norm ρ is a semi-norm if it enjoys the additional property that $\rho(\lambda x) = |\lambda|\rho(x)$ for each scalar λ and each vector x. Just as the continuous semi-norms determine the topology of a locally convex space, the topology of an arbitrary topological vector space is determined by the continuous quasi-norms on the space. More precisely, whenever V is a topological vector space, a base of open neighbourhoods of the origin is given by the family of all sets of the form $\{x \in V : \rho(x) < \varepsilon\}$ where ρ is a continuous quasi-norm and ε is a positive real number. [8, page 76].

In Theorem 1.5, Corollary 1.6 and Corollary 1.7 no local convexity conditions are imposed on any of the vector spaces.

THEOREM 1.5. *Let X be a Frechet space, Y an Hausdorff topological vector space and $T : X \to Y$ a linear map. If, for each continuous quasi-norm ρ on Y, the quasi-norm ρT on X is $\mathcal{BP}(X)$-measurable then T is a continuous linear operator.*

Let ρ be a continuous quasi-norm on Y. Clearly ρT is subadditive and, since $\rho T(\frac{1}{n}x) = \rho(\frac{1}{n}Tx)$, ρT is Archimedean. So, by Theorem 1.2, T is continuous.

COROLLARY 1.6. *Let X be a Frechet space, Y an Hausdorff topological vector space and $T : X \to Y$ a linear map. If, for each closed separable subspace F of X and for each continuous quasi-norm ρ on Y, the restriction of ρT to F is $\mathcal{BP}(F)$-measurable then T is a continuous linear operator.*

We recall that T is continuous if, and only if, whenever (x_n) $(n = 1, 2, \ldots)$ is a sequence on X which converges to 0 then $\lim Tx_n = 0$.

COROLLARY 1.7. *Let X be a Frechet space and Y a Suslin topological vector space. Let $T : X \to Y$ be a linear map such that Graph T is a Borel subset of $X \times Y$. Then T is a continuous linear operator.*

It suffices to prove the result when X is separable. Then, by [14, page 107], since X is Polish, Y is Suslin and Graph T is a Borel set, T is a Borel measurable map. Hence, by Theorem 1.5, T is continuous.

The Borel Graph Theorem of Schwartz [15] and [14, page 160] is an immediate consequence of the above.

COROLLARY 1.8. *Let X be an ultrabornological space and Y a locally convex Suslin space. Let $T : X \to Y$ be a linear map such that Graph T is a Borel subset of $X \times Y$. Then T is a continuous operator.*

Since X is ultrabornological there exists a family of Banach space X_γ and linear maps $u_\gamma : X_\gamma \to X$ such that T is continuous only if $Tu_\gamma : X_\gamma \to Y$ is continuous for each γ.

For each γ, Graph Tu_γ is the inverse of the Borel set Graph T with respect to the continuous map $(x, y) \to (u_\gamma(x), y)$ of $X_\gamma \times Y$ into $X \times Y$. So, by Corollary 1.7, Tu_γ is continuous. Thus T is continuous.

§2 APPLICATIONS IN ZF + DC + BP

Let BP be the proposition: *Every subset of a complete separable metric space has the Baire property.* Then, see [16], when BP is interpreted in Solovay's model of set theory it becomes a true statement. So, provided Solovay's model exists, any theorem of ZF + DC + BP is consistent with ZF + DC.

In this section all arguments are in the system ZF + DC + BP .

THEOREM 2.1. *Let X be a Frechet space and let $\rho : X \to \mathbb{R}$ be an Archimedean subadditive functional. Then ρ is continuous at the origin.*

This is an immediate consequence of Corollary 1.3 and BP .

THEOREM 2.2. *Let X be any Frechet space and Y any Hausdorff topological vector space. Let $T : X \to Y$ be a linear map. Then T is continuous.*

This is an immediate consequence of Corollary 1.6 and BP .

At the price of introducing local convexity we have the following Corollary.

COROLLARY 2.3. *Let X be an ultrabornological space and Y a locally convex space. Let $T : X \to Y$ be a linear map. Then T is continuous.*

By a familiar argument, see the proof of Corollary 1.7, it is sufficient to prove this in the special situation where X is a Banach space, but this has already been established by Theorem 2.2.

REFERENCES

1. Ajtai, M. (to appear). On the boundedness of definable linear operators.
2. Beltrami, E. (1866). Risoluzione del problema di riportare i punti di una superficie supra un piano in modo che le linee geodetiche vengano rappresentate da linee rette. Annali di Matematica 7, 185-204.
3. ———— (1868). Saggio di interpretazione della geometria non euclidea. Giornale di Matematiche 6, 284-312.
4. ———— (1868). Teoria fondamentale degli spazii di curvature costante. Annali di Matematica (2) 2, 232-255.
5. Coxeter, H. S. M. (1947). Non-Euclidean Geometry. (University of Toronto Press).
6. Garnir, H. G., De Wilde, M. and Schmets, J. (1968). Théorie constructive des espaces linéaires à semi-normes. (Volume 1, Théorie générale), Birkhäuser Verlag (Basel-Stuttgart).
7. Garnir, H. G. Solovay's axiom and functional analysis. Proc. Madras Conference Functional Analysis. Springer Lecture notes, No. 399, 189-204.
8. Hewitt, E. and Ross, K. A. (1963). Abstract Harmonic Analysis. (Springer).
9. Kipling, J. R. (London 1921) Rudyard Kipling's Verse Inclusive Edition 1885-1918. (Hodder and Stoughton).
10. Klein, F. (1928). Vorlesungen über Nicht-Euklidische Geometrie. (Berlin.)
11. Kuratowski, C. (1948). Topologie (Vol. I) Monagrafie Matematyczne 20. (Warsaw).
12. de Molière, J. B. P. (London 1962). The Miser and other Plays. (Penguin).
13. Robertson, A. P. and W. (1964). Topological vector spaces. (Cambridge University Press).
14. Schwartz, L. (1973). Radon measures on arbitrary topological spaces and cylindrical measures. (Oxford University Press).
15. ———— (1966). Extensions du théorème du graphe fermé. C. R.Acad. Sci. Paris, Sér. A-B 263, A602-A605.
16. Solovay, R. M. (1970). A model of set theory in which every set of reals is Lebesgue measurable. Ann. of Math. 92, 1-56.
17. Wright, J. D. M. (1973). All operators on a Hilbert space are bounded. Bull. Amer. Math. Soc. 79, 1247-1250.
18. ———— (1975). On the continuity of mid-point convex functions. Bull. London Math. Soc. 7, 89-92.